Stochastic Processes and Operator Calculus on Quantum Groups

Mathematics and Its Applications

Volume 490

Stochastic Processes and Operator Calculus on Quantum Groups

by

Uwe Franz

Institut für Mathematik und Informatik,
Ernst-Moritz-Arndt-Universität Greifswald,
Greifswald, Germany

and

René Schott

Institut Elie Cartan and Loria,
Université Henri Poincaré-Nancy 1,
Vandoeuvre-lès-Nancy,
France

KLUWER ACADEMIC PUBLISHERS
DORDRECHT / BOSTON / LONDON

A C.I.P. Catalogue record for this book is available from the Library of Congress.

ISBN 978-90-481-5290-2

Published by Kluwer Academic Publishers,
P.O. Box 17, 3300 AA Dordrecht, The Netherlands.

Sold and distributed in North, Central and South America
by Kluwer Academic Publishers,
101 Philip Drive, Norwell, MA 02061, U.S.A.

In all other countries, sold and distributed
by Kluwer Academic Publishers,
P.O. Box 322, 3300 AH Dordrecht, The Netherlands.

Printed on acid-free paper

Contents

Preface

Quantum groups have been investigated rather deeply in mathematical physics over the last decade. Among the most prominent contributions in this area let us mention the works of V.G. Drinfeld, S.L. Woronowicz, S. Majid. Probability theory on quantum groups has developed in several directions (see works of P. Biane, R.L. Hudson and K.R. Partasarathy, P.A. Meyer, M. Schürmann, D. Voiculescu). The aim of this book is to present several new aspects related to quantum groups: operator calculus, dual representations, stochastic processes and diffusions, Appell polynomials and systems in connection with evolution equations. Much of the material is scattered throughout available literature, however, we have nowhere found in accessible form all of this material collected. The presentation of representation theory in connection with Appell systems is original with the authors. Stochastic processes (example: Brownian motion, diffusion processes, Lévy processes) are investigated and several examples are presented.

As a text the work is intended to be accessible to graduate students and researchers not specialised in quantum probability.

We would like to acknowledge our colleagues P. Feinsilver, R. Lenzceswki, D. Neuenschwander, and M. Schürmann for allowing us to use material which has been the object of joint works. We would like as well to acknowledge our colleagues from Clausthal (Arnold Sommerfeld Institute), Greifswald (Institut für Mathematik und Informatik), Nancy (Institut Elie Cartan and LORIA), Strasbourg (IRMA) for numerous discussions on the topic of this book. Finally we express gratitude to our families and to all our friends. Their patience and encouragement made this project possible.

Chapter 1

Introduction

Quantum probability is now a quite active research area motivated by applications in physics. Recent books of Biane [Bia93], Meyer [Mey93], and Parthasarathy [Par92] are recommended as introductions to this field. For quantum probability on Hopf algebras see Schürmann [Sch93], or also [Mey93, Chapter VII] and Majid [Maj95b, Chapter 5].

In this book we present some recent developments in the theory of stochastic processes and operator calculus on quantum groups.

The organisation of this book is as follows:

We begin in Chapter 2 with some preliminaries on Lie groups and related topics: representation theory, construction of stochastic processes, Appell systems. This chapter is intended to provide some intuition that might be helpful later on. Many of our results for quantum groups were motivated by their counterparts on Lie groups and still contain them as a special case or as "classical limit".

Chapter 3 provides the necessary background on the algebraic structures that we are going to use in the following chapters.

Even though no generally accepted definition seems to exist to this day, most authors agree that a quantum group should at least be a bialgebra or a Hopf algebra. For the definition of Lévy processes in Chapter 4, it is sufficient to have an involutive bialgebra structure.

First, in Section 3.1, we give the basic definitions of bialgebras and Hopf algebras. Duality and dual representations are also already introduced. In the next section (Section 3.2) we present several examples of bialgebras and Hopf algebras. We start with a few classical examples arising from groups or semi-groups and continue then with some standard examples of non-commutative and non-cocommutative Hopf algebras. In Section 3.3, we study dual representations. We show that the definition of dual representations for dually paired bialgebras extends that of dual representations for Lie algebras introduced in Section 2.3, show that similar techniques can be used for calculating them, and present the explicit calculations for several examples. To underline the similarity to the theory on Lie groups further we introduce an analogue of the composition law in Section 3.4. It is a consequence of duality, but written in terms of the dual pairing it has the same form as the composition (or group) law on Lie groups. In Section 3.6 we introduce matrix elements for dual representations of quantum groups along the line of Section 2.6. In Section 3.7 we introduce braided tensor categories, an important concept that is intimately related to the representation theory of quantum groups. In the next two Sections (Sections 3.8 and 3.9) we define braided bialgebras and braided Hopf algebras and present several examples. In Section 3.10 we concentrate on a special class of braided Hopf algebras which are primitively generated, called braided spaces.

In Chapter 4 we come to our main topic, the study of stochastic processes on quantum groups. First, we recall the basic terminology of non-commutative or quantum probability. In Section 4.2 we introduce the notion of independence for quantum random variables. The next section (Section 4.3) contains the definition of quantum stochastic processes with independent and stationary increments, i.e. Lévy processes. As in the classical situation, these processes are completely characterised by their convolution semi-group or by their generators. In Sections 4.4 and 4.5 we describe two constructions of realisations of Lévy processes. The first starts from the generator and gives a realisation on a Bose Fock space. In order to do so, one constructs the so-called Schürmann triple of the generator, which appears as coefficients in the quantum stochastic differential equation of the process. The second requires a convolution semi-group of normalised functionals as input and uses an inductive limit procedure. The resulting realisation is poorer from the analytical point of view, but it has the advantage that it can be applied in a more general situation, even if the semi-group is not positive. In Section 4.6, we construct convolution semi-groups on quantum groups from classical Lévy processes and in Section 4.7 we prove an analogue of the Feynman -Kac formula for these semi-groups. The last section (Section 4.8) deals with duality and time-reversal.

In Chapter 5 we show that Lévy processes have a natural quantum Markov structure and use this to deal with the question which quantum stochastic processes admit classical versions, i.e. for what families of operators $(X_t)_{t\in I}$ (of the form $X_t = j_t(x)$, where (j_t) is a Lévy process), there exists a classical stochastic process $(\tilde{X}_t)_{t\in I}$ on some probability space $(\Omega, \mathcal{F}, P)$ such that all time-ordered moments agree, i.e.

$$\Phi(X_{t_1}^{k_1} \cdots X_{t_n}^{k_n}) = \mathbb{E}(\tilde{X}_{t_1}^{k_1} \cdots \tilde{X}_{t_n}^{k_n}), \tag{1.1}$$

for all $n, k_1, \ldots, k_n \in \mathbb{N}$, $t_1 \leq \ldots \leq t_n \in I$, for the (vacuum) state Φ.

A famous example of a classical version of a quantum Lévy process is the Azéma martingale. It is well known that on a commutative algebra the quantum Markov property is sufficient for the existence of classical versions. Therefore it is sufficient to find commutative $*$-subalgebras such that the restriction of our process is still Markovian. We show that we can obtain quantum Markov processes on commutative subalgebras from quantum Lévy processes that are not commutative, and that there exist also commutative processes that are not Markovian. We also show that this approach leads to a powerful tool for the explicit calculation of the classical generators or measures.

The next two chapters investigate the relation between Lévy processes and evolution equations. In Chapter 6, we define and construct diffusions on braided spaces, and we show that their shifted moments sequences are polynomials (the so-called Appell polynomials). These polynomials satisfy equations of the form $(\partial_t - L)u = 0$, where L is an operator consisting of linear and quadratic terms in the braided-partials ∂^i. In Section 6.3 we give a brief discussion of how density functions can be introduced.

In Chapter 7 we consider stochastic processes on quantum groups that are related to evolution equations of the form

$$\partial_t u = Lu,$$

with some difference-differential operator L. For the equations considered in Section 7.1, u is an element of a quantum or braided group $\mathcal{A}$. We recall that solutions of these equations can be given as Appell systems or shifted moments of the associated process, and show how these can be calculated explicitly on the q-affine group and on a braided analogue of the Heisenberg-Weyl group. In Section 7.2, we define a Wigner map from functionals on a quantum group or braided group to a "Wigner" density on the undeformed space. We prove that the densities associated in this way to (pseudo-) Lévy processes satisfy a Fokker-Planck type equation. In the one-dimensional case these coincide with the evolution equations of Section 7.1, but in the general case we get new equations.

In Chapter 8, we turn to the characterisation of certain probability laws and convolution semi-groups on nilpotent quantum groups and on nilpotent braided groups. In Section 8.1 we determine the functionals which satisfy an analogue of the Bernstein property, i.e. that the sum and difference of independent random variables are also independent, on several braided groups. This extends results obtained on Lie groups by D. Neuenschwander et al., cf. [Neu93], [NRS97]. As for Lie groups this class turns out to be too small to constitute a satisfactory definition of Gaussianity. Therefore we turn to convolution semi-groups. We show the uniqueness of embedding into continuous convolution semi-groups on nilpotent quantum groups and nilpotent braided groups in Section 8.2. In Section 8.3 we define Gaussian convolution semi-groups in the sense of Bernstein and calculate their generators on several nilpotent braided groups. We show that quadratric generators on Hopf algebrs (as defined by Schürmann[Sch93]) generate semi-groups that are Gaussian in the sense of Bernstein.

Chapter 9 presents the results of Franz, Neuenschwander, and Schott[FNS97b] for the problem of phase retrieval on nilpotent quantum groups and nilpotent braided groups: Given the symmetrisation $\mu * \overline{\mu}$ and the first moments of a unital functional μ on $\mathcal{A}$, when is it possible to retrieve the original functional μ from these data? The somewhat surprising answer is that in this framework, the retrieval is always possible (provided that the quantum or braided group is "sufficiently" non-commutative, e.g. if q is not a root of unity). By definition functionals on quantum groups have all moments and are uniquely determined by them. So it will suffice to show that the moments of the symmetrisation and the first moments of μ together allow to calculate all moments of μ recursively. Observe that one can not expect to be able to remove the condition of knowledge of the first moments of μ, since already on the classical real line, in the best possible case, μ can be determined by its symmetrisation only up to a shift. The situation on nilpotent quantum groups and nilpotent braided groups is in sharp contrast to the classical case of simply connected nilpotent Lie groups, where the moments can not be retrieved, as we see in Section 9.1.

In Chapter 10 we present several limit theorems. In Section 10.1, the general results for limit theorems on bialgebras due to Schürmann [Sch93] is presented. Then a randomised q-central (or q-commutative) limit theorem on a family of bialgebras with one complex parameter is shown in Section 10.2. This result is due to U. Franz [Fra98]. Section 10.3 is devoted to Woronowicz results [Wor87] on convergence of convolution products of probability measures to the Haar functional on compact quantum groups. A q-central limit theorem for $U_q(su(2))$ has been proved by R. Lenczewski [Len93, Len94], this result is stated in Section 10.4 as well as a weak law of large numbers which derives easily from the central limit theorem. Then we recall D. Neuenschwander and R. Schott's (cf. [NS97]) results on domains of attraction for q-transformed random variables in Section 10.5.

Chapter 2

Preliminaries on Lie groups

We begin by recalling some basic definitions. The theory of Lie groups and Lie algebras can be found in many textbooks, see e. g. [HN91, Hum87, SW86, Ser92, Var84]. We list several examples of Lie algebras in section 2.2.

In the following sections we develop some representation theory that we will need later to construct stochastic processes on Lie groups . The methods that we apply are taken from e. g. [GDF82, FS90, FKS93].

First we consider the actions of a Lie algebra on its universal enveloping algebra defined by multiplication from the left or from the right (the master representation of [GDF82]). We then show how vector field realisations of the Lie algebra arise as the duals of these actions.

Next we show how these realisations can be used to compute decomposition formulas, i. e. formulas that represent a group element as product of N one-parameter subgroups, where N is the dimension of the corresponding Lie algebra. We show how the group law looks in these parameters, and define the matrix elements of the representation of the group on the universal enveloping algebra. These will be special functions. Their properties, such as summation theorems and recurrence relations, can also be derived from the approach presented here.

2.1 BASIC DEFINITIONS

Let G be a group and an analytic $\mathbb{K}$-manifold, where $\mathbb{K}$ stands for the field of complex or real numbers. G is called a *Lie group*, if the multiplication

$$\begin{aligned} m \quad : \quad & G \times G \to G \\ & (g_1, g_2) \mapsto g_1 g_2 \end{aligned}$$

and the inverse

$$\begin{aligned} i \quad : \quad & G \to G \\ & g \mapsto g^{-1} \end{aligned}$$

are analytical maps. A *Lie algebra* over $\mathbb{K}$ is a $\mathbb{K}$-vector space $\mathfrak{g}$ equipped with a $\mathbb{K}$-bilinear map $[\cdot,\cdot] : \mathfrak{g} \times \mathfrak{g} \to \mathfrak{g}$ that satisfies

- $[X,Y] = -[Y,X]$ for all $X, Y \in \mathfrak{g}$ (anti-symmetry),
- $[X,[Y,Z]] + [Y,[Z,X]] + [Z,[X,Y]] = 0$ for all $X, Y, Z \in \mathfrak{g}$ (Jacobi identity).

Example: Let $(A, \cdot)$ be an associative $\mathbb{K}$-algebra. Then $A_L = (A, [\cdot,\cdot])$, with the Lie bracket $[\cdot,\cdot]$ defined by

$$[X,Y] = X \cdot Y - Y \cdot X \qquad \text{for } X, Y \in A,$$

is a $\mathbb{K}$-Lie algebra.

For each $\mathbb{K}$-Lie group G there exists a unique $\mathbb{K}$-Lie algebra $L(G)$ whose elements can be considered as left-invariant vector fields on G, and for every $\mathbb{K}$-Lie algebra $\mathfrak{g}$ there exists a simply connected $\mathbb{K}$-Lie group G such that $L(G) = \mathfrak{g}$.

One of the most important tools for the study of Lie groups and Lie algebras is the *exponential map* $\exp : L(G) \to G$. For nilpotent Lie algebras and for linear Lie groups (i. e. $G \subseteq Gl(n, \mathbb{K})$ and $L(G) \subseteq M(n, \mathbb{K})$) it can be defined via the exponential series

$$\exp(X) = \sum_{k=0}^{\infty} \frac{X^k}{k!}.$$

In general it is defined by

$$\exp(X) = \gamma_X(1),$$

where $\gamma_X(t)$ is the integral curve of the vector field X with initial value $\gamma_X(0) = e$ (the unit element of G). With the exponential map we can write the action of $X \in L(G)$ as a vector field on G as

$$(Xf)(g) = \left.\frac{\mathrm{d}}{\mathrm{d}t}\right|_{t=0} f(g \exp(tX)) \qquad \text{for } f \in C^\infty(G).$$

We will now introduce the universal enveloping algebra $U(\mathfrak{g})$ of a Lie algebra $\mathfrak{g}$. Let $\mathcal{T}(\mathfrak{g})$ be the tensor algebra of $\mathfrak{g}$, and J the ideal generated by

$$\{X \otimes Y - Y \otimes X - [X,Y]; X, Y \in \mathfrak{g}\}.$$

Then the *universal enveloping algebra* $U(\mathfrak{g})$ is defined as the quotient

$$U(\mathfrak{g}) = \mathcal{T}(\mathfrak{g})/J.$$

We identify $\mathfrak{g}$ with its image $\sigma(\mathfrak{g})$ under the canonical mapping $\sigma : \mathfrak{g} \to U(\mathfrak{g})$. If $\{X_1, \ldots, X_n\}$ is a basis of $\mathfrak{g}$, then, by the *Poincaré-Birkhoff-Witt theorem*, the set

$$\{X_1^{\mu_1} \cdots X_n^{\mu_k}; \mu = (\mu_1, \ldots, \mu_n) \in \mathbb{Z}_+^n\}$$

is a basis of $U(\mathfrak{g})$. The universal enveloping algebra is characterised by the universal property:

Let A be an associative algebra and let $\varphi : \mathfrak{g} \to A_L$ be a Lie algebra homomorphism. Then φ extends to a unique algebra homomorphism $\tilde{\varphi} : U(\mathfrak{g}) \to A$ such that $\varphi(X) = \tilde{\varphi}(X)$ for all $X \in \mathfrak{g}$.

A *representation* ρ of a (Lie) algebra A (resp. $\mathfrak{g}$) on a vector space V is a linear mapping

$$\begin{aligned} \rho \ &: \ A \to \mathrm{Hom}(V,V) \\ \Big(\text{resp. } \rho \ &: \ \mathfrak{g} \to \mathrm{Hom}(V,V)\Big) \end{aligned}$$

such that

$$\begin{aligned} \rho(a \cdot b) &= \rho(a) \circ \rho(b) \\ \Big(\text{resp. } \rho([X,Y]) &= \rho(X) \circ \rho(Y) - \rho(Y) \circ \rho(X)\Big) \end{aligned}$$

for all $a, b \in A$ (resp. $X, Y \in \mathfrak{g}$). The universality property implies that the representation theory of $\mathfrak{g}$ is equivalent to that of $U(\mathfrak{g})$.

For every algebra (A, m, e) we can define its opposite $A^{\mathrm{op}} = (A, m^{\mathrm{op}}, e)$ with the multiplication m^{op} defined by

$$m^{\mathrm{op}}(a,b) = m(b,a), \qquad \text{for all } a, b \in A.$$

For Lie algebras the anti-symmetry implies that $S_{\mathfrak{g}} : \mathfrak{g} \to \mathfrak{g}^{\mathrm{op}}$, $S(X) = -X$ for all $X \in \mathfrak{g}$ is a Lie algebra homomorphism, i. e.

$$S_{\mathfrak{g}}([X,Y]) = [S_{\mathfrak{g}}(X), S_{\mathfrak{g}}(Y)]^{\mathrm{op}} \qquad \Big(\ = [S_{\mathfrak{g}}(Y), S_{\mathfrak{g}}(X)] \ \Big).$$

This map extends to a homomorphism between the enveloping algebras $S : U(\mathfrak{g}) \to U(\mathfrak{g}^{\mathrm{op}}) \cong U(\mathfrak{g})^{\mathrm{op}}$. Note that since the algebras $U(\mathfrak{g})$ and $U(\mathfrak{g})^{\mathrm{op}}$ are defined on the same space, we can also consider S as an algebra anti-homomorphism from $U(\mathfrak{g})$ to itself.

2.2 EXAMPLES

We now introduce the Lie algebras that we will use in the following.

2.2.1 *The affine group and its Lie algebra*

The group of affine motions on the real line is equal to $Aff(1) = \{(a,b) \in \mathbb{R}^2; a > 0\}$. Its group law and its action on the real line are given by

$$\begin{aligned} (A,B) \circ (a,b) &= (Aa, aB + b), \\ (a,b)(x) &= ax + b. \end{aligned}$$

$Aff(1)$ is a Lie group. A basis of the tangent space at the unit element $(1,0)$ is given by $X_1 = \partial_a|_{(a,b)=(1,0)}$ and $X_2 = \partial_b|_{(a,b)=(1,0)}$. These tangent vectors, when extended as left-invariant vector fields, become $X_1 = a\partial_a + b\partial_b$ and $X_2 = \partial_b$. They satisfy $[X_1, X_2] = -X_2$. In the following calculations we choose the basis $X = -X_1$ and $Y = X_2$ with the commutation relations

$$[X, Y] = Y$$

as a basis for the Lie algebra $aff(1)$. Note that this is the only noncommutative two-dimensional Lie algebra.

2.2.2 *N-step nilpotent Lie algebras*

The Lie algebra $\mathfrak{g}$ with basis $X_0, X_1, \ldots, X_N$ and commutation relations

$$\begin{aligned}
[X_0, X_N] &= 0, \\
[X_0, X_i] &= X_{i+1}, \qquad 1 \le i \le N-1, \\
[X_i, X_j] &= 0, \qquad 1 \le i, j \le N,
\end{aligned}$$

is nilpotent, with $\mathfrak{g}^2 = [\mathfrak{g}, \mathfrak{g}] = \text{span}\{X_2, \ldots, X_N\}, \ldots, \mathfrak{g}^i = [\mathfrak{g}, \mathfrak{g}^{i-1}]$ $= \text{span}\{X_i, \ldots, X_N\}$, $(2 \le i \le N)$, and $\mathfrak{g}^{N+1} = [\mathfrak{g}, \mathfrak{g}^N] = \{0\}$. For $N = 2$ this algebra is called the *Heisenberg-Weyl algebra*.

2.2.3 *The finite-difference algebra*

Consider the operators T, Δ, and X defined by

$$\begin{aligned}
Tf(x) &= f(x+1), \\
\Delta f(x) &= f(x+1) - f(x), \\
Xf(x) &= xf(x).
\end{aligned}$$

These operators are a basis of the Lie algebra with basic commutation relations

$$\begin{aligned}
[T, X] &= T, \\
[\Delta, X] &= T, \\
[T, \Delta] &= 0.
\end{aligned}$$

This Lie algebra is solvable, $\mathfrak{g}^{(1)} = \mathfrak{g}$, $\mathfrak{g}^{(2)} = [\mathfrak{g}, \mathfrak{g}] = \text{span}\{T\}$, and $\mathfrak{g}^{(3)} = [\mathfrak{g}^{(2)}, \mathfrak{g}^{(2)}] = \{0\}$.

2.2.4 *The rotation group and its Lie algebra*

The real Lie algebra generated by X_1, X_2, X_3 with the commutation relations

$$[X_1, X_2] = X_3, \qquad +\text{cycl.}^{1}$$

[1] where "+cycl." means that the relations obtained by cyclic permutation of the indices are also true, i. e. $[X_2, X_3] = X_1$ and $[X_3, X_1] = X_2$. Because of the anti-symmetry this determines the Lie bracket completely.

is generally denoted as $so(3)$. It is useful to notice that the complexification of $so(3)$ is isomorphic to the Lie algebra $sl(2, \mathbb{C})$. If we set $H = 2iX_3$, $X_+ = iX_1 - X_2$, $X_- = iX_1 + X_2$, we get the standard $sl(2)$ commutation relations

$$\begin{aligned} [H, X_\pm] &= \pm 2X_\pm, \\ [X_+, X_-] &= H. \end{aligned}$$

The real Lie algebra generated by H, X_+, X_- is denoted by $sl(2, \mathbb{R})$.

The corresponding groups are

$$\begin{aligned} SO(3) &= \{A \in \mathbb{R}^{3\times 3}; A^T A = 1 \text{ and } \det(A) = 1\} \\ SL(2, \mathbb{R}) &= \{A \in \mathbb{R}^{2\times 2}; \det(A) = 1\} \\ SL(2, \mathbb{C}) &= \{A \in \mathbb{C}^{2\times 2}; \det(A) = 1\} \end{aligned}$$

with the usual matrix multiplication (where we followed the tradition of denoting the algebra by lowercase letters and the corresponding group by uppercase letters).

2.3 DUAL REPRESENTATIONS

The multiplication from the right and from the left gives actions

$$\rho_L, \ \rho_R : U(\mathfrak{g}) \to \mathrm{Hom}(U(\mathfrak{g}), U(\mathfrak{g}))$$

of $U(\mathfrak{g})$ on itself,

$$\begin{aligned} \rho_L(X)Y &= XY, \\ \rho_R(X)Y &= YX, \end{aligned}$$

which satisfy

$$\begin{aligned} \rho_L(X) \circ \rho_L(Y) &= \rho_L(XY), \\ \rho_R(X) \circ \rho_R(Y) &= \rho_R(YX), \end{aligned}$$

i. e. ρ_L is a representation fo $U(\mathfrak{g})$, and ρ_R is a representation of $U(\mathfrak{g})^{\mathrm{op}}$ called left and right representation, respectively. This implies that $\rho_R \circ S$ is a representation of $U(\mathfrak{g})$ (with S as defined on page 7). Finite-dimensional representations of $U(\mathfrak{g})$ can be obtained from ρ_L by taking the quotient with respect to invariant subspaces $I \subset U(\mathfrak{g})$ whose codimension is finite (a subspace $I \subset U(\mathfrak{g})$ is called (ρ_L-) invariant, if for all $Y \in I$ and all $X \in U(\mathfrak{g})$: $\rho_L(X)Y \in I$). If we fix a basis $\{\psi_n\}$ for $U(\mathfrak{g})$, then we can define the matrix elements $m_n^N(X)$ of $\rho_L(X)$ (with respect to this basis), by

$$\rho_L(X)\psi_n = \sum_{n+N\geq 0} m_n^N(X)\psi_N, \tag{2.1}$$

for $X \in U(g)$ (n and N are generally multi-indices). Another representation, the *adjoint representation*, is given by

$$\mathrm{ad}(X)Y = XY - YX, \qquad \text{for } X, Y \in U(\mathfrak{g}).$$

Example: We look at the right and left actions for the affine algebra. Fix the basis $\{\psi_{nm} = X^nY^m\}$ of $U(aff(1))$. Then the matrix elements of the right and left action can be read from the following relations,

$$\begin{aligned}
\rho_L(X)\psi_{nm} &= X \cdot X^nY^m = X^{n+1}Y^m \\
&= \psi_{n+1,m} \\
\rho_L(Y)\psi_{nm} &= Y \cdot X^nY^m = (X-1)^nY^m \\
&= \sum_{\nu=0}^{n} \binom{n}{\nu} (-1)^{n-\nu}\psi_{\nu m} \\
\rho_R(X)\psi_{nm} &= \psi_{n+1,m} + m\psi_{nm} \\
\rho_R(Y)\psi_{nm} &= \psi_{n,m+1}.
\end{aligned}$$

We will show one example, how quotient representations can be calculated. Let I be the left ideal (i. e. ρ_L-invariant subspace) generated by $X-\alpha$, i. e. $I = \{x \cdot (X-\alpha); x \in U(\mathfrak{g})\}$. Then $\{\tilde{\psi}_m = [Y^m]\}$ is a basis of $U(\mathfrak{g})/I$, and we get

$$\begin{aligned}
\tilde{\rho}_L(X)\tilde{\psi}_m &= [X \cdot Y^m] = [Y^m(X+m)] = [Y^m(\alpha+m) + Y^m(X-\alpha)] \\
&= (\alpha+m)\tilde{\psi}_m \\
\tilde{\rho}_L(Y)\tilde{\psi}_m &= \tilde{\psi}_{m+1}.
\end{aligned}$$

We will now show how the elements of $\mathfrak{g}$ can be realized as vector fields. Fix a basis $\{X_1, \ldots, X_n\}$ of $\mathfrak{g}$. Then

$$g(a_1, \ldots, a_n; X_1, \ldots, X_n) = \exp(a_1X_1)\cdots\exp(a_nX_n)$$

for $a = (a_1, \ldots, a_n) \in U \subseteq \mathbb{R}^n$ in an appropriate neighbourhood of $0 \in \mathbb{R}^n$ defines a coordinate system for a neighbourhood of the unit element $e \in G$. These coordinates are called *coordinates of the second kind.* Coordinates of the first kind are introduced in Section 2.4.

Note that $g(a_1, \ldots, a_n; X_1, \ldots, X_n)$ can also be interpreted as a *formal pairing* or *dual pairing* $\langle a, X\rangle$ between the algebra $\mathbb{K}[a_1, \ldots, a_n]$ of polynomials in the commuting variables $a_1, \ldots, a_n$, and the universal enveloping algebra $U(\mathfrak{g})$. Choose $\{c_m(a) = \frac{a^m}{m!} = \frac{a_1^{m_1}\cdots a_n^{m_n}}{m_1!\cdots m_n!}; m = (m_1, \ldots, m_n) \in \mathbb{Z}_+^n\}$ and $\{\psi_m(X) = X^m = X_1^{m_1}\cdots X_n^{m_n}; m \in \mathbb{Z}_+^n\}$ as bases of $\mathbb{K}[a_1, \ldots, a_n]$ and $U(\mathfrak{g})$, respectively. Then

$$\langle a, X\rangle = \sum_{m\in\mathbb{N}^n} c_m(a)\psi_m(X) = g(a_1, \ldots, a_n; X_1, \ldots, X_n).$$

This interpretation will also be useful when we later generalise our procedure to quantum groups.

Using the relation $e^XYe^{-X} = e^{\mathrm{ad}(X)}Y$, we can determine the matrices $\pi^{\dagger}_{ij}(a)$ and $\pi^{*}_{ij}(a)$ such that

$$X_i^{\dagger}g(a;X) = X_ig(a;X)$$

$$
\begin{aligned}
&= \sum_{j=1}^{n} \pi^{\dagger}_{ij}(a) \frac{\partial}{\partial a_j} g(a_1, \ldots, a_n; X_1, \ldots, X_n) \\
X_i^* g(a;X) &= g(a;X)X_i \\
&= \sum_{j=1}^{n} \pi^{*}_{ij}(a) \frac{\partial}{\partial a_j} g(a_1, \ldots, a_n; X_1, \ldots, X_n).
\end{aligned}
$$

The maps $\rho^\dagger : U(\mathfrak{g}) \to \mathrm{Hom}(\mathbb{K}[a], \mathbb{K}[a])$ and $\rho^* : U(\mathfrak{g}) \to \mathrm{Hom}(\mathbb{K}[a], \mathbb{K}[a])$ defined above are called the left and right *dual representation*, respectively. From the viewpoint of the dual pairing they are introduced by requiring

$$
\begin{aligned}
\langle \rho^\dagger(Y)X; a\rangle &= \langle a, \rho_L(Y)X\rangle = \langle a, YX\rangle, \\
\langle \rho^*(Y)X; a\rangle &= \langle a, rho_R(Y)X\rangle = \langle a, XY\rangle,
\end{aligned}
$$

i. e. they are the duals of the left and right multiplication. They satisfy

$$
\begin{aligned}
\rho^\dagger(X) \circ \rho^\dagger(Y) &= \rho^\dagger(YX), \\
\rho^*(X) \circ \rho^*(Y) &= \rho^*(XY),
\end{aligned}
$$

for $X, Y \in U(\mathfrak{g})$, i. e. ρ^* is a representation of $U(\mathfrak{g})$, and $\rho^\dagger$ is a representation of $U(\mathfrak{g})^{\mathrm{op}}$. The *dual adjoint representation* is given by

$$
X' = X^\dagger - X^*. \tag{2.2}
$$

We can also introduce dual representations for quotient representations, see e. g. the following example on the affine group.

2.3.1 *Example: the affine group*

We define the group elements with respect to coordinates of the second kind by

$$
g(a, b; X, Y) = \exp(aX)\,\exp(bY).
$$

If we interpret the group element $g(a, b; X, Y)$ as a formal pairing, this corresponds to choosing $\{c_{nm} = \frac{a^n b^m}{n!m!}\}$ and $\{\psi_{nm} = X^n Y^m\}$ as dual bases of the universal enveloping algebra $U(aff(1))$ and the algebra A of polynomials in the two commuting variables a and b. We obtain vector field realisations corresponding to the left and right multiplication using the formulas

$$
\begin{aligned}
e^{\mathrm{ad}aX} Y &= Y + a[X,Y] + \frac{a^2}{2}[X,[X,Y]] + \cdots \\
&= e^a Y \\
e^{\mathrm{ad}bY} X &= X + b[Y,X] + \frac{b^2}{2}[Y,[Y,X]] + \cdots \\
&= X + bY.
\end{aligned}
$$

Then

$$\begin{aligned} X^{\dagger}g &= Xe^{aX}a^{bY} = \partial_a g, \\ X^{*}g &= e^{aX}e^{bY}X = e^{aX}e^{bY}Ye^{-bY}e^{bY} \\ &= e^{aX}e^{\mathrm{ad}(bY)}Xe^{bY} = e^{aX}(X - bY)e^{bY} \\ &= (\partial_a - b\partial_b)g, \end{aligned}$$

and so on. For the adjoint representation $X'g = Xg - gX = (X^{\dagger} - X^{*})g = b\partial_b g$.

Proposition 2.3.1 *For the action on the left*

$$\begin{aligned} X^{\dagger} &= \partial_a, \\ Y^{\dagger} &= e^{-a}\partial_b, \end{aligned}$$

for the action on the right

$$\begin{aligned} X^{*} &= \partial_a - b\partial_b. \\ Y^{*} &= \partial_b, \end{aligned}$$

and for the adjoint representation

$$\begin{aligned} X' &= b\partial_b, \\ Y' &= (a^{-a} - 1)\partial_b. \end{aligned}$$

We also give a list of realisations that can be obtained from quotient representations:

$$\begin{aligned} X\Omega \equiv \alpha\Omega : &\qquad X^{\dagger} = b\partial_b + \alpha, \\ &\qquad Y^{\dagger} = \partial_b, \\ Y\Omega \equiv \beta\Omega : &\qquad X^{\dagger} = \partial_a, \\ &\qquad Y^{\dagger} = \beta e^{-a}, \\ \Omega X \equiv \alpha\Omega : &\qquad X^{*} = \alpha - b\partial_b, \\ &\qquad Y^{*} = \partial_b, \\ \Omega Y \equiv \beta\Omega : &\qquad X^{*} = \partial_a, \\ &\qquad Y^{*} = \beta e^{a}. \end{aligned}$$

2.3.2 *Example: nilpotent groups*

The defining commutation relations imply

$$\begin{aligned} e^{\mathrm{ad}\, tX_0}X_j &= X_j + tX_{j+1} + \frac{t^2}{2}X_{j+2} + \cdots + \frac{t^{N-j}}{(N-j)!}X_N \\ &= \sum_{i=0}^{N-j}\frac{t^i}{i!}X_{i+j}, \qquad (1 \le j \le N) \\ e^{\mathrm{ad}\, tX_j}X_0 &= X_0 - tX_{j+1}. \end{aligned}$$

With these formulas we can derive the dual representations, if we set

$$g = g(a;X) = g(a_0,\dots,a_N;X_0,\dots,X_N) = e^{a_N X_N} e^{a_{N-1}X_{N-1}} \cdots e^{a_0 X_0}.$$

Proposition 2.3.2 (Dual representations) *The left dual representation is given by*

$$\begin{aligned} X_0^\dagger &= \partial_0 + \sum_{i=1}^{N-1} a_i \partial_{i+1}, \\ X_i^\dagger &= \partial_i, \qquad (1 \le i \le N). \end{aligned}$$

The right dual representation is given by

$$\begin{aligned} X_0^* &= \partial_0, \\ X_i^* &= \sum_{j=0}^{N-i} \frac{a_0^j}{j!} \partial_{i+j}, \qquad (1 \le i \le N). \end{aligned}$$

One easily verifies that

$$\begin{aligned} \hat{X}_0 &= \partial_x, \\ \hat{X}_1 &= \alpha \frac{x^{(N-1)}}{(N-1)!}, \\ &\vdots \\ \hat{X}_i &= \alpha \frac{x^{N-i}}{(N-i)!}, \\ &\vdots \\ \hat{X}_N &= \alpha, \end{aligned}$$

is also a representation.

2.3.3 *Example: the finite-difference group*

We define the *finite-difference group* via the finite-difference algebra (see paragraph 2.2.3) by

$$g(a,b,c) = g(a,b,c;T,\Delta,X) = \exp(aT)\exp(b\Delta)\exp(cX),$$

and calculate the right and left dual representations.

Proposition 2.3.3 *The left dual representation is given by*

$$\begin{aligned} T^\dagger &= \partial_a, \\ \Delta^\dagger &= \partial_b, \\ X^\dagger &= -(a+b)\partial_a + \partial_c. \end{aligned}$$

The right dual representations are given by

$$\begin{aligned} T^* &= e^{-c}\partial_a, \\ \Delta^* &= \left(e^{-c} - 1\right)\partial_a + \partial_b, \\ X^* &= \partial_c. \end{aligned}$$

2.3.4 *Example: so(3) and sl(2)*

We define the group elements by

$$\begin{aligned} g(a,b,c;X_1,X_2,X_3) &= \sum_{n,m,r=0}^{\infty} c_{nmr}(a,b,c)\psi_{nmr}(X_1,X_2,X_3) \\ &= \sum_{n,m,r=0}^{\infty} \frac{a^n b^m c^r X_1^n X_2^m X_3^r}{n!m!r!} = e^{aX_1}e^{bX_2}e^{cX_3}. \end{aligned}$$

From the defining commutation relations of X_1, X_2, X_3 one deduces

$$\begin{aligned} e^{aX_1}X_2e^{-aX_1} &= e^{\mathrm{ad}(aX_1)}(X_2) \\ &= X_2 + aX_3 - \frac{a^2}{2}X_2 - \frac{a^3}{3!}X_3 + - \ldots \\ &= X_2\cos(a) + X_3\sin(a), \end{aligned}$$

and therefore

$$\begin{aligned} e^{aX_1}X_2 &= (X_2\cos(a) + X_3\sin(a))\, e^{aX_1}, \\ X_2e^{aX_1} &= e^{aX_1}X_2(\cos a)^{-1} - X_3\tan(a)e^{aX_1} \end{aligned}$$

(+ cycl.). With these relations the dual left and right realisations can be calculated, e. g. ($g = g(a,b,c;X_1,X_2,X_3)$)

$$\begin{aligned} X_1^\dagger g &= X_1e^{aX_1}e^{bX_2}e^{cX_3} \\ &= \partial_a e^{aX_1}e^{bX_2}e^{cX_3}, \\ X_1^* g &= e^{aX_1}e^{bX_2}e^{cX_3}X_1 \\ &= e^{aX_1}e^{bX_2}(X_1\cos(c) + X_2\sin(c))e^{cX_3} \\ &= e^{aX_1}\left(\frac{\cos(c)}{\cos(b)}X_1e^{bX_2} - \cos(c)\tan(b)e^{bX_2}X_3\right)e^{cX_3} \\ &\quad + e^{aX_1}e^{bX_2}\sin(c)X_2e^{cX_3} \\ &= \left(\frac{\cos(c)}{\cos(b)}\partial_a + \sin(c)\partial_b - \cos(c)\tan(b)\partial_c\right)e^{aX_1}e^{bX_2}e^{cX_3}. \end{aligned}$$

The realisations of X_2 and X_3 are calculated analogously.

Proposition 2.3.4 (Realisations of $so(3)$)

1. *The dual left realisation is given by*

$$\begin{aligned} X_1^\dagger &= \partial_a, \\ X_2^\dagger &= \sin(a)\tan(b)\partial_a + \cos(a)\partial_b - \frac{\sin(a)}{\cos(b)}\partial_c, \\ X_3^\dagger &= -\cos(a)\tan(b)\partial_a + \sin(a)\partial_b + \frac{\cos(a)}{\cos(b)}\partial_c. \end{aligned}$$

2. *The dual right realisation is given by*

$$
\begin{aligned}
X_1^* &= \frac{\cos(c)}{\cos(b)}\partial_a + \sin(c)\partial_b - \cos(c)\tan(b)\partial_c, \\
X_2^* &= -\frac{\sin(c)}{\cos(b)}\partial_a + \cos(c)\partial_b + \sin(c)\tan(b)\partial_c, \\
X_3^* &= \partial_c
\end{aligned}
$$

Notice that

$$\hat{X}_1 = \cos(x)\partial_x, \qquad \hat{X}_2 = \sin(x)\partial_x, \qquad \hat{X}_3 = \partial_x,$$

is also a realisation of $so(3)$.

For $sl(2)$ we set

$$g(a, b, c; X_+, H, X_-) = e^{aX_+}e^{bH}e^{cX_-}.$$

Proposition 2.3.5 (Realisation of $sl(2)$)

1. *The dual left realisation is given by*

$$
\begin{aligned}
H^\dagger &= 2a\partial_a + \partial_b, \\
X_+^\dagger &= \partial_a, \\
X_-^\dagger &= -a^2\partial_a - a\partial_b + e^b\partial_c.
\end{aligned}
$$

2. *The dual right realisation is given by*

$$
\begin{aligned}
H^* &= \partial_b + 2c\partial_c, \\
X_+^* &= e^{-b}\partial_a - c\partial_b - c^2\partial_c, \\
X_-^* &= \partial_c.
\end{aligned}
$$

Notice that

$$\hat{X}_+ = z^2\partial_z, \qquad \hat{H} = 2z\partial_z, \qquad \hat{X}_- = \partial_z,$$

is also a realisation of $sl(2)$.

2.4 THE SPLITTING LEMMA

We have already introduced one canonical coordinate system for Lie groups, the coordinates of the second kind. As the name suggests, there are also *coordinates of the first kind.* They are defined by

$$(\alpha_1, \ldots, \alpha_n) \mapsto g_1(\alpha; X) = \exp(\alpha_1 X_2 + \cdots + \alpha_n X_n).$$

These coordinates can be defined for any Lie group for some neighbourhood of the unit element. The following lemma shows how the transformation between these two coordinate systems can be calculated. It is called *splitting lemma*, because it allows to "split" the one-parameter subgroup $\exp(tX)$ generated by $X = \sum_{i=1}^{n} \alpha_i X_i$ into a product $\exp(a_1 X_1) \cdots \exp(a_n X_n)$.

Lemma 2.4.1 (The Splitting Lemma) *Let* $X = \sum_{i=1}^{n} \alpha_i X_i$, *and let*

$$\exp(tX) = \exp(a_1(t)X_1) \cdots \exp(a_n(t)X_n) \tag{2.3}$$

be the one-parameter subgroup generated by X. *Let* $\tilde{X}_i$ *denote either the left or the right dual representation, with the matrix* $\tilde{\pi}_{ij}(a)$ *defined by*

$$\tilde{X}_i = \sum_{j=1}^{n} \tilde{\pi}_{ij}(a) \frac{\partial}{\partial a_j}.$$

Then the coordinates $a_1(t), \ldots, a_n(t)$ *are determined by the differential equations*

$$\dot{a}_j = \sum_{i=1}^{n} \alpha_i \tilde{\pi}_{ij}(a), \qquad j = 1, \ldots, n$$

with the initial conditions $a_1(0) = \cdots = a_n(0) = 0$ *for* t *sufficiently small.*

Proof: (Cf. [FKS93]) Differentiating (2.3) with respect to t we get

$$\begin{aligned} X \exp(tX) &= \sum_{j=1}^{n} \left(\prod_{i=1}^{j-1} e^{a_i(t)X_i} \right) \dot{a}_j(t) X_j e^{a_j(t)X_j} \left(\prod_{i=j+1}^{n} e^{a_i(t)X_i} \right) \\ &= \sum_{j=1}^{n} \dot{a}_j(t) \frac{\partial}{\partial a_j} g(a(t); X) \end{aligned} \tag{2.4}$$

For the left-hand side

$$\begin{aligned} X \exp(tX) &= \sum_{i=1}^{n} \alpha_i X_i g(a; X) \\ &= \sum_{i=1}^{n} \alpha_i X_i^{\dagger} g(a; X) \\ &= \sum_{i,j=1}^{n} \alpha_i \pi_{ij}^{\dagger}(a) \frac{\partial}{\partial a_j} g(a; X) \end{aligned} \tag{2.5}$$

or, similarly,

$$\begin{aligned} X \exp(tX) &= \exp(tX) X \\ &= \sum_{i,j=1}^{n} \alpha_i \pi_{ij}^{*}(a) \frac{\partial}{\partial a_j} g(a; X). \end{aligned} \tag{2.6}$$

Comparing (2.4) with (2.5) (or (2.6)) yields the result. ∎

2.4.1 *Example: the affine group*

Proposition 2.4.2

$$\begin{aligned} e^{\alpha X+\beta Y} &= e^{\alpha X} e^{-\frac{\beta}{\alpha}(e^{-\alpha}-1)Y}, \\ e^{aX}e^{bY} &= e^{aX-\frac{ab}{e^{-a}-1}Y}. \end{aligned}$$

Proof: We solve one of the two equivalent systems of ODEs

$$\begin{aligned} \dot{a}(t) &= \alpha, \\ \dot{b}(t) &= \beta e^{-a(t)}, \end{aligned} \qquad \begin{aligned} \dot{a}(t) &= \alpha, \\ \dot{b}(t) &= -\alpha b(t)+\beta. \end{aligned}$$

with the initial conditions $a(0) = b(0) = 0$, and then set $t = 1$ to get the first formula. The second formula follows immediately. ∎

2.4.2 *Example: nilpotent groups*

Proposition 2.4.3 (The Splitting Lemma) *The relation*

$$e^{t\sum_{i=0}^{N}\alpha_i X_i} = e^{a_N(t)X_N}\cdots e^{a_1(t)X_1}e^{a_0(t)X_0}$$

is satisfied by the functions

$$\begin{aligned} a_0(t) &= \alpha_0 t, \\ a_1(t) &= \alpha_1 t, \\ a_2(t) &= \frac{\alpha_0\alpha_1}{2}t^2+\alpha_2 t, \\ a_3(t) &= \frac{\alpha_0^2\alpha_1}{6}t^3+\frac{\alpha_0\alpha_2}{2}t^2+\alpha_3 t, \\ &\vdots \\ a_i(t) &= \sum_{\gamma=0}^{i-1}\frac{\alpha_0^\gamma\,\alpha_{i-\gamma}}{(\gamma+1)!}t^{\gamma+1}, \qquad (1\le i\le N) \end{aligned}$$

Proof: We have to solve the system of differential equations

$$\begin{aligned} \dot{a}_0 &= \alpha_0, \\ \dot{a}_1 &= \alpha_1, \\ \dot{a}_2 &= \alpha_0 a_1+\alpha_2, \\ &\vdots \\ \dot{a}_i &= \alpha_0 a_{i-1}+\alpha_i, \end{aligned}$$

with the initial conditions

$$a_0(0) = a_1(0) = \cdots = a_N(0) = 0.$$

∎

2.4.3 *Example: the finite-difference group*

Proposition 2.4.4 *We have the following splitting formula*

$$\exp[t(\alpha T + \beta\Delta + \gamma X)] = \exp(a(t)T)\exp(b(t)\Delta)\exp(c(t)X),$$

where the functions $a(t)$, $b(t)$, and $c(t)$ are given by

$$\begin{aligned} a(t) &= -\frac{\alpha+\beta}{\gamma}\exp(-\gamma t) - \beta t + \frac{\alpha+\beta}{\gamma}, \\ b(t) &= \beta t, \\ c(t) &= \gamma t. \end{aligned}$$

Proof: Apply lemma 2.4.1. ∎

Remark: For $\gamma = 0$ we get $a(t) = \alpha t$.

2.4.4 *Example: $so(3)$ and $sl(2)$*

Proposition 2.4.5 (Splitting Lemma) *The functions $a(t), b(t), c(t)$, such that*

$$e^{t(\alpha X_1+\beta X_2+\gamma X_3)} = e^{a(t)X_1}e^{b(t)X_2}e^{c(t)X_3}$$

holds, are uniquely determined by the differential equations

$$\begin{aligned} \dot a &= \alpha + \beta\sin(a)\tan(b) - \gamma\cos(a)\tan(b), \\ \dot b &= \beta\cos(a) + \gamma\sin(a), \\ \dot c &= -\beta\frac{\sin(a)}{\cos(b)} + \gamma\frac{\cos(a)}{\cos(b)}, \end{aligned}$$

(or equivalently

$$\begin{aligned} \dot a &= \alpha\frac{\cos(c)}{\cos(b)} - \beta\frac{\sin(c)}{\cos(b)}, \\ \dot b &= \alpha\sin(c) + \beta\cos(c), \\ \dot c &= -\alpha\cos(c)\tan(b) + \beta\sin(c)\tan(b) + \gamma,) \end{aligned}$$

and the initial conditions

$$a(0) = b(0) = c(0) = 0.$$

Proof: We apply the splitting lemma 2.4.1. ∎

Proposition 2.4.6 (Splitting Lemma for $sl(2)$; see [FS93] Chapter 1, 4.3.10)

$$e^{s(X_+ + aH + bX_-)} = e^{v(s)X_+}e^{h(s)H}e^{bv(s)X_-},$$

where

$$\begin{aligned} \delta^2 &= a^2 - b, \\ h(s) &= \log\left(\frac{\delta \operatorname{sech} \delta s}{\delta - a \operatorname{sech} \delta s}\right), \\ v(s) &= \frac{\tanh \delta s}{\delta - a \tanh \delta s}. \end{aligned}$$

We will now derive the splitting lemma for $so(3)$, using the adjoint representation.

Proposition 2.4.7 *The adjoint representation of $so(3)$ with respect to the basis $\{X_1, X_2, X_3\}$ is given by*

$$\begin{aligned} \operatorname{ad}(X_1) &= \begin{pmatrix} 0 & 0 & 0 \\ 0 & 0 & -1 \\ 0 & 1 & 0 \end{pmatrix}, & e^{\operatorname{ad} tX_1} &= \begin{pmatrix} 1 & 0 & 0 \\ 0 & \cos t & -\sin t \\ 0 & \sin t & \cos t \end{pmatrix}, \\ \operatorname{ad}(X_2) &= \begin{pmatrix} 0 & 0 & 1 \\ 0 & 0 & 0 \\ -1 & 0 & 0 \end{pmatrix}, & e^{\operatorname{ad} tX_2} &= \begin{pmatrix} \cos t & 0 & \sin t \\ 0 & 1 & 0 \\ -\sin t & 0 & \cos t \end{pmatrix}, \\ \operatorname{ad}(X_3) &= \begin{pmatrix} 0 & -1 & 0 \\ 1 & 0 & 0 \\ 0 & 0 & 0 \end{pmatrix}, & e^{\operatorname{ad} tX_3} &= \begin{pmatrix} \cos t & -\sin t & 0 \\ \sin t & \cos t & 0 \\ 0 & 0 & 1 \end{pmatrix}. \end{aligned} \tag{2.7}$$

Now we can calculate $e^{\operatorname{ad} aX_1} e^{\operatorname{ad} bX_2} e^{\operatorname{ad} cX_3}$. On the other hand we find for $Y = \alpha X_1 + \beta X_2 + \gamma X_3$

$$\begin{aligned} \operatorname{ad}(Y) &= \begin{pmatrix} 0 & -\gamma & \beta \\ \gamma & 0 & -\alpha \\ -\beta & \alpha & 0 \end{pmatrix}, \\ \operatorname{ad}(Y)^2 &= \begin{pmatrix} \beta^2+\gamma^2 & -\alpha\beta & -\alpha\gamma \\ -\alpha\beta & \alpha^2+\gamma^2 & -\beta\gamma \\ -\alpha\gamma & -\beta\gamma & \beta^2+\gamma^2 \end{pmatrix}, \\ \operatorname{ad}(Y)^3 &= -(\alpha^2+\beta^2+\gamma^2)\begin{pmatrix} 0 & -\gamma & \beta \\ \gamma & 0 & -\alpha \\ -\beta & \alpha & 0 \end{pmatrix} \\ &= -(\alpha^2+\beta^2+\gamma^2)\operatorname{ad}(Y), \\ &\;\vdots \\ \operatorname{ad}(Y)^{2n+1} &= \left\{-(\alpha^2+\beta^2+\gamma^2)\right\}^n \operatorname{ad}(Y), \\ \operatorname{ad}(Y)^{2n+2} &= \left\{-(\alpha^2+\beta^2+\gamma^2)\right\}^n \operatorname{ad}(Y)^2. \end{aligned}$$

With these equations we can compute $e^{\operatorname{ad}(Y)}$.

Proposition 2.4.8 *Let $Y = \alpha X_1 + \beta X_2 + \gamma X_3$ and $t^2 = \alpha^2 + \beta^2 + \gamma^2$. Then*

$$e^{\operatorname{ad}(Y)} = \mathbb{1} + \frac{\sin t}{t}\begin{pmatrix} 0 & -\gamma & \beta \\ \gamma & 0 & -\alpha \\ -\beta & \alpha & 0 \end{pmatrix} + \frac{\cos t - 1}{t^2}\begin{pmatrix} \beta^2+\gamma^2 & -\alpha\beta & -\alpha\gamma \\ -\alpha\beta & \alpha^2+\gamma^2 & -\beta\gamma \\ -\alpha\gamma & -\beta\gamma & \beta^2+\gamma^2 \end{pmatrix}.$$

We can now give a different form of the splitting lemma for $so(3)$.

Proposition 2.4.9 (Splitting Lemma) *Let α, β, γ be chosen such that $\alpha^2 + \beta^2 + \gamma^2 = 1$ and let $Y = \alpha X_1 + \beta X_2 + \gamma X_3$. Then (for t sufficiently small) the functions $a(t), b(t), c(t)$ in*

$$e^{tY} = e^{a(t)X_1} e^{b(t)X_2} e^{c(t)X_3}$$

can be determined by

$$\begin{aligned} \sin b &= \beta \sin t - \alpha\gamma(\cos t - 1), \\ \sin a &= \frac{1}{\cos b} \{\alpha \sin t + \beta\gamma(\cos t - 1)\}, \\ \sin c &= \frac{1}{\cos b} \{\gamma \sin t + \alpha\beta(\cos t - 1)\}, \end{aligned}$$

where a, b, c have to be chosen as continuous functions with $a(0) = b(0) = c(0) = 0$.

Proof: Compare $e^{\mathrm{ad}(Y)}$ and $e^{\mathrm{ad}\, aX_1} e^{\mathrm{ad}\, bX_2} e^{\mathrm{ad}\, cX_3}$. ∎

2.5 THE COMPOSITION LAW

We will now look at a composition law for formal pairings. Since the formal pairings can be interpreted as group elements this will be the group multiplication in coordinates of the second kind. If $g(a; X)$ and $g(b; X)$ are two group elements, then their product $g(a; X)g(b; X)$ is also a group element. We will use the notation

$$g(a \odot b; X) = g(a; X)g(b; X)$$

for the product. The map $\odot$ is called composition law and is in general uniquely defined only for a sufficiently small neighbourhood of 0. We give the explicit form of the composition law for several groups.

2.5.1 *Example: the affine group*

Proposition 2.5.1 1. *In terms of coordinates of the second kind the composition law is*

$$g(A, B; X, Y)\, g(a, b; X, Y) = g(A + a, e^{-a}B + b; X, Y).$$

2. *In terms of coordinates of the first kind the composition law is*

$$e^{AX+BY}\, e^{aX+bY} = \exp\left((A + a)X + \frac{aB(a^{-A} - 1) - Ab(e^{a} - 1)}{aA(e^{-(A+a)} - 1)} Y \right).$$

Proof: *1.* From $Ye^{aX} = e^{aX} e^{\mathrm{ad}(-aX)}(Y) = e^{aX} e^{-a} Y$ follows $Y^n e^{aX} = e^{aX}\, (e^{a}Y)^n$ and thus $e^{BY} e^{aX} = e^{aX} e^{e^{-a}BY}$. Therefore

$$e^{AX} e^{BY} e^{aX} e^{bY} = e^{(A+a)X} e^{(e^{-a}B+b)Y}.$$

2. follows from *1.* with proposition 2.4.2. ∎

2.5.2 *Example: nilpotent groups*

We will need the formula

$$\begin{aligned} e^{a_0 X_0} e^{A_i X_i} &= e^{a_0 X_0} e^{A_i X_i} e^{-a_0 X_0} e^{a_0 X_0} \\ &= \exp\left(A_i e^{a_0 \mathrm{ad} X_0} X_i\right) e^{a_0 X_0} \\ &= \exp\left(A_i \sum_{j=0}^{N-i} \frac{a_0^j}{j!} X_{i+j}\right) e^{a_0 X_0} \\ &= e^{A_i \frac{a_0^{N-i}}{(N-i)!} X_N} \cdots e^{A_i \frac{a_0^j}{j!} X_{i+j}} \cdots e^{A_i X_i} e^{a_0 X_0}, \end{aligned} \tag{2.8}$$

($1 \leq i \leq N$). Applying this formula N times and reordering the resulting terms then gives the following proposition.

Proposition 2.5.2 *We have* $g(a)g(A) =$

$$g\left(a_0 + A_0, a_1 + A_1, \ldots, a_i + \sum_{j=0}^{i-1} \frac{a_0^j}{j!} A_{i-j}, \ldots, a_N + \sum_{j=0}^{N-1} \frac{a_0^j}{j!} A_{N-j}\right)$$

where $a = (a_0, \ldots, a_N)$, $A = (A_0, \ldots, A_N)$.

2.5.3 *Example: the finite-difference group*

Proposition 2.5.3 *The composition law with respect to coordinates of the second kind is given by*

$$g(a,b,c)g(A,B,C) = g(a + Ae^{-c} + B(e^{-c} - 1), b + B, c + C).$$

2.5.4 *Example: $so(3)$ and $sl(2)$*

Proposition 2.5.4 (Composition Law for $sl(2)$; see [FS93] Chapter 1, 3.4.1) *For* $g(a,b,c) = e^{aX_+} b^H e^{cX_-}$, *we have the composition law*

$$g(a,b,c)g(A,B,C) = g\left(a + \frac{Ab^2}{1 - Ac}, \frac{bB}{1 - Ac}, C + \frac{cB^2}{1 - Ac}\right)$$

One can calculate the group law of $so(3)$ from this group law and the splitting lemmas of the previous section. Or we can again use the adjoint representation. The second method gives the following expressions for the group law.

Proposition 2.5.5 (Composition Law for $so(3)$) *If we denote the group elements of $so(3)$ by*

$$g(a,b,c) = e^{aX_1} e^{bX_2} e^{cX_3},$$

then the composition law can be written as

$$g(a,b,c)g(a',b',c') = g(A,B,C),$$

where

$$
\begin{aligned}
\sin B &= \cos b \cos c \sin b' + \cos b \sin c \sin a' \cos b' + \sin b \cos a' \cos b', \\
\sin A &= \frac{1}{\cos B}\Big\{ -(\sin a \sin b \cos c + \cos a \sin c) \sin b' \\
&\quad +(\cos a \cos c - \sin a \sin b \sin c) \sin a' \cos b' \\
&\quad + \sin a \cos b \cos a' \cos b' \Big\}, \\
\sin C &= \frac{1}{\cos B}\Big\{ -(\sin a \sin c - \cos a \sin b \cos c) \sin b' \\
&\quad +(\sin a \cos c + \cos a \sin b \sin c) \sin a' \cos b' \\
&\quad + \cos a \cos b \cos a' \cos b' \Big\}
\end{aligned}
$$

Proof: Calculate $e^{\mathrm{ad}\, aX_1} e^{\mathrm{ad}\, bX_2} e^{\mathrm{ad}\, cX_3} e^{\mathrm{ad}\, a'X_1} e^{\mathrm{ad}\, b'X_2} e^{\mathrm{ad}\, c'X_3}$ and $e^{\mathrm{ad}\, AX_1} e^{\mathrm{ad}\, BX_2} e^{\mathrm{ad}\, CX_3}$ and compare the matrix elements. ∎

Remark: The preceding proposition is only good for sufficiently small values of a, b, c and a', b', c'. If, for example, we get $B = \frac{\pi}{2}$, then A and C are not uniquely determined.

2.6 MATRIX ELEMENTS

Exponentiating the representation ρ_L of $U(\mathfrak{g})$ on $U(\mathfrak{g})$ we get a representation of G on $U(\mathfrak{g})$. To simplify notation we will identify $\rho_L(X)$ and X. We define the *matrix elements* $M_m^M(a)$ of the representation G on $U(g)$ by

$$g(a; X)\psi_m(X) = \sum_{m+M \geq 0} M_m^M(a)\psi_{m+M}(X),$$

where $\{\psi_m(X); m \in \mathbb{Z}_+^n\}$ is the basis of $U(\mathfrak{g})$ introduced on page 10. These matrix elements are generally well-known special functions, such as hypergeometric polynomials, etc. For a complete discussion of how their special function properties can be derived with many interesting examples see [FS90]. We present here only a strongly abbreviated version. We will sometimes switch to the subscript notation defined by $M_{kl}(a) = M_k^{l-k}(a)$, i. e.

$$g(a; X)\psi_k(X) = \sum_{l \in \mathbb{Z}_+^n} M_{kl}(a)\psi_l(X).$$

The following proposition gives a useful formula for calculating the matrix elements.

Proposition 2.6.1 (Principal formula)

$$M_m^M(a) = \psi_m(X^*)c_{m+M}(a)$$

Proof: We can write the product of two group elements $g(a;X)$ and $g(b;X)$ as

$$\begin{aligned} g(a;X)g(b;X) &= g(a,X)\sum_m c_m(b)\psi_m(X) \\ &= \sum_m c_m(b)g(a;X)\psi_m(X) \\ &= \sum_{m,M} c_m(b)M_m^M(a)\psi_{m+M}(X), \end{aligned}$$

since the a's and b's commute. On the other hand, we can also write

$$\begin{aligned} g(a;X)g(b;X) &= g(b,X^*)g(a,X) \\ &= \sum_{k,l} c_k(b)\psi_k(X^*)c_l(a)\psi_l(X). \end{aligned}$$

Comparing these two expressions leads to the desired formula. ∎
Many interesting relations for the matrix elements can now be deduced from the group law, and the relations of the operators X^*.

Addition theorems: Writing the group law as

$$g(a;X)g(b;X) = \sum c_m(b)M_m^M(a)\psi_{m+M}(X)$$

and as

$$g(a\odot b;X) = \sum c_m(a\odot b)\psi_m(X),$$

we can read off the transformation formula

$$c_m(a\odot b) = \sum c_k(b)M_k^{m-k}(a).$$

Similarly,

$$g(a;X)g(b;X)\psi_m(X) = g(a\odot b;X)\psi_m(X)$$

yields the addition theorem

$$M_m^M(a\odot b) = \sum_{M'+M''=M} M_{m+M'}^{M''}(a)M_m^{M'}(b).$$

Differential recurrence relations: We use the right dual representation.

$$\begin{aligned} X_i^*M_m^M(a) &= X_i^*\psi_m(X^*)c_{m+M}(a) \\ &= \sum m_m^k(X_i)\psi_{m+k}(X^*)c_{m+M}(a) \\ &= \sum m_m^k(X_i)M_{m+k}^{M-k}(a), \end{aligned}$$

where $m_m^k(X_i)$ are the matrix elements of $\rho_L(X_i)$, cf. equation (2.1).

2.6.1 *Example: the affine group*

The matrix elements are given by the principal formula (proposition 2.6.1)

$$M_{nm}^{NM}(a,b) = \psi_{nm}(X^*,Y^*)c_{n+N,m+M}(a,b).$$

This gives us

$$\begin{aligned} M_{nm}^{NM} &= (\partial_a - b\partial_b)^n(\partial_b)^m \frac{a^{N+n}b^{M+m}}{(N+n)!(M+m)!} \\ &= \sum_{\nu=0}^{n} \binom{n}{\nu} \partial_a^{n-\nu}(-b\partial_b)^\nu \frac{a^{N+n}b^M}{(N+n)!M!} \\ &= \sum_{\nu=0}^{n} \binom{n}{\nu} \frac{(-M)^\nu a^{N+\nu}b^M}{(N+\nu)!M!} \\ &= \frac{a^N b^M}{N!M!}\, {}_1F_1(-n;N+1;aM), \end{aligned}$$

where ${}_1F_1(a;b;x) = \sum_{\nu=0}^{\infty} \frac{(a)_\nu x^\nu}{(b)_\nu \nu!}$ and $(a)_\nu = a(a+1)\cdots(a+\nu-1)$.

Proposition 2.6.2 *The matrix elements are given by*

$$M_{nm}^{NM}(a,b) = \frac{a^N b^M}{N!M!}\, {}_1F_1(-n;N+1;aM).$$

We get the following differential recurrence relations:

$$(\partial_a - b\partial_b)M_{nm}^{NM}(a,b) = M_{n+1,m}^{N-1,M}(a,b)$$

or, after some simplification,

$$\frac{a}{N}\partial_a\, {}_1F_1(-n;N+1;aM) = \left(\frac{aM}{N} - 1\right) {}_1F_1(-n;N+1;aM) + {}_1F_1(-n-1;N;aM),$$

and

$$\partial_b M_{nm}^{NM}(a,b) = \sum_{\nu=0}^{n}(-1)^\nu \binom{n}{\nu} M_{n-\nu,m+1}^{N+\nu,M-1}(a,b)$$

or

$${}_1F_1(-n;N+1;aM) = \sum_{\nu=0}^{n} \frac{(-n)_\nu a^\nu}{(N+1)_\nu \nu!}\, {}_1F_1(-n+\nu;N+\nu+1;a(M-1)).$$

2.7 STOCHASTIC PROCESSES ON LIE GROUPS

In this section we study stochastic processes on Lie groups. Subsection 2.7.1 shows how, for a given continuous stochastic process X_t on a Euclidean space, an analogue $\tilde{X}_t$ on a Lie group G can be constructed. To do so we first embed the given process linearly in the Lie algebra $\mathfrak{g}$ of G. This embedding depends on the choice of a basis for $\mathfrak{g}$ and uses only the vector space structure of g. This process could then be mapped to the group via the exponential map, but the resulting process would not behave similarly with respect to the multiplication of G as the original process does with respect to the addition of the Euclidean space. Therefore we decompose the original process into increments, i. e. differences $X_{t+\Delta t} - X_t$ corresponding to small time intervals, map these to the group, and then multiply them back together, using the group multiplication. The process $\tilde{X}_t$ on the group is then defined as the limit for $\Delta t \to 0$. We see that the generator of $\tilde{X}_t$ is also directly related to the generator of X_t.

The next section contains an application of these processes. They can be used to define solutions of generalised heat equations as Appell polynomials. These polynomials are also related to limit theorems, and have nice special function properties (see also [FS92]).

Finally, in the last section of this chapter, we calculate "by hand" certain solutions of heat equations related to the affine group and to several nilpotent groups.

2.7.1 *A construction of stochastic processes on Lie groups*

We now give an explicit construction of stochastic processes on a Lie group G. The approach we follow can be traced back to McKean's multiplicative integrals (cf. [McK69]), our presentation is taken from [FS89a, FS89b]. We start with a process $W(t) = (W_1(t), \ldots, W_d(t))$ on $\mathbb{R}^d$ and embed it linearly in the linear space of the Lie algebra $\mathfrak{g} = L(G)$ ($\dim \mathfrak{g} = n \geq d$). Then we use the exponential mapping that is defined by the Lie algebra structure of $\mathfrak{g}$ to map $W(t)$ to G. We suppose that $W(t)$ is a continuous stochastic process with stationary and independent increments and with $W(0) = 0$, and that G is an exponential group (e. g. a nilpotent group), i. e. that $\exp : L(G) \to G$ is a bijective diffeomorphism. Furthermore, we identify $W(t)$ and its image in $\mathfrak{g}$, $W(t) = \sum_{i=1}^n W_i(t) X_i$, where the last $n - d$ coordinates are equal to zero.

We can view $W(t)$ on $\mathbb{R}^d$ as being built from its increments $w^{(k)}(n)$. Fix $t \in \mathbb{R}_+$ and $k \in \mathbb{N}$, and let

$$w^{(k)}(n) = W\left(\frac{(n+1)t}{2^k}\right) - W\left(\frac{nt}{2^k}\right), \qquad n = 0, \ldots, 2^k - 1.$$

Then obviously

$$W(t) = \sum_{n=0}^{2^k-1} w^{(k)}(n). \tag{2.9}$$

In the limit $k \to \infty$, this sum becomes a stochastic integral,

$$W(t) = \int_0^t \mathrm{d}W(s).$$

To generalise the increment property (2.9) to processes on a group G we replace the sum by the group product. But for non-commutative groups this fails for the process $g(W(t);X)$, i. e. in general

$$g(W(t);X) \neq g(w^{(k)}(0);X)\cdots g(w^{(k)}(n);X)\cdots g(w^{(k)}(2^k-1);X). \tag{2.10}$$

A way out is to change $W(t)$. We start with the right hand side of (2.10) and define it as stochastic process $g(\tilde{W}^{(k)}(t);X)$ on G, i. e.

$$g^{(k)}(t) = g(\tilde{W}^{(k)}(t);X) = g(w^{(k)}(0);X)\cdots g(w^{(k)}(n);X)\cdots g(w^{(k)}(2^k-1);X).$$

This defines $\tilde{W}^{(k)}(t)$ uniquely, because G is an exponential group. We set

$$\tilde{W}(t) = \lim_{k\to\infty} \tilde{W}^{(k)}(t).$$

The expression obtained for $\tilde{W}(t)$ will again involve stochastic integrals.

Remarks:

- To see how the convergence of $g^{(k)}(t)$ can be proved in general, consult [FS89b, Fei78].
- For non-exponential groups one needs techniques to patch together the processes across different coordinate patches.
- Using semi-martingales one can define a stochastic exponential also for more general processes as considered here [HDL86]. It is introduced as the solution of a stochastic differential equation of the form $\mathrm{d}X_t = X_t \mathrm{d}M_t$.

Assume now that $W(t)$ is a d-dimensional Brownian motion. Then the expectation of an increment $g(w^{(k)}(n);X)$ is

$$\begin{aligned}\mathbb{E}\Big(g(w^{(k)}(n);X)\Big) &= \mathbb{E}\Big(\exp(w_1^{(k)}(n)X_1)\cdots\exp(w_d^{(k)}(n)X_d)\Big)\\ &= \exp\left(\frac{tX_1^2}{2\cdot 2^k}\right)\cdots\exp\left(\frac{tX_1^2}{2\cdot 2^k}\right)\end{aligned}$$

and therefore, since the increments are independent,

$$\mathbb{E}\Big(g(\tilde{W}^{(k)}(t);X)\Big) = \left(\exp\left(\frac{tX_1^2}{2\cdot 2^k}\right)\cdots\exp\left(\frac{tX_1^2}{2\cdot 2^k}\right)\right)^{2^k}.$$

Using *Trotter's product formula*

$$\lim_{N\to\infty}\left(\exp\left(\frac{X_1}{N}\right)\cdots\exp\left(\frac{X_d}{N}\right)\right)^N = \exp(X_1+\cdots+X_d)$$

we find

$$\mathbb{E}\Big(g(\tilde{W}(t);X)\Big) = \exp\frac{t}{2}\left(X_1^2+\cdots X_d^2\right).$$

This result remains true, if $W(t)$ is a process with stationary and independent increments, and with independent components. Let $L(x) = \sum_{i=1}^{d} L_i(x_i)$ be the generator of $W(t)$, i. e.

$$\mathbb{E}\Big(\exp(W(t)x)\Big) = \exp tL(x). \qquad \text{for } x \in \mathbb{R}^d,$$

then we find

$$\mathbb{E}\Big(g(\tilde{W}(t); X)\Big) = \exp tL(X) = \exp t \sum_{i=1}^{d} L_i(X_i). \tag{2.11}$$

Remark: Equation (2.11) can be considered as a generalised *Feynman-Kac formula*, as is shown in [FS89a, FS89b]. To give the limits involved in this equation a meaning, we have to replace the Lie algebra elements by operators and use the operator topology. Trotter's product formula can be applied in different contexts, e. g. if the operators are bounded or unitary.

Example: the affine group

Let A_t, B_t be two stochastic processes on $\mathbb{R}^2$ and $g(\tilde{A}_t, \tilde{B}_t; X, Y)$ the corresponding process on the group in terms of coordinates of the second kind. As in the preceding discussion we choose a partition[2] $\{0 = t_0, t_1, \ldots, t_N, t_{N+1} = t\}$ with width $\Delta t = t_{i+1} - t_i$ $(i = 0, \ldots, N)$ of the interval $[0, t]$. This corresponds to approximating the stochastic process by a discretised process. We have $A_{t_N} = \sum_{n=0}^{N-1} a_{t_n}$, $A_{t_{N+1}} = A_{t_N} + a_{t_N}$, $B_{t_N} = \sum_{n=0}^{N-1} b_{t_n}$, $B_{t_{N+1}} = B_{t_N} + b_{t_N}$ and

$$\begin{aligned} g(\tilde{A}^{(N)}_{t_{N+1}}, \tilde{B}^{(N)}_{t_{N+1}}; X, Y) &= g(a_{t_N}, b_{t_N}; X, Y) g(\tilde{A}^{(N)}_{t_N}, \tilde{B}^{(N)}_{t_N}; X, Y) \\ &= g(\tilde{A}^{(N)}_{t_N} + a_{t_N}, e^{-\tilde{A}^{(N)}_{t_N}} b_{t_N} + \tilde{B}^{(N)}_{t_N}; X, Y), \end{aligned}$$

therefore

$$\begin{aligned} \tilde{A}^{(N)}_t &= \sum_{n=0}^{N} a_{t_n} = A_{t_N}, \\ \tilde{B}^{(N)}_t &= \sum_{n=0}^{N} e^{-\tilde{A}^{(N)}_{t_n}} b_{t_n} = \sum_{n=0}^{N} e^{-A_{t_n}} b_{t_n}. \end{aligned}$$

In the continuous time limit ($N \to \infty$, $\Delta t \to 0$), this becomes

$$\begin{aligned} \tilde{A}_t &= A_t, \\ \tilde{B}_t &= \int_0^t e^{-A_s} \mathrm{d}B_s, \end{aligned}$$

where the integral is the usual Itô integral. With these arguments we have proven the following proposition.

[2] For convenience we again assume that this is an equi-distant partition, but this is not necessary.

Proposition 2.7.1 *For a continuous time stochastic process with components (A_t, B_t) on $\mathbb{R}^2$ the corresponding process on the affine group in terms of coordinates of the second kind is given by*

$$g(A_t, \int_0^t \exp(-A_s)\mathrm{d}B_s; X, Y).$$

If we again denote by $\mathbb{E}(\cdot)$ the expectation value, and take standard Brownian motion on $\mathbb{R}^2$ for (A_t, B_t), we get the following result.

Proposition 2.7.2 *For Brownian motion on the affine group we have the relation*

$$e^{tH} = \mathbb{E}\left(g(A_t, \int_0^t \exp(-A_s)\mathrm{d}B_s; X, Y)\right),$$

where $H = \frac{1}{2}(X^2 + Y^2)$.

Remark: This is just a special case of equation (2.11).
Proof: In the discrete time approximation

$$\begin{aligned}\mathbb{E}\left(g(A_{t_N}, \sum_{n=0}^{N-1} e^{-A_{t_n}} b_{t_n}; X, Y)\right) &= \mathbb{E}\left(\prod_{n=0}^{N-1} e^{a_{t_n} X} e^{b_{t_n} Y}\right) \\ &= \prod_{n=0}^{N-1} e^{\frac{\Delta t}{2} X^2} e^{\frac{\Delta t}{2} Y^2}.\end{aligned}$$

In the limit $\Delta t \to 0$, $N \to \infty$, $(N \cdot \Delta t = t)$, the Trotter product formula

$$\lim_{k\to\infty}\left(\exp\left(\frac{X_1}{k}\right)\exp\left(\frac{X_2}{k}\right)\right)^k = \exp(X_1 + X_2)$$

yields the proposition. ∎

This means that for every (sufficiently well-behaved) realisation ρ of $Aff(1)$ (see e. g. proposition 2.3.1), we have two equivalent ways to express the time evolution of (sufficiently well-behaved) solutions $f(x,t)$ of the equation

$$\partial_t f(x,t) = \rho(H) f(x,t),$$

namely

$$f(x,t) = e^{t\rho(H)} f(x,0),$$

and

$$f(x,t) = \mathbb{E}\left(g(A_t, \int_0^t e^{-A_s}\mathrm{d}B_s; \rho(X), \rho(Y)) f(x,0)\right).$$

But in general it might not be easy to evaluate either of these expressions (see also 2.7.3).

Example: nilpotent groups

We state the analogous results for the nilpotent groups that correspond to the nilpotent Lie algebras introduced in paragraph 2.3.2. They can be proven along the same lines as propositions 2.7.1 and 2.7.2.

Proposition 2.7.3 *Let $A_t = (A_t^0, \ldots, A_t^N)$ be a continuous time stochastic process on $\mathbb{R}^{N+1}$. Then the corresponding process $g(\tilde{A}_t) = g(\tilde{A}_t^0, \ldots, \tilde{A}_t^N)$ on the N-step nilpotent Lie group is given by*

$$\begin{aligned}
\tilde{A}_t^0 &= A_t^0, \\
\tilde{A}_t^1 &= A_t^1, \\
\tilde{A}_t^2 &= A_t^2 + \int_0^t A_s^0 \mathrm{d}A_s^1, \\
&\vdots \\
\tilde{A}_t^i &= A_t^i + \int_0^t A_s^0 \mathrm{d}A_s^{i-1} + \cdots + \frac{1}{(i-1)!}\int_0^t \left(A_s^0\right)^{i-1} \mathrm{d}A_s^1, \\
&\vdots \\
\tilde{A}_t^N &= \sum_{j=0}^{N-1} \frac{1}{j!}\int_0^t \left(A_s^0\right)^j \mathrm{d}A_s^{N-j}.
\end{aligned}$$

Proposition 2.7.4 *If we take two-dimensional Brownian motion $B_t = (B_t^0, B_t^1, 0, \ldots, 0)$ we find*

$$\mathbb{E}\Big(g(\tilde{B}_t; X)\Big) = \exp\left(\frac{t}{2}(X_0^2 + X_1^2)\right)$$

With the realisation given after proposition 2.3.2, we get

$$\begin{aligned}
&\mathbb{E}\left(g(B_t^0, B_t^1, \int_0^t B_s^0 \mathrm{d}B_s^1, \ldots, \frac{1}{(N-1)!}\int_0^t \left(B_s^0\right)^{N-1} \mathrm{d}B_s^1; \hat{X})\right) \\
&= \exp\frac{t}{2}\left(\partial_x^2 + \alpha^2 \frac{x^{2N-2}}{[(N-1)!]^2}\right),
\end{aligned}$$

It would be of great interest to find simpler expressions for the action of $\exp\frac{t}{2}(\partial_x^2 + c^2 x^{2N-2})$. Let for example $f(x,t)$ be defined by

$$f(x,t) = \exp\frac{t}{2}\left(\partial_x^2 + \alpha^2 \frac{x^{2N-2}}{[(N-1)!]^2}\right) 1,$$

i. e. $f(x,t)$ is the solution of

$$\partial_t \psi(x,t) = \frac{1}{2}\left(\partial_x^2 + \alpha^2 \frac{x^{2N-2}}{[(N-1)!]^2}\right)\psi(x,t) \tag{2.12}$$

with the initial condition $\psi(x,0) = 1$ for all x. Then

$$\mathbb{E}\left(\exp\left(\alpha\sum_{i=1}^{N}\frac{x^{N-i}}{(N-i)!}\int_0^t \left(B_s^0\right)^{i-1} \mathrm{d}B_s^1\right)\right) = f(x,t)$$

would allow us to find the characteristic function of the process given by the stochastic integrals (see also 2.7.3).

Example: the finite-difference group

For the finite-difference group we can prove the following proposition similarly as proposition 2.7.1.

Proposition 2.7.5 *If (A_t, B_t, C_t) is a stochastic process on $\mathbb{R}^3$, then the corresponding process $(\tilde{A}_t, \tilde{B}_t, \tilde{C}_t)$ on the finite-difference group with respect to coordinates of the second kind is given by*

$$\begin{aligned}\tilde{A}_t &= \int_0^t e^{-C_s}\mathrm{d}A_s + \int_0^t (e^{-C_s}-1)\mathrm{d}B_s,\\ \tilde{B}_t &= B_t,\\ \tilde{C}_t &= C_t.\end{aligned}$$

2.7.2 *Appell systems*

Appell polynomials $\{h_n(x); n \in \mathbb{N}\}$ on $\mathbb{R}$ are usually characterised by the two conditions

- $h_n(x)$ is a polynomial of degree n,
- $\frac{\mathrm{d}}{\mathrm{d}x}h_n(x) = nh_{n-1}(x)$.

Interesting examples are furnished by the shifted moment sequences

$$h_n(x) = \int_{-\infty}^{\infty} (x+y)^n p(\mathrm{d}y), \tag{2.13}$$

where p is a probability measure on $\mathbb{R}$ with all moments finite. This includes in particular the Hermite polynomials

$$H_n(x) = \frac{1}{\sqrt{2\pi}}\int_{-\infty}^{\infty} (x+iy)^n e^{-y^2/2}\mathrm{d}y$$

for the Gaussian case. P. Feinsilver and R. Schott [FS92] have used the probabilistic interpretation of Appell systems to define their analoga on Lie groups (where, in general, they are no longer polynomials). We will present this definition and some of the basic properties of Appell systems on Lie groups, before we look at several examples.

We are here interested in Appell systems, because they are polynomial solutions of generalised heat equations (see proposition 2.7.6). Naturally, they can also be used to give more general solutions in the form of infinite series. In probability theory they are also used to obtain non-central limit theorems (see references in [FS92]).

The left and right *Appell polynomials* (corresponding to the random variable Z) on a Lie group are defined as

$$\begin{aligned} h_n^L(a) &= \sum_{j+J=n} c_j(a)\mathbb{E}(M_j^J(Z)), \\ h_n^R(a) &= \sum_{j+J=n} \mathbb{E}\Big(c_j(Z)\Big)M_j^J(a). \end{aligned}$$

The convolution formulas

$$\begin{aligned} h_n^L(a) &= \mathbb{E}\Big(c_n(Z \odot a)\Big) \\ h_n^R(a) &= \mathbb{E}\Big(c_n(a \odot Z)\Big) \end{aligned} \tag{2.14}$$

show the analogy to formula (2.13). To derive these formulas we have to expand the product $g(Z;X)g(a;X)$ in two ways:

$$\begin{aligned} \mathbb{E}\Big(g(Z;X)g(a;X)\Big) &= \mathbb{E}\Big(g(Z \odot a;X)\Big) \\ &= \sum \mathbb{E}\Big(c_n(Z \odot a)\Big)\psi_n(X) \end{aligned}$$

and

$$\begin{aligned} \mathbb{E}\Big(g(Z;X)g(a;X)\Big) &= \mathbb{E}\Big(g(Z;X)\sum c_j(a)\psi_j(X)\Big) \\ &= \mathbb{E}\Big(\sum c_j(a)M_j^J(Z)\psi_{j+J}(X)\Big) \\ &= \sum h_n^L(a)\psi_n(X). \end{aligned}$$

Equation (2.14) follows, since $\{\psi_n(X)\}$ is a basis for $U(\mathfrak{g})$. Starting with $< g(a;X)g(Z;X) >$, we get the corresponding formula for the right Appell polynomials.

Of particular interest are the Appell systems related to the process $\tilde{W}(t)$ constructed in the previous section.

Proposition 2.7.6 *For the case of the process* $\mathbb{E}\Big(g(\tilde{W}(t);X)\Big) = e^{tL(X)}$ *we have*

$$\begin{aligned} h_n^L(a,t) &= e^{tL(X^\dagger)}c_n(a), \\ h_n^R(a,t) &= e^{tL(X^*)}c_n(a), \end{aligned}$$

i. e. $h_n^L(a,t)$ *(resp.* $h_n^R(a,t)$*) is the solution of the evolution equation*

$$\begin{aligned} \frac{\partial}{\partial t}f(a,t) &= L(X^\dagger)f(a,t) \\ \Big(\mathit{resp.}\ \frac{\partial}{\partial t}f(a,t) &= L(X^*)f(a,t)\Big) \end{aligned}$$

with initial condition $f(a,0) = c_n(a)$.

Proof: We show the first equation, the second follows similarly.

$$\begin{aligned}\sum h_n^L(a,t)\psi_n(X) &= \sum \mathbb{E}\Big(c_n(\tilde{W}(t)\odot a)\Big)\psi_n(t)\\ &= \mathbb{E}\Big(g(\tilde{W}(t);X)g(a;X)\Big)\\ &= \mathbb{E}\Big(g(\tilde{W}(t);X^\dagger)\Big)\sum c_n(a)\psi_n(X)\\ &= \sum e^{tL(X^\dagger)}c_n(a)\psi_n(X).\end{aligned}$$

∎

Example: the affine group

We use here the results of Section 2.7.3. Using the group law (see 2.5.1)

$$g(A,B)g(a,b) = g\Big((A,B)\odot(a,b)\Big) = g(A+a, e^{-a}B+b)$$

and the convolution formulas (2.14), we can calculate the left Appell polynomials with the formula

$$\begin{aligned}h_{nm}^L(a,b) &= \mathbb{E}\left(c_{nm}\left((\tilde{A}_t,\tilde{B}_t)\odot(a,b)\right)\right)\\ &= \frac{1}{n!m!}\mathbb{E}\left((\tilde{A}_t+a)^n(e^{-a}\tilde{B}_t+b)^m\right),\end{aligned} \tag{2.15}$$

e. g.

$$\begin{aligned}h_{0,2}^L(a,b) &= \mathbb{E}\left(c_{0,2}\left((\tilde{A}_t,\tilde{B}_t)\odot(a,b)\right)\right) = \frac{1}{2}\mathbb{E}\left((e^{-a}\tilde{B}_t+b)^2\right)\\ &= \frac{1}{2}\mathbb{E}\left(e^{-2a}\tilde{B}_t^2+2e^{-a}\tilde{B}_t b+b^2\right)\\ &= \frac{1}{4}\left(e^{2t}-1\right)e^{-2a}+\frac{b^2}{2},\end{aligned}$$

where we have used $\mathbb{E}(\tilde{A}_t)=\mathbb{E}(\tilde{B}_t)=0$, $\mathbb{E}\left(\tilde{B}_t^2\right)=\frac{1}{2}\left(e^{2t}-1\right)$ (see 2.7.3). We see that in this case the Appell polynomials are only for $t=0$ polynomials. We list some more left Appell "polynomials".

$$\begin{aligned}h_{0,0}^L(a,b) &= 1\\ h_{1,0}^L(a,b) &= a\\ h_{0,1}^L(a,b) &= b\\ h_{2,0}^L(a,b) &= \frac{1}{2}\left(a^2+t\right)\\ h_{1,1}^L(a,b) &= ab\\ h_{0,2}^L(a,b) &= \frac{1}{4}\left(e^{2t}-1\right)e^{-2a}+\frac{b^2}{2}\end{aligned}$$

$$
\begin{aligned}
h^L_{3,0}(a,b) &= \frac{a^3}{6} + \frac{at}{2} \\
h^L_{2,1}(a,b) &= \frac{1}{2}\left(a^2 b + tb\right) \\
h^L_{1,2}(a,b) &= \frac{ab^2}{2} + \frac{1}{4}\left(e^{2t}-1\right) a e^{-2a} + \frac{1}{4}\left((1-2t)e^{2t}-1\right) e^{-2a} \\
h^L_{0,3}(a,b) &= \frac{1}{4}\left(e^{2t}-1\right) e^{-2a} b + \frac{b^3}{6} \\
&\vdots
\end{aligned}
$$

All these functions are solutions of the following partial differential equation

$$
\partial_t h^L_{nm}(a,b;t) = \frac{1}{2}\left(X^{\dagger 2} + Y^{\dagger 2}\right) h^L_{nm}(a,b;t) = \frac{1}{2}\left(\partial_a^2 + e^{-2a}\partial_b^2\right) h^L_{nm}(a,b;t).
$$

that arises from the left dual representation (see 2.3.1). For solution of the heat equation related to the right dual representation

$$
\partial_t h^R_{nm}(a,b;t) = \frac{1}{2}\left(X^{*2} + Y^{*2}\right) h^R_{nm}(a,b;t) = \frac{1}{2}\left((\partial_a - b\partial_b)^2 + \partial_b^2\right) h^R_{nm}(a,b;t)
$$

we need the right Appell system. To get a formula for the right Appell polynomials we have to exchange $(\tilde{A}_t, \tilde{B}_t)$ and (a,b) in equation (2.15). This gives

$$
h^R_{nm}(a,b) = \frac{1}{n!m!}\mathbb{E}\left((\tilde{A}_t + a)^n (e^{-\tilde{A}_t} b + \tilde{B}_t)^m\right).
$$

We calculate for example

$$
\begin{aligned}
h^R_{0,0}(a,b) &= 1 \\
h^R_{1,0}(a,b) &= a \\
h^R_{0,1}(a,b) &= e^{t/2} b \\
h^R_{2,0}(a,b) &= \frac{1}{2}\left(a^2 + t\right) \\
h^R_{1,1}(a,b) &= e^{t/2}(ab - tb) \\
h^R_{0,2}(a,b) &= \frac{1}{2} e^{2t} b^2 + \frac{1}{4}\left(e^{2t}-1\right) \\
&\vdots
\end{aligned}
$$

Example: nilpotent groups

Here we get

$$
\begin{aligned}
& h^L_{n_0 n_1 \cdots n_N}(a_0, a_1, \ldots, a_N) \\
= \; & \frac{1}{n!}\mathbb{E}\left((\tilde{B}^0_t + a_0)^{n_0} (\tilde{B}^1_t + a_1)^{n_1} \cdots \left(\tilde{B}^N_t + \sum_{j=0}^{N-1} \frac{1}{j!} \left(\tilde{B}^0_t\right)^j a_{N-j} \right)^{n_N} \right)
\end{aligned}
$$

$$
\begin{aligned}
&h^R_{n_0n_1\cdots n_N}(a_0,a_1,\ldots,a_N)\\
&= \frac{1}{n!}\mathbb{E}\left((\tilde{B}^0_t+a_0)^{n_0}(\tilde{B}^1_t+a_1)^{n_1}\cdots\left(\tilde{B}^N_t+\sum_{j=0}^{N-1}\frac{1}{j!}\tilde{B}^{N-j}_t a_0^j\right)^{n_N}\right)
\end{aligned}
$$

with equation (2.14) and 2.5.2. In 2.7.3 we will show how the moments

$$E_{n_0n_1\cdots n_N} = \mathbb{E}\left(\left(\tilde{B}^0_t\right)^{n_0}\left(\tilde{B}^1_t\right)^{n_1}\cdots\left(\tilde{B}^N_t\right)^{n_N}\right)$$

can be calculated.

2.7.3 *Explicit calculations of solutions of the heat equation*

In the previous sections we have developed a constructive approach to stochastic processes on Lie groups, and have shown the relations of these processes to special functions and solutions of evolution equations (heat equations). In some cases, e. g. for the groups of type H, this yields a method for calculating the polynomial solutions of the corresponding evolution equations (see [FKS93]). But in many cases this last step is not complete, because there is not enough information about the stochastic processes involved. We will use the fact that the correspondence works in both ways, i. e. we will try to solve the evolution equation for a particularly simple realisation, and use the results to calculate the moments of the stochastic processes, its characteristic function, or determine its asymptotic behaviour, etc. This information can then, in turn, be used to study the evolution equations of other realisations. The results in this section have yet to be completed.

Heat polynomials for the affine group

In this section we study the polynomial solutions of a particular generalised heat equation related to the affine group.

For the realisations that we obtained in proposition 2.3.1, it is not difficult to calculate

$$g(A_t,\int_0^t \exp(-A_s)\mathrm{d}B_s;\rho(X),\rho(Y))f(x,0)$$

for e.g. analytic functions f. The hard part is to compute the expectation value of this expression, because it requires the knowledge of the expectation values

$$E_{nm} = \mathbb{E}\left(A_t^n\left(\int_0^t \exp(-A_s)\mathrm{d}B_s\right)^m\right).$$

On the other hand it does not seem easy to calculate $e^{t\rho(H)}f(x,0)$ for, say, analytic functions either.

We will calculate the action of $e^{t\rho(H)}$ on polynomials for one realisation by hand, and use the results to compute the E_{nm}.

Let $\rho(X)=\alpha-x\partial_x$, $\rho(Y)=\partial_x$ (see eq.s (2.3),(2.3)). Then

$$\rho(H)=\frac{1}{2}\left((\alpha-x\partial_x)^2+\partial_x^2\right)=\frac{1}{2}\left(\alpha^2+(1-2\alpha)x\partial_x+(x^2+1)\partial_x^2\right).$$

We find

$$H^m x^n = \sum_{\mu=0}^{m\wedge[\frac{n}{2}]} C^n_{m\mu} x^{n-2\mu},$$

where the coefficients $C^n_{m\mu}$ satisfy the following recursion relation

$$C^n_{m+1,\mu} = \frac{(n-2\mu-\alpha)^2}{2} C^n_{m,\mu} + \frac{(n-2\mu+2)(n-2\mu+1)}{2} C^n_{m,\mu-1}. \tag{2.16}$$

In particular we get $C^n_{m,0} = \frac{(n-\alpha)^{2m}}{2^m}$, and $C^n_{m\mu} = 0$ if $\mu > [\frac{n}{2}] \wedge m$. By induction follows (with $a = \max(2, |\alpha|)$)

$$|C^n_{m\mu}| \leq (n+a)^{2m}, \tag{2.17}$$

since $C^n_{00} = 1$.

$$\begin{aligned} \exp(t\rho(H))x^n &= \sum_{m=0}^{\infty} \frac{t^m}{m!} H^m x^n = \sum_{m=0}^{\infty} \frac{t^m}{m!} \sum_{\mu=0}^{m\wedge[\frac{n}{2}]} C^n_{m\mu} x^{n-2\mu} \\ &= \sum_{\mu=0}^{[\frac{n}{2}]} \sum_{m=\mu}^{\infty} \frac{C^n_{m\mu} t^m}{m!} x^{n-2\mu} = \sum_{\mu=0}^{[\frac{n}{2}]} f^n_\mu(t) x^{n-2\mu}, \end{aligned}$$

if we define

$$f^n_\mu(t) = \sum_{m=\mu}^{\infty} C^n_{m\mu} \frac{t^m}{m!}.$$

Equation (2.17) shows that the series converges

$$f^n_\mu(t) \leq \sum_{m=\mu}^{\infty} (n+a)^{2m} \frac{|t|^m}{m!} \leq \exp\left[(n+a)^2 |t|\right].$$

Obviously

$$f^n_0(t) = \exp\left(\frac{(n-\alpha)^2}{2} t\right) \tag{2.18}$$

and $f^n_\mu = 0$ for $\mu > [\frac{n}{2}]$. Note also that from the definition of f^n_μ follows

$$\begin{aligned} \frac{\mathrm{d}^\nu}{\mathrm{d}t^\nu} f^n_\mu(0) &= 0, \qquad \text{if } 0 \leq \nu < \mu \\ \frac{\mathrm{d}^\mu}{\mathrm{d}t^\mu} f^n_\mu(0) &= C^n_{\mu\mu}. \end{aligned}$$

From the recursion relation (2.16) we get a recursion relation for the functions f^n_μ:

$$\frac{\mathrm{d}}{\mathrm{d}t} f^n_\mu(t) = \frac{(n-2\mu-\alpha)^2}{2} f^n_\mu(t) + \frac{(n-2\mu+2)(n-2\mu+1)}{2} f^n_{\mu-1}(t) \tag{2.19}$$

for $n \in N, 0 \leq \mu \leq [\frac{n}{2}]$. Solving this differential equation for $\mu = 1, n \geq 2$, we get

$$f_1^n(t) = \begin{cases} \frac{n(n-1)}{2} t e^{\frac{t}{2}} & \alpha = n-1, \\ \frac{n(n-1)}{4(1+\alpha-n)} \left(e^{\frac{(n-2-\alpha)^2}{2} t} - e^{\frac{(n-\alpha)^2}{2} t} \right) & \text{else.} \end{cases} \tag{2.20}$$

In general, for $\alpha \notin \{n, n-1, \ldots, n-2\mu+1\}$, the function f_μ^n can be calculated with the help of the following proposition.

Proposition 2.7.7 *Let $n, \mu \in N$, $n \geq 2\mu$, and $\alpha \notin \{n, n-1, \ldots, n-2\mu+1\}$. Then*

$$f_\mu^n(t) = \sum_{\nu=0}^{\mu} A_{\mu\nu}^n e^{\frac{(n-2\nu-\alpha)^2}{2} t}, \tag{2.21}$$

where the $A_{\mu\nu}^n$ are determined by the $\mu + 1$ equations

$$\sum_{\nu=0}^{\mu} \left[\frac{(n-2\nu-\alpha)^2}{2} \right]^\rho A_{\mu\nu}^n = 0, \qquad 0 \leq \rho < \mu, \tag{2.22}$$

$$\sum_{\nu=0}^{\mu} \left[\frac{(n-2\nu-\alpha)^2}{2} \right]^\mu A_{\mu\nu}^n = \frac{n!}{2^\mu (n-2\mu)!}. \tag{2.23}$$

Proof: Notice first that from (2.16) follows

$$\begin{aligned} C_{\mu\mu}^n &= \frac{(n-2\mu+2)(n-2\mu+1)}{2} C_{\mu-1,\mu-1}^n \\ &= \frac{n(n-1)\cdots(n-2\mu+2)(n-2\mu+1)}{2^\mu} \\ &= \frac{n!}{2^\mu (n-2\mu)!}, \end{aligned}$$

and that, if f_μ^n has the form given in eq. (2.21), then eq.s (2.22),(2.23) follow from eq.s (2.19),(2.19). Since f_0^n and f_1^n have the form stated in the proposition (see (2.18), (2.20)), we only need to verify that the solutions given in (2.21) satisfy the recursion relation (2.19). Assume now that $f_{\mu-1}^n$ has the form of (2.21). Then f_μ^n has to satisfy the differential equation

$$\frac{\mathrm{d}}{\mathrm{d}t} f_\mu^n(t) = \frac{(n-2\mu-\alpha)^2}{2} f_\mu^n(t) + \frac{(n-2\mu+2)(n-2\mu+1)}{2} \sum_{\nu=0}^{\mu-1} A_{\mu-1,\nu}^n e^{\frac{(n-2\nu-\alpha)^2}{2} t},$$

where the hypotheses of the proposition guarantee

$$\begin{aligned} n-2\nu-\alpha &\neq 0 & 0 \leq \nu < \mu, \\ (n-2\mu+2)(n-2\mu+1) &\neq 0, & \\ (n-2\mu-\alpha)^2 &\neq (n-2\nu-\alpha)^2, & 0 \leq \nu < \mu, \end{aligned}$$

and therefore f_μ^n has the form given in eq. (2.21). Since then equations (2.22)(2.23) completely determine f_μ^n, this completes the proof. ∎

Nonetheless it is useful to look more closely at the conditions on the $A^n_{\mu\nu}$ that follow from (2.19). After some simplification we get the recursion relation

$$4(\mu-\nu)(n-\nu-\mu-\alpha)A^n_{\mu\nu} = (n-2\mu+2)(n-2\mu+1)A^n_{\mu-1,\nu}$$

for $0 \le \nu < \mu$, which, if we add $A^n_{\mu\mu} = -\sum_{\nu=0}^{\mu-1} A^n_{\mu\nu}$ and notice $A^n_{00} = 1$, allows us to calculate all required $A^n_{\mu\nu}$.

Proposition 2.7.8 *Let $n, \mu \in N$, $n \ge 2\mu$ and $\alpha \notin \{n, n-1, \ldots, n-2\mu+1\}$. Then the coefficients in proposition 2.7.7 are determined by the recurrence formulas*

$$\begin{aligned} A^n_{\mu\nu} &= \frac{(n-2\mu+2)(n-2\mu+1)}{4(\mu-\nu)(n-\nu-\mu-\alpha)} A^n_{\mu-1,\nu}, \qquad 0 \le \nu < \mu, \\ A^n_{\mu\mu} &= -\sum_{\nu=0}^{\mu-1} A^n_{\mu\nu}, \\ A^n_{00} &= 1. \end{aligned}$$

Assume now that we have calculated the f^n_μ. Then

$$\begin{aligned} e^{t\rho(H)}x^n &= \sum_{\mu=0}^{[\frac{n}{2}]} f^n_\mu(t) x^{n-2\mu} & (2.24) \\ &= \mathbb{E}\left(g(A_t, \int_0^t \exp(-A_s)\mathrm{d}B_s; \alpha - x\partial_x, \partial_x)x^n\right) & (2.25) \\ &= \mathbb{E}\left(e^{\alpha A_t}\left(e^{-A_t}x + \int_0^t e^{-A_s}\mathrm{d}B_s\right)^n\right) & (2.26) \\ &= \sum_{\nu=0}^{n} \binom{n}{\nu} x^\nu \mathbb{E}\left(e^{(\alpha-\nu)A_t}\left(\int_0^t e^{-A_s}\mathrm{d}B_s\right)^{n-\nu}\right), & (2.27) \end{aligned}$$

and by comparing the coefficients of x^ν and α^k in both expressions we get the values $< A^n_t \left(\int_0^t \exp(-A_s)\mathrm{d}B_s\right)^m >$. For $n = 2$, $\alpha \ne 1$, follows

$$\begin{aligned} \mathbb{E}\left(e^{(\alpha-2)A_t}\right) &= e^{\frac{(\alpha-2)^2}{2}t}, \\ \mathbb{E}\left(e^{(\alpha-1)A_t}\int_0^t e^{-A_s}\mathrm{d}B_s\right) &= 0, \\ \mathbb{E}\left(e^{\alpha A_t}\left(\int_0^t e^{-A_s}\mathrm{d}B_s\right)^2\right) &= \frac{1}{2(\alpha-1)}\left(e^{\frac{\alpha^2}{2}t} - e^{\frac{(\alpha-2)^2}{2}t}\right). \end{aligned}$$

If we set $\alpha = 0$ in the last equation, we get

$$\mathbb{E}\left(\left(\int_0^t e^{-A_s}\mathrm{d}B_s\right)^2\right) = \frac{e^{2t}-1}{2}.$$

In general we can deduce from (2.24)-(2.27) by comparing the x^0 terms

$$\mathbb{E}\left(e^{\alpha A_t}\left(\int_0^t e^{-A_s}\mathrm{d}B_s\right)^n\right) = \begin{cases} f^n_{n/2}(t) & \text{if } n \text{ is even,} \\ 0 & \text{if } n \text{ is odd.} \end{cases} \tag{2.28}$$

Expanding both sides in power series with respect to α now yields the values of $E_{nm} = \mathbb{E}\left(A_t^n\left(\int_0^t \exp(-A_s)\mathrm{d}B_s\right)^m\right)$. This can easily be done with programs such as Mathematica or Maple. We get for example

$$\begin{aligned} f_1^2(t) &= \frac{-1+e^{2t}}{2} + \frac{(-1+e^{2t}-2e^{2t}t)\,\alpha}{2} \\ &+\frac{(-2+2e^{2t}-t-3e^{2t}t+4e^{2t}t^2)\,\alpha^2}{4} \\ &+\frac{\left(-1+e^{2t}-\frac{t}{2}-\frac{3e^{2t}t}{2}+e^{2t}t^2-\frac{4e^{2t}t^3}{3}\right)\alpha^3}{2} \\ &+\frac{\left(-1+e^{2t}-\frac{t}{2}-\frac{3e^{2t}t}{2}-\frac{t^2}{8}+\frac{9e^{2t}t^2}{8}-\frac{e^{2t}t^3}{3}+\frac{2e^{2t}t^4}{3}\right)\alpha^4}{2}+\cdots \end{aligned}$$

$$\begin{aligned} f_2^4(t) &= \frac{3e^{(\alpha-4)^2t/2}}{4(2-\alpha)(3-\alpha)} - \frac{3e^{(\alpha-2)^2t/2}}{2(1-\alpha)(3-\alpha)} + \frac{3e^{\alpha^2t/2}}{4(1-\alpha)(2-\alpha)} \\ &= \left[\frac{3}{8}-\frac{e^{2t}}{2}+\frac{e^{8t}}{8}\right]+\left[\frac{9}{16}-\frac{2e^{2t}}{3}+\frac{5e^{8t}}{48}+e^{2t}t-\frac{e^{8t}t}{2}\right]\alpha \\ &+\left[\frac{3\,(7+2t)}{32}-\frac{e^{2t}\,(26-39t+36t^2)}{36}+\frac{e^{8t}\,(19-102t+288t^2)}{288}\right]\alpha^2 \\ &+\left[\frac{9\,(5+2t)}{64}+\frac{e^{8t}\,(65-366t+1008t^2-2304t^3)}{1728}\right. \\ &\left.-\frac{e^{2t}\,(40-60t+45t^2-36t^3)}{54}\right]\alpha^3 \\ &+\left[\frac{3\,(31+14t+2t^2)}{128}-\frac{e^{2t}\,(968-1452t+1089t^2-504t^3+432t^4)}{1296}\right. \\ &\left.+\frac{e^{8t}\,(211-1218t+3474t^2-6336t^3+13824t^4)}{10368}\right]\alpha^4+\cdots \end{aligned}$$

In this way we can obtain the values of E_{nm}, e. g.

$$\begin{aligned} E_{02} &= \frac{1}{2}(e^{2t}-1) \\ E_{12} &= \frac{1}{2}((1-2t)e^{2t}-1) \\ E_{22} &= \frac{1}{2}((4t^2-3t+2)e^{2t}-2-t) \end{aligned}$$

$$\begin{aligned} &\vdots \\ E_{04} &= \frac{1}{8}\left(e^{8t} - 4e^{2t} + 3\right) \\ E_{14} &= \frac{1}{48}\left((5-24t)e^{8t} + (48t-32)e^{2t} + 27\right) \\ &\vdots \end{aligned}$$

With these values the time evolution of the polynomial[3] solutions of

$$\rho(H)f(x,t) = \partial_t f(x,t)$$

in other realisations can be determined.

One can use (2.28) also to give an expression for the Fourier transform of the density of A_t and $\int_0^t \exp(-A_s)\mathrm{d}B_s$. Let α be purely imaginary, $\alpha = ix$, then

$$\mathbb{E}\left(\exp\left(ixA_t + iy\int_0^t e^{-A_s}\mathrm{d}B_s\right)\right) = \sum_{n=0}^{\infty}(-1)^n \frac{y^{2n}}{(2n)!} f_n^{2n}(t), \tag{2.29}$$

where the dependence on $x = \frac{\alpha}{i}$ is hidden in the f_n^{2n},

A heat equation for the nilpotent groups

In previous calculations concerning Brownian motion on nilpotent groups we encountered the following partial differential equation (2.12):

$$\partial_t\psi(x,t) = \frac{1}{2}\left(\partial_x^2\psi(x,t) + \beta x^n\psi(x,t)\right).$$

To solve this equation we consider the corresponding time-independent equation

$$\partial_x^2 f_\lambda(x) + \beta x^n f_\lambda(x) = \lambda f_\lambda(x).$$

This equation is found if we make the separation ansatz $\psi(x,t) = f_\lambda(x)\exp\left(\frac{\lambda t}{2}\right)$. It is equivalent to the system

$$\begin{pmatrix} Y_1(x) \\ Y_2(x) \end{pmatrix}' = \begin{pmatrix} 0 & 1 \\ \lambda - \beta x^n & 0 \end{pmatrix}\begin{pmatrix} Y_1(x) \\ Y_2(x) \end{pmatrix}.$$

If we denote

$$V(x) = \left(v_{ij}(x)\right)_{i,j=1,2} = \begin{pmatrix} 0 & 1 \\ \lambda - \beta x^n & 0 \end{pmatrix},$$

then we can write the solution as

$$Y(x) = U(x)Y(0),$$

[3]More precisely, where $f(x,0)$ is a polynomial in the $x_1,\ldots,x_k$, $(x = (x_1,\ldots,x_k))$.

where the matrix $U(x)$ is given by

$$U(x) = \mathbb{1} + \int_0^x V(x_1)\mathrm{d}x_1 + \cdots + \int_0^x \cdots \int_0^{x_{n-1}} V(x_1)\cdots V(x_n)\mathrm{d}x_n \cdots \mathrm{d}x_1 + \cdots.$$

From this equation we can derive

$$f_\lambda(x) = a\sum_{k=0}^{\infty} g_k(x) + b\sum_{k=0}^{\infty} h_k(x),$$

where

- the constants a and b are determined by the initial values

$$\begin{aligned} a &= f_\lambda(0), \\ b &= f_\lambda{}'(0), \end{aligned}$$

- and the functions g_k and h_k are determined by the recursion relations

$$\begin{aligned} g_{k+1}(x) &= \int_0^x \int_0^{x_1} (\lambda - \beta x_2^n) g_k(x_2)\mathrm{d}x_2\mathrm{d}x_1, \qquad g_0(x) = 1, \\ h_{k+1}(x) &= \int_0^x \int_0^{x_1} (\lambda - \beta x_2^n) h_k(x_2)\mathrm{d}x_2\mathrm{d}x_1, \qquad h_0(x) = x. \end{aligned}$$

These expressions allow us to find an upper bound for $f(x)$. We get

$$\begin{aligned} |g_k(x)| &= \left| \int_0^x \int_0^{x_1} \cdots \int_0^{x_{2k-1}} (\lambda - \beta x_2^n) \cdots (\lambda - \beta x_{2k}^n)\mathrm{d}x_{2k} \cdots \mathrm{d}x_2\mathrm{d}x_1 \right| \\ &\le \left(\max_{|\xi|\le|x|} |\lambda - \beta\xi^n| \right) \frac{|x|^{2k}}{(2k)!} \\ |h_k(x)| &= \left| \int_0^x \int_0^{x_1} \cdots \int_0^{x_{2k-1}} (\lambda - \beta x_2^n) \cdots (\lambda - \beta x_{2k}^n) x_{2k}\mathrm{d}x_{2k} \cdots \mathrm{d}x_2\mathrm{d}x_1 \right| \\ &\le \left(\max_{|\xi|\le|x|} |\lambda - \beta\xi^n| \right) \frac{|x|^{2k+1}}{(2k+1)!}, \end{aligned}$$

and therefore

$$\begin{aligned} |f_\lambda(x)| &\le |a| \cosh\left[|x| \sqrt{\max_{|\xi|\le|x|} |\lambda - \beta\xi^n|} \right] + |b| \frac{\sinh\left[|x| \sqrt{\max_{|\xi|\le|x|} |\lambda - \beta\xi^n|} \right]}{\sqrt{\max_{|\xi|\le|x|} |\lambda - \beta\xi^n|}} \\ &\underset{x\to\infty}{\approx} C \exp\left(|x|^{\frac{n}{2}+1} \right). \end{aligned}$$

For the special case $\lambda = 0$ we get the stationary solution

$$\begin{aligned} \psi_0(x,t) &= a\left(1 - \frac{\beta x^{n+2}}{(n+1)(n+2)} + \frac{\beta^2 x^{2n+4}}{(n+1)(n+2)(2n+3)(2n+4)} - + \cdots \right) \\ &+ b\left(x - \frac{\beta x^{n+3}}{(n+2)(n+3)} + \frac{\beta^2 x^{2n+5}}{(n+2)(n+3)(2n+4)(2n+5)} - + \cdots \right). \end{aligned}$$

Solutions of the time-dependent equation can now be constructed as linear combinations of the form

$$\psi(x,t)=\sum_n c_n \exp\left(\frac{\lambda t}{2}\right) f_\lambda(x).$$

Let us now see how we can use these solutions to calculate the moments of Brownian motion on nilpotent groups.

$N=2$: Here we have $\beta=\alpha^2$ and $n=2$, and the two stationary solutions are

$$\begin{aligned}\psi_1(x) &= 1-\frac{\alpha^2x^4}{12}+\frac{\alpha^4x^8}{672}-\frac{\alpha^6x^{12}}{88704}+-\cdots\\ \psi_2(x) &= x-\frac{\alpha^2x^5}{20}+\frac{\alpha^4x^9}{1440}-\frac{\alpha^6x^{13}}{224640}+-\cdots.\end{aligned}$$

From the leading terms of $< g(A=B_t^0, B=B_t^1, C=\int_0^t B_s^0 \mathrm{d}B_s^1;\hat{X})\psi_i(x) >$ we calculate, using the Feynman-Kac formula

$$\mathbb{E}\Big(g(A,B,C;\hat{X})\psi_i(x)\Big)=e^{tH}\psi_i(x)=\psi_i(x),$$

the expectation of the coefficients of the expansion of $r_i=g(A,B,C;\hat{X})\psi_i(x)$,

$$\begin{aligned}r_1 &= 1+[\mathrm{C}+B\,x]\,\alpha\\ &+\left[\left(\frac{-A^4}{12}+\frac{\mathrm{C}^2}{2}\right)+\left(\frac{-A^3}{3}+B\,\mathrm{C}\right)x+\frac{(-A^2+B^2)\,x^2}{2}-\frac{A\,x^3}{3}-\frac{x^4}{12}\right]\alpha^2\\ &+\cdots\\ r_2 &= A+x+\left[A\,\mathrm{C}+(A\,B+\mathrm{C})\,x+B\,x^2\right]\alpha\\ &+\left[\left(\frac{-A^5}{20}+\frac{A\,\mathrm{C}^2}{2}\right)+\left(\frac{-A^4}{4}+A\,B\,\mathrm{C}+\frac{\mathrm{C}^2}{2}\right)x\right.\\ &\left.+\left(\frac{-A^3}{2}+\frac{A\,B^2}{2}+B\,\mathrm{C}\right)x^2+\frac{(-A^2+B^2)\,x^3}{2}-\frac{A\,x^4}{4}-\frac{x^5}{20}\right]\alpha^2\\ &+\cdots\end{aligned}$$

We know that $\mathbb{E}(A^nB^m)=0$ if n or m is odd, and $\mathbb{E}(A^{2n}B^{2m})=\frac{(2n)!(2m)!}{2^{n+m}n!m!}t^{n+m}$. Then $\mathbb{E}(r_i)=\psi_i$ yields

$$\begin{aligned}\mathbb{E}(C^2) &= \frac{t^2}{2}\\ \mathbb{E}(AC) &= < BC >= 0\\ \mathbb{E}(AC^2) &= 0\\ \mathbb{E}(ABC) &= \frac{t^2}{2}.\end{aligned}$$

$N=3:$ Here $\beta = \frac{\alpha^2}{4}$, $n=4$, and

$$\begin{aligned}
\psi_1(x) &= 1 - \frac{\alpha^2 x^6}{120} + \frac{\alpha^4 x^{12}}{63360} - + \cdots \\
\psi_2(x) &= x - \frac{\alpha^2 x^7}{168} + \frac{\alpha^4 x^{13}}{104832} - + \cdots .
\end{aligned}$$

We set $A = B_t^0$, $B = B_t^1$, $C = \int_0^t B_s^0 \mathrm{d}B_s^1$, $D = \int_0^t \left(B_s^0\right)^2 \mathrm{d}B_s^1$, and

$$r_i = g(A, B, C, D; \hat{X}\psi_i(x),$$

and get the expansions

$$\begin{aligned}
r_1 &= 1 + \left[\frac{D}{2} + C\,x + \frac{B\,x^2}{2}\right]\alpha \\
&+ \left[\left(\left(\frac{-A^6}{120} + \frac{D^2}{8}\right) + \left(\frac{-A^5}{20} + \frac{C\,D}{2}\right)x + \left(\frac{-A^4}{8} + \frac{C^2}{2} + \frac{B\,D}{4}\right)x^2\right.\right. \\
&+ \left.\left(\frac{-A^3}{6} + \frac{B\,C}{2}\right)x^3 + \frac{\left(-A^2+B^2\right)x^4}{8} - \frac{A\,x^5}{20} - \frac{x^6}{120}\right]\alpha^2 \\
&+ \cdots
\end{aligned} \tag{2.30}$$

$$\begin{aligned}
r_2 &= A + x + \left[\frac{A\,D}{2} + \left(A\,C + \frac{D}{2}\right)x + \left(\frac{A\,B}{2} + C\right)x^2 + \frac{B\,x^3}{2}\right]\alpha \\
&+ \left[\left(\frac{-A^7}{168} + \frac{A\,D^2}{8}\right) + \left(\frac{-A^6}{24} + \frac{A\,C\,D}{2} + \frac{D^2}{8}\right)x\right. \\
&+ \left(\frac{-A^5}{8} + \frac{A\,C^2}{2} + \frac{A\,B\,D}{4} + \frac{C\,D}{2}\right)x^2 \\
&+ \left(\frac{-5\,A^4}{24} + \frac{A\,B\,C}{2} + \frac{C^2}{2} + \frac{B\,D}{4}\right)x^3 \\
&+ \left(\frac{-5\,A^3}{24} + \frac{A\,B^2}{8} + \frac{B\,C}{2}\right)x^4 \\
&+ \left.\frac{\left(-A^2+B^2\right)x^5}{8} - \frac{A\,x^6}{24} - \frac{x^7}{168}\right]\alpha^2 \\
&+ \cdots
\end{aligned} \tag{2.31}$$

From this we deduce

$$\begin{aligned}
\mathbb{E}(D^2) &= t^3 \\
\mathbb{E}(AD) &= \mathbb{E}(CD) = 0 \\
\mathbb{E}(BD) &= \frac{t^2}{2} \\
\mathbb{E}(ABD) &= 0 \\
\mathbb{E}(ACD) &= \frac{t^3}{4}.
\end{aligned}$$

Higher moments can be calculated using more terms of the expansion, and including time-dependent solutions.

Some calculations for $SO(3)$

We show another method to calculate the expectation value of certain functions of the coordinates of Brownian motion on a group. We will consider the group $SO(3)$, and use finite-dimensional representations. This method is not as powerful as the methods presented in the previous two sections.

It is well-known that for every integer $n \geq 1$ there exists an n-dimensional representation of $sl(2)$ on the vector space $V = \text{span}\{|j,j>, |j,j-2>, \ldots, |j,-j>\}$, where $j = n-1$. The representation is defined by

$$\begin{aligned} \rho(H)|j,m> &= m|j,m> \\ \rho(X_+)|j,m> &= \frac{1}{2}\sqrt{(j-m)(j+m+2)}|j,m+2> \\ \rho(X_-)|j,m> &= \frac{1}{2}\sqrt{(j-m+2)(j+m)}|j,m-2> \end{aligned}$$

for $m = j, j-2, \cdots, -j$. Using the isomorphism between (the complexifications of) $sl(2)$ and $so(3)$

$$X_1 = \frac{1}{2i}(X_+ + X_-), \qquad X_2 = \frac{1}{2}(-X_+ + X_-), \qquad X_3 = \frac{1}{2i}H,$$

this gives a representation of $so(3)$. For the Casimir operator $C = 2\mathcal{H} = X_1^2 + X_2^2 + X_3^3$ we get in this representation

$$\rho(C) = \frac{1}{4}j(j+2)\mathbb{1}.$$

If we now calculate $g(\tilde{A}_t, \tilde{B}_t, \tilde{C}_t; \rho X) = e^{\tilde{A}_t\rho(X_1)}e^{\tilde{B}_t\rho(X_2)}e^{\tilde{C}_t\rho(X_3)}$, then we can use the Feynman-Kac formula

$$\mathbb{E}\left(e^{\tilde{A}_t\rho(X_1)}e^{\tilde{B}_t\rho(X_2)}e^{\tilde{C}_t\rho(X_3)}\right) = e^{t\mathcal{H}} = e^{tj(j+2)/8}\mathbb{1} \qquad (2.32)$$

to determine the expectation values of the matrix elements of $g(\tilde{A}_t, \tilde{B}_t, \tilde{C}_t; \rho X)$.

The following lemma is useful for calculating the exponential of a matrix. If M is an $n \times n$ matrix, then we know that there exists a polynomial p of degree $r \leq n$ such that $p(M) = 0$. We can assume that the coefficient of M^r is equal to 1, i. e. that

$$0 = p(M) = M^r + a_{r-1}M^{r-1} + \cdots a_1 M + a_0\mathbb{1}.$$

Lemma 2.7.9 *The matrix elements $F(t)_{ij}$ of $F(t) = \exp tM$ satisfy the differential equation*

$$F^{(r)}(t)_{ij} = -\sum_{\rho=0}^{r-1} a_\rho F^{(\rho)}(t)_{ij}.$$

The initial values are given by

$$F^{(\rho)}(0)_{ij} = (M^\rho)_{ij}.$$

Proof: The proof is straightforward if we observe

$$M \exp tM = \sum \frac{t^k M^{k+1}}{k!} = \frac{\mathrm{d}}{\mathrm{d}t} \exp tM.$$

This implies

$$\begin{aligned} F^{(r)}(t) &= M^r F(t) = -\sum_{\rho=0}^{r-1} a_\rho M^\rho F(t) \\ &= -\sum_{\rho=0}^{r-1} a_\rho F^{(\rho)}(t), \end{aligned}$$

and shows thus the first part of the lemma. For the second part

$$F^{(\rho)}(0) = M^\rho \exp tM|_{t=0} = M^\rho.$$

■

Equation (2.32) gives for example for $j = 1$:

$$\begin{aligned} \mathbb{E}\left(e^{-i\tilde{C}_t/2}(\cos(\tilde{A}_t/2)\cos(\tilde{B}_t/2) - i\sin(\tilde{A}_t/2)\sin(\tilde{B}_t/2))\right) &= e^{-3t/8} \\ \mathbb{E}\left(e^{i\tilde{C}_t/2}(-i\cos(\tilde{B}_t/2)\sin(\tilde{A}_t/2) - \cos(\tilde{A}_t/2)\sin(\tilde{B}_t/2))\right) &= 0 \\ \mathbb{E}\left(e^{-i\tilde{C}_t/2}(-i\cos(\tilde{B}_t/2)\sin(\tilde{A}_t/2) + \cos(\tilde{A}_t/2)\sin(\tilde{B}_t/2))\right) &= 0 \\ \mathbb{E}\left(e^{i\tilde{C}_t/2}(\cos(\tilde{A}_t/2)\cos(\tilde{B}_t/2) + i\sin(\tilde{A}_t/2)\sin(\tilde{B}_t/2))\right) &= e^{-3t/8} \end{aligned}$$

and for $j = 2$:

$$\begin{aligned} e^{-t} &= \mathbb{E}\left(e^{-i\tilde{C}_t}\left(\cos(\frac{\tilde{A}_t}{2})^2\cos(\frac{\tilde{B}_t}{2})^2 + \sin(\frac{\tilde{A}_t}{2})^2\sin(\frac{\tilde{B}_t}{2})^2 - \frac{iI\,\sin(\tilde{A}_t)\sin(\tilde{B}_t)}{2}\right)\right) \\ 0 &= \mathbb{E}\left(i\,\sin(\tilde{A}_t - \tilde{B}_t) + 2\,\sin(\tilde{B}_t) + i\,\sin(\tilde{A}_t + \tilde{B}_t)\right) \\ 0 &= \mathbb{E}\left(e^{i\tilde{C}_t}\left(-\cos(\frac{\tilde{B}_t}{2})^2\sin(\frac{\tilde{A}_t}{2})^2 - \cos(\frac{\tilde{A}_t}{2})^2\sin(\frac{\tilde{B}_t}{2})^2 + \frac{iI\,\sin(\tilde{A}_t)\sin(\tilde{B}_t)}{2}\right)\right) \\ 0 &= \mathbb{E}\left(e^{-i\tilde{C}_t}\left(-\sin(\tilde{A}_t - \tilde{B}_t) - i\,\sin(\tilde{A}_t + \tilde{B}_t)\right)\right) \\ e^{-t} &= \mathbb{E}\left(\cos(\tilde{A}_t)\,\cos(\tilde{B}_t)\right) \\ 0 &= \mathbb{E}\left(e^{i\tilde{C}_t}\left(-i\,\sin(\tilde{A}_t - \tilde{B}_t) - \sin(\tilde{A}_t + \tilde{B}_t)\right)\right) \\ 0 &= \mathbb{E}\left(e^{-i\tilde{C}_t}\left(-1 + i - i\,\cos(\tilde{A}_t - \tilde{B}_t) + \cos(\tilde{A}_t + \tilde{B}_t)\right)\right) \\ 0 &= \mathbb{E}\left(-i\,\sin(\tilde{A}_t - \tilde{B}_t) + 2\,\sin(\tilde{B}_t) - i\,\sin(\tilde{A}_t + \tilde{B}_t)\right) \\ e^{-t} &= \mathbb{E}\left(e^{i\tilde{C}_t}\left(1 - i + \cos(\tilde{A}_t - \tilde{B}_t) - i\,\cos(\tilde{A}_t + \tilde{B}_t)\right)\right). \end{aligned}$$

Chapter 3

Hopf algebras, quantum groups and braided spaces

This chapter provides the necessary background on the algebraic structures that we are going to use in the following chapters.

Quantum groups have received a lot of attention from the physics and mathematics community since the early eighties. Even though no generally accepted definition seems to exist to this day, most authors agree that a quantum group should at least be a bialgebra or a Hopf algebra. For the definition of Lévy processes, the main class of processes that we shall be concerned with, it is sufficient to have an involutive bialgebra structure, see Chapter 4.

Standard textbooks on bialgebras and Hopf algebras are [Swe69, Abe80], but see also [DHL91, Fuc92, FK93, Lus93, Mon93, SS93, CP95, Gui95, Kas95, Maj95b] and the references therein.

This chapter is organised as follows:

First, in Section 3.1, we give the basic definitions of bialgebras and Hopf algebras. Duality and dual representations are also already introduced. In the next section (Section 3.2) we present several examples of bialgebras and Hopf algebras. We start with a few classical examples arising from groups or semi-groups and continue then with some standard examples of non-commutative and non-cocommutative Hopf algebras. In Section 3.3, we study dual representations. We show that the definition of dual representations for dually paired bialgebras extends that of dual representations for Lie algebras introduced in Section 2.3, show that similar techniques can be used for calculating them, and present the explicit calculations for several examples. To underline the similarity to the theory on Lie groups further we introduce an analogue of the composition law in Section 3.4. It is a consequence of duality, but written in terms of the dual pairing it has the same form as the composition (or group) law on Lie groups. Formulas of the form (3.12) will play a central role in the proof of the Feynman-Kac formula in Section 4.7. In Section 3.6 we introduce matrix elements for dual representations of quantum groups along the line of Section 2.6.

In Section 3.7 we introduce braided tensor categories, an important concept that is intimately related to the representation theory of quantum groups. In the next two Sections (Sections 3.8 and 3.9) we defined braided bialgebras and braided Hopf algebras and present several examples. In Section 3.10 we concentrate on a special class of braided Hopf algebras which are primitively generated, called braided spaces.

3.1 COALGEBRAS, BIALGEBRAS AND HOPF ALGEBRAS

An associative algebra $\mathcal{A}$ over a field $\mathbb{K}$ is a $\mathbb{K}$-vector space equipped with a bilinear mapping $m : \mathcal{A} \times \mathcal{A} \to \mathcal{A}$ such that $m(a, m(b, c)) = m(m(a, b), c))$ for all $a, b, c \in \mathcal{A}$ or, equivalently, with a linear mapping $m : \mathcal{A} \otimes \mathcal{A} \to \mathcal{A}$ that satisfies

$$m \circ (m \otimes \mathrm{id}) = m \circ (\mathrm{id} \otimes m) \qquad \text{(associativity)}.$$

We will always assume that our algebras are unital, i. e. that an element $e \in \mathcal{A}$ exists such that

$$m(a \otimes e) = m(e \otimes a) = a \qquad \text{for all } a \in \mathcal{A}.$$

Such a unit element e can be identified with the mapping $e : \mathbb{K} \to \mathcal{A}$ defined by $e(\lambda) = \lambda e$. The tensor product $\mathcal{A} \otimes \mathcal{A}$ has again an associative algebra structure given by

$$\begin{aligned} e^{\otimes} &= e \otimes e, \\ m^{\otimes} &= (m \otimes m) \circ (\mathrm{id} \otimes \tau \circ \mathrm{id}), \end{aligned}$$

where $\tau : \mathcal{A} \otimes \mathcal{A} \to \mathcal{A} \otimes \mathcal{A}$ is the *flip automorphism* defined by $\tau(a \otimes b) = b \otimes a$, $\forall a, b \in A$.

An algebra $(\mathcal{A}, m)$ is called *commutative*, if $m(a \otimes b) = m(b \otimes a)$ for all $a, b \in \mathcal{A}$, i.e. $m = m \circ \tau$.

Assume for the moment that $\mathcal{A}$ is finite dimensional. The algebra structure of $\mathcal{A}$, given by m and e, defines a dual structure $m^* : \mathcal{A}^* \to (\mathcal{A} \otimes \mathcal{A})^* \cong \mathcal{A}^* \otimes \mathcal{A}^*$, $e^* : \mathcal{A}^* \to \mathbb{K}$ on the dual space $\mathcal{A}^* = \{\varphi : \mathcal{A} \to \mathbb{K}; \varphi \text{ linear}\}$ by

$$\begin{aligned} m^*(\varphi)(a \otimes b) &= \varphi(m(a \otimes b)), \\ e^*(\varphi) &= \varphi(e), \end{aligned} \tag{3.1}$$

for $\varphi \in \mathcal{A}^*$, $a, b \in \mathcal{A}$. The algebra axioms imply that m^* and e^* satisfy

$$\begin{aligned} (\mathrm{id} \otimes m^*) \circ m^* &= (m^* \otimes \mathrm{id}) \circ m^*, \\ (e^* \otimes \mathrm{id}) \circ m^* &= (\mathrm{id} \otimes e^*) \circ m^* = \mathrm{id}. \end{aligned}$$

We therefore define a *coalgebra* over the field $\mathbb{K}$ abstractly as a triple $(\mathcal{C}, \Delta, \epsilon)$ of a $\mathbb{K}$-vector space $\mathcal{C}$ (not necessarily finite dimensional) and $\mathbb{K}$-linear maps $\Delta : \mathcal{C} \to \mathcal{C} \otimes \mathcal{C}$, $\epsilon : \mathcal{C} \to \mathbb{K}$ that satisfy the coalgebra axioms

$$\begin{aligned} (\Delta \otimes \mathrm{id}) \circ \Delta &= (\mathrm{id} \otimes \Delta) \circ \Delta && \text{(coassociativity)} \\ (\epsilon \otimes \mathrm{id}) \circ \Delta &= (\mathrm{id} \otimes \epsilon) \circ \Delta = \mathrm{id} && \text{(counit)} \end{aligned}$$

The dual of an infinite-dimensional algebra is not always a coalgebra, because for infinite vector spaces $(\mathcal{A}\otimes\mathcal{A})^* \not\cong \mathcal{A}^*\otimes\mathcal{A}^*$, but only $\mathcal{A}^*\otimes\mathcal{A}^* \subset (\mathcal{A}\otimes\mathcal{A})^*$ (in a canonical way), and the image of the map m^* defined in Equation (3.1) will in general not lie inside $\mathcal{A}^*\otimes\mathcal{A}^*$. But the dual of a coalgebra is always an algebra with the multiplication $\Delta^* : \mathcal{C}^*\otimes\mathcal{C}^* \to \mathcal{C}^*$ and the unit $\varepsilon : \mathbb{K} \to \mathcal{C}^*$ defined in the obvious way.

Similarly as for algebras the maps

$$\begin{aligned}\Delta^{\otimes} &: \mathcal{C}\otimes\mathcal{C} \to (\mathcal{C}\otimes\mathcal{C})\otimes(\mathcal{C}\otimes\mathcal{C}),\\ \Delta^{\otimes} &= (\mathrm{id}\otimes\tau\otimes\mathrm{id})\circ(\Delta\otimes\Delta),\\ \epsilon^{\otimes} &: \mathcal{C}\otimes\mathcal{C} \to \mathbb{K},\\ \epsilon^{\otimes} &= \epsilon\otimes\epsilon\end{aligned}$$

define a coalgebra $(\mathcal{C}\otimes\mathcal{C}, \Delta^{\otimes}, \epsilon^{\otimes})$.

A coalgebra $(\mathcal{C}, \Delta)$ is called *cocommutative*, if $\Delta = \tau\circ\Delta$.

Since Δc is an element of $\mathcal{C}\otimes\mathcal{C}$, it is a finite sum of the form $\Delta c = \sum_i c_i^{(1)}\otimes c_i^{(2)}$. It has become customary to omit the summation index and/or the summation symbol, e.g.

$$\Delta c = c^{(1)}\otimes c^{(2)}, \tag{3.2}$$

this kind of short-hand notation is generally referred to as *Sweedler's notation*. If we apply id$\otimes\Delta$ or $\Delta\otimes$id to Δc, we get the same result (due to the coassociativity), in Sweedler's notation this is expressed in the form

$$c^{(1)}\otimes c^{(2)(1)}\otimes c^{(2)(2)} = c^{(1)(1)}\otimes c^{(1)(2)}\otimes c^{(2)} =: c^{(1)}\otimes c^{(2)}\otimes c^{(3)}.$$

A *sub-coalgebra* $\mathcal{C}_0$ of a coalgebra $\mathcal{C}$ is a sub-space such that $\Delta\mathcal{C}_0 \subseteq \mathcal{C}_0\otimes\mathcal{C}_0$. A *right (left) coideal* $\mathcal{I}$ of $\mathcal{C}$ is a sub-space such that $\Delta\mathcal{I} \subseteq \mathcal{C}\otimes\mathcal{I}$ ($\Delta\mathcal{I} \subseteq \mathcal{I}\otimes\mathcal{C}$, resp.). For a *coideal* one requires $\Delta\mathcal{I} \subseteq \mathcal{C}\otimes\mathcal{I} + \mathcal{I}\otimes\mathcal{C}$ and $\varepsilon(\mathcal{I}) = 0$, these are exactly the conditions that are necessary so that Δ and ε factor through to the maps $\Delta_{\mathcal{C}/\mathcal{I}} : \mathcal{C}/\mathcal{I} \to \mathcal{C}/\mathcal{I}\otimes\mathcal{C}/\mathcal{I}$ and $\varepsilon_{\mathcal{C}/\mathcal{I}} : \mathcal{C}/\mathcal{I} \to \mathbb{K}$. The triple $(\mathcal{C}/\mathcal{I}, \Delta_{\mathcal{C}/\mathcal{I}}, \varepsilon_{\mathcal{C}/\mathcal{I}})$ is again a coalgebra, the so-called *quotient coalgebra*.

The following useful result is a consequence of the fact that the algebraic tensor product is used in the definition of coalgebras.

Theorem 3.1.1 (Fundamental theorem of coalgebras) *In a coalgebra $\mathcal{C}$, every finite-dimensional sub-space is contained in a finite-dimensional sub-coalgebra.*

For the proof see [Swe69].

Definition 3.1.2 *A* bialgebra *over $\mathbb{K}$ is a 5-tuple $(\mathcal{A}, m, e, \Delta, \epsilon)$ where*

- *$(\mathcal{A}, m, e)$ is an algebra over $\mathbb{K}$,*
- *$(\mathcal{A}, \Delta, \epsilon)$ is a coalgebra over $\mathbb{K}$,*
- *and the algebra structure and the coalgebra structure are compatible, i. e.*

  $\Delta : \mathcal{A} \to \mathcal{A}\otimes\mathcal{A}$ and $\epsilon : \mathcal{A} \to \mathbb{K}$ are algebra homomorphisms

or, equivalently,

$$m : \mathcal{A} \otimes \mathcal{A} \to \mathcal{A} \text{ and } e : \mathbb{K} \to \mathcal{A} \text{ are coalgebra homomorphisms.}$$

The compatibility conditions can also be written as

$$\begin{aligned} \Delta \circ m &= m^{\otimes} \circ (\Delta \otimes \Delta) = (m \otimes m) \circ \Delta^{\otimes}, \\ \Delta \circ e &= e \otimes e, \\ \epsilon \circ m &= \epsilon \otimes \epsilon, \\ \epsilon \circ e &= \mathrm{id}_{\mathbb{K}}. \end{aligned}$$

For a Hopf algebra we need one more ingredient, the antipode.

Definition 3.1.3 *Let $(\mathcal{A}, m, e, \Delta, \epsilon)$ be a bialgebra. A linear map $S : \mathcal{A} \to \mathcal{A}$ that satisfies*

$$m \circ (\mathrm{id} \otimes S) \circ \Delta = m \circ (S \otimes \mathrm{id}) \circ \Delta = e \circ \epsilon$$

is called an antipode, *and $(\mathcal{A}, m, e, \Delta, \epsilon, S)$ is called a* Hopf algebra.

If an antipode exists, it is unique, and an algebra anti-homomorphism, i. e. $m \circ (S \otimes S) = S \circ m \circ \tau$, or $S(a)S(b) = S(ba)$ for all $a, b \in \mathcal{A}$, and a coalgebra anti-homomorphism, i.e. $\Delta \circ S = \tau \circ (S \otimes S) \circ \Delta$.

A $*$-*bialgebra* is a bialgebra $\mathcal{A}$ (over a field with involution, e.g. $\mathbb{C}$) with an involution $* : \mathcal{A} \to \mathcal{A}$ such that $(\mathcal{A}, m, e, *)$ is a $*$-algebra (i.e. $(e(\lambda))^* = e(\overline{\lambda})$, $* \circ m = m \circ \tau \circ (* \otimes *)$, $* \circ * = \mathrm{id}$), and Δ and ε are $*$-algebra homomorphisms (i.e. $(* \otimes *) \circ \Delta = \Delta \circ *$ and $\varepsilon(a^*) = \overline{(\varepsilon(a))}$ for all $a \in A$). For a $*$-*Hopf algebra* one requires in addition $S \circ * \circ S \circ * = \mathrm{id}$. Note that this implies $* \circ S \circ * \circ S = \mathrm{id}$ and therefore that S^{-1} exists and is given by $S^{-1} = * \circ S \circ *$.

We say that two $\mathbb{K}$-bialgebras $(\mathcal{A}_1, m_1, e_1, \Delta_1, \epsilon_1)$ and $(\mathcal{A}_2, m_2, e_2, \Delta_2, \epsilon_2)$ are *dually paired* (form a *dual pair*), if there exists a non-degenerate bilinear map $\langle \cdot, \cdot \rangle : \mathcal{A}_1 \times \mathcal{A}_2 \to \mathbb{K}$, such that

$$\begin{aligned} \langle m_1(a_1 \otimes b_1), c_2 \rangle &= \langle a_1 \otimes b_1, \Delta_2(c_2) \rangle_{\otimes}, \\ \langle c_1, m_2(a_2 \otimes b_2) \rangle &= \langle \Delta_1(c_1), a_2 \otimes b_2 \rangle_{\otimes}, \\ \langle e_1, a_2 \rangle &= \epsilon_2(a_2), \\ \langle a_1, e_2 \rangle &= \epsilon_1(a_1), \end{aligned}$$

for all $a_1, b_1, c_1 \in \mathcal{A}_1$, $a_2, b_2, c_2 \in \mathcal{A}_2$. For a dual pairing between two Hopf algebras $(\mathcal{A}_1, m_1, e_1, \Delta_1, \epsilon_1, S_1)$ and $(\mathcal{A}_2, m_2, e_2, \Delta_2, \epsilon_2, S_2)$ we also require

$$\langle S_1(a_1), a_2 \rangle = \langle a_1, S_2(a_2) \rangle$$

to be satisfied for all $a_1 \in \mathcal{A}_1$, $a_2 \in \mathcal{A}_2$.

If $\mathcal{A}_1$ and $\mathcal{A}_2$ are $*$-Hopf algebras, then there is an additional condition, namely

$$\langle a_1^*, a_2 \rangle = \overline{\langle a_1, (S(a_2))^* \rangle}$$

for all $a_1 \in \mathcal{A}_1$, $a_2 \in \mathcal{A}_2$.

Whenever we have a dual pair, we also have actions of each Hopf algebra on the other one, the so-called right and left *regular* or *dual representation.* They are defined by $\rho_R^*, \rho_L^* : \mathcal{A}_1 \to \mathrm{Hom}(\mathcal{A}_2, \mathcal{A}_2)$, $\rho_R^*(X) = (\mathrm{id} \otimes X) \circ \Delta_2$ and $\rho_L^*(X) = (X \otimes \mathrm{id}) \circ \Delta_2$, respectively, and they satisfy

$$\rho_R^*(XY) = \rho_R^*(X)\rho_R^*(Y), \qquad \rho_L^*(XY) = \rho_L^*(Y)\rho_L^*(X),$$

for all $X, Y \in \mathcal{A}_1$.

These representations have a 'nice' behaviour with respect to the multiplication in $\mathcal{A}_2$. Since the coproduct in $\mathcal{A}_1$ is the dual of the multiplication in $\mathcal{A}_2$, we get the following *Leibniz formulas.* Let $\Delta_1(X) = \sum X_i^{(1)} \otimes X_i^{(2)}$, then

$$\begin{aligned} \rho_L^*(X)(ab) &= \sum(\rho_L^*(X_i^{(1)})a)(\rho_L^*(X_i^{(2)})b) \\ \rho_R^*(X)(ab) &= \sum(\rho_R^* X_i^{(1)})a)(\rho_R^*(X_i^{(2)})b) \end{aligned} \tag{3.3}$$

for all $X \in \mathcal{A}_1$, $a, b \in \mathcal{A}_2$. For *primitive* elements X (i.e. $\Delta X = X \otimes 1 + 1 \otimes X$) this is the usual Leibniz formula $\rho_{R,L}^*(X)(ab) = (\rho_{R,L}^*(X)a)b + a(\rho_{R,L}^*(X)b)$. Note that the primitive elements of a bialgebra form a Lie algebra, since for X, Y with $\Delta X = X \otimes 1 + 1 \otimes X$ and $\Delta Y = Y \otimes 1 + 1 \otimes Y$, we have $\Delta[X, Y] = [X \otimes 1 + 1 \otimes X, Y \otimes 1 + 1 \otimes Y] = XY \otimes 1 + X \otimes Y + Y \otimes X + 1 \otimes XY - YX \otimes 1 - X \otimes Y - Y \otimes X - 1 \otimes YX = [X, Y] \otimes 1 + 1 \otimes [X, Y]$.

For *group-like* elements g (i.e. $\Delta g = g \otimes g$ and $\varepsilon(g) = 1$) the Leibniz formula reads $\rho_{R,L}^*(g)(ab) = (\rho_{R,L}^*(X)a)(\rho_{R,L}^*(X)b)$. The group-like elements of a bialgebra form a unital semi-group, since $\Delta 1 = 1 \otimes 1$ and since $\Delta g_1 = g_1 \otimes g_1$ and $\Delta g_2 = g_2 \otimes g_2$ implies $\Delta g_1 g_2 = g_1 g_2 \otimes g_1 g_2$. In a Hopf algebra they even form a group. To see this note that the definition of the antipode implies $S(g) = g^{-1}$ for group-like elements.

Let us also introduce some basic cohomology language that we will meet again in Section 4.4. If $\mathcal{A}$ is an associative algebra and $\mathcal{M}$ an $\mathcal{A}$-bimodule, with the left and right actions $\rho_l : \mathcal{A} \times \mathcal{M} \to \mathcal{M}$ and $\rho_r : \mathcal{M} \times \mathcal{A} \to \mathcal{M}$, then we define the space of (ρ_l, ρ_r)-*n-cochains* as $C^n = \mathrm{Hom}_{\mathbb{K}}(\mathcal{A}^{\otimes n}, \mathcal{M})$ and we define $\mathbb{K}$-bimodule maps

$$\begin{aligned} \partial &: C^n \to C^{n+1} \\ \partial\alpha(a_1 \otimes \cdots \otimes a_{n+1}) &= \rho_l(a_1)\alpha(a_2 \otimes \cdots \otimes a_{n+1}) \\ &\quad + \sum_{i=1}^{n}(-1)^i \alpha(a_1 \otimes \cdots \otimes a_i a_{i+1} \otimes \cdots \otimes a_{n+1}) \\ &\quad + (-1)^{n+1}\alpha(a_1 \otimes \cdots \otimes a_n)\rho_r(a_{n+1}) \end{aligned}$$

for a (ρ_l, ρ_r)-n-cochain $\alpha \in C^n$ and for $a_1, \ldots, a_{n+1} \in \mathcal{A}$. An (ρ_l, ρ_r)-n-cochain α is called an (ρ_l, ρ_r)-*n-cocycle*, if

$$\partial\alpha = 0$$

and a (ρ_l, ρ_r)-*n-coboundary* (of β) if there exists an (ρ_l, ρ_r)-$(n-1)$-cochain $\beta \in C^{n-1}$ such that

$$\partial\beta = \alpha.$$

We denote by $Z^n(\mathcal{A},\mathcal{M})$ the space of (ρ_l,ρ_r)-n-cocycles and by $B^n(\mathcal{A},\mathcal{M})$ the space of (ρ_l,ρ_r)-n-coboundaries. Since $\partial^2=0$, we have $B^n(\mathcal{A},\mathcal{M})\subseteq Z^n(\mathcal{A},\mathcal{M})$, i.e. every coboundary is also a cocycle, and we can define the *cohomology group* $H^n(\mathcal{A},\mathcal{M})=B^n(\mathcal{A},\mathcal{M})/Z^n(\mathcal{A},\mathcal{M})$.

There also exists a cohomology theory for coalgebras that is essentially the dual version of that for associative algebras. If $\mathcal{N}$ is a $\mathcal{C}$-bicomodule, with the left and right coactions $\gamma_l:\mathcal{N}\to\mathcal{C}\otimes\mathcal{N}$ and $\gamma_r:\mathcal{N}\to\mathcal{N}\otimes\mathcal{C}$, then the space of n-cochains is defined as $D^n=\mathrm{Hom}_{\mathbb{K}}(\mathcal{N},\mathcal{C}^{\otimes n})$. The bicomodule maps $\partial:D^n\to D^{n+1}$ are now defined as

$$\begin{aligned}
\partial &: D^n\to D^{n+1}\\
\partial\alpha &= (\mathrm{id}_{\mathcal{C}}\otimes\alpha)\circ\gamma_l\\
&\quad+\sum_{i=1}^{n}(-1)^i(\underbrace{\mathrm{id}_{\mathcal{C}}\otimes\cdots\otimes\mathrm{id}_{\mathcal{C}}}_{i-1\text{ times}}\otimes\Delta\otimes\underbrace{\mathrm{id}_{\mathcal{C}}\otimes\cdots\otimes\mathrm{id}_{\mathcal{C}}}_{n-i-1\text{ times}})\circ\alpha\\
&\quad+(-1)^{n+1}(\alpha\otimes\mathrm{id}_{\mathcal{C}})\circ\gamma_r,
\end{aligned}$$

for $\alpha\in D^n$.

As before we can define (γ_l,γ_r)-n-coboundaries, (γ_l,γ_r)-n-cocycles, and the cohomology group.

If $\mathcal{A}$ is actually a Hopf algebra, then both constructions can be combined to define Hopf algebra cohomology theory (see, e.g., [CP95, Paragraph 6.1.B]), but we shall not need it here.

3.2 EXAMPLES OF BIALGEBRAS AND HOPF ALGEBRAS

3.2.1 *The bialgebras* $\mathbb{K}G$ *and* $(\mathbb{K}G)^*$

Let $(G,\circ,e)$ be a finite unital semi-group. The algebra $A=\mathrm{Fun}\,G=\mathbb{K}^G=\{\varphi:G\to\mathbb{K}\}$ of functions on G and the group algebra $a=\mathbb{K}G$ can be turned into a dual pair of $\mathbb{K}$-bialgebras. The set of the characteristic functions (indicator functions) $\{\mathbb{1}_g;g\in G\}$ with

$$\mathbb{1}_g(g')=\begin{cases}1 & \text{if } g=g',\\ 0 & \text{else,}\end{cases}$$

is a basis of $\mathrm{Fun}\,G$. For the dual space we will use the dual basis $\{b_g;g\in G\}$. The b_g are defined by

$$\langle b_g,\mathbb{1}_{g'}\rangle=b_{g'}(\mathbb{1}_g)=\begin{cases}1 & \text{if } g=g',\\ 0 & \text{else,}\end{cases}$$

The multiplication in $\mathrm{Fun}\,G$ is defined as pointwise multiplication

$$(\varphi\cdot\psi)(g)=\varphi(g)\psi(g),\qquad \varphi,\psi\in\mathrm{Fun}\,G,\quad g\in G.$$

In terms of the basis introduced above this becomes

$$m_A(\lambda \otimes \mu) = m_A\left(\sum \lambda_g \mathbb{1}_g \otimes \sum \mu_g \mathbb{1}_g\right) = \sum \lambda_g \mu_g \mathbb{1}_g.$$

(We use the subscripts a and A to distinguish between the operations in $A = \operatorname{Fun} G$ and $a = \mathbb{K}G$). The unit element (with respect to m_A) is

$$e_A = \sum \mathbb{1}_g,$$

i. e. the constant function with value 1. The coproduct of $\operatorname{Fun} G$ is induced by the multiplication of G. It is given by

$$(\Delta_A \varphi)(g_1, g_2) = \varphi(g_1 \circ g_2), \qquad \varphi \in \mathbb{K}G, \;\; g_1, g_2 \in G.$$

The counit $\epsilon_A : \operatorname{Fun} G \to \mathbb{K}$ is defined by

$$\epsilon_A \varphi = \varphi(e).$$

Again using the basis $\{\mathbb{1}_g; g \in G\}$, we can rewrite these relations as

$$\begin{aligned} \Delta_A(\lambda) &= \Delta_A\left(\sum \lambda_g \mathbb{1}_g\right) = \sum_{g,g'} \lambda_{g\circ g'} \mathbb{1}_g \otimes \mathbb{1}_{g'}, \\ \epsilon_A \lambda &= \epsilon_A\left(\sum \lambda_g \mathbb{1}_g\right) = \lambda_e. \end{aligned}$$

We now use the duality between $\operatorname{Fun} G$ and $\mathbb{K}G$ to define a bialgebra structure on $\mathbb{K}G$. We set $m_a = \Delta_A^*$ and $\Delta_a = m_A^*$, i. e. we define m_a and Δ_a by requiring

$$\begin{aligned} \langle m_A(\lambda \otimes \mu), x\rangle &= \langle \lambda \otimes \mu, \Delta_a(x)\rangle_\otimes, \\ \langle \Delta_A(\lambda), x \otimes y\rangle_\otimes &= \langle \lambda, m_a(x \otimes y)\rangle, \end{aligned}$$

to be satisfied for all $\lambda, \mu \in \operatorname{Fun} G$ and all $x, y \in \mathbb{K}G$. This gives

$$\begin{aligned} m_a(x \otimes y) &= m_a\left(\sum x_g b_g \otimes \sum y_g b_g\right) = \sum_g \left(\sum_{g' \circ g'' = g} x_{g'} y_{g''}\right) b_g, \\ \Delta_a(x) &= \Delta_a\left(\sum x_g b_g\right) = \sum x_g b_g \otimes b_g, \end{aligned}$$

and we see that we really recovered the multiplication of the group algebra, i.e. the multiplication defined by setting $b_g b_{g'} = b_{gg'}$. We see that the basis elements b_g are group-like. Furthermore, they are the only group-like elements and the semi-group formed by them is obviously isomorphic to G.

The unit e_a and counit ϵ_a of $\mathbb{K}G$ are given by

$$\begin{aligned} e_a &= b_e, \\ \epsilon_a(x) &= \epsilon_a\left(\sum x_g b_g\right) = \sum x_g. \end{aligned}$$

They satisfy the relations

$$\begin{aligned} \langle \lambda, e_a \rangle &= \epsilon_A(\lambda), \\ \langle e_A, x \rangle &= \epsilon_a(x), \end{aligned}$$

for all $\lambda \in \mathrm{Fun}\, G$, $x \in \mathbb{K}G$, i. e. $\mathrm{Fun}\, G$ and $\mathbb{K}G$ are dually paired.

If G is a group, then we can also define antipodes on $\mathrm{Fun}\, G$ and $\mathbb{K}G$, namely

$$\begin{aligned} S_A(\mathbb{1}_g) &= \mathbb{1}_{g^{-1}} \\ S_a(b_g) &= b_{g^{-1}}, \end{aligned}$$

and $\mathrm{Fun}\, G$ and $\mathbb{K}G$ are also dually paired as Hopf algebras, i. e.

$$\langle S_a(x), \lambda \rangle = \langle x, S_A(\lambda) \rangle \qquad \text{for all } \lambda \in \mathrm{Fun}\, G, x \in \mathbb{K}G.$$

If the field $\mathbb{K}$ has an involution (e.g., if $\mathbb{K} = \mathbb{C}$), then $\mathrm{Fun}\, G$ is an involutive algebra with the pointwise involution,

$$\lambda^*(g) = \overline{\lambda(g)}, \qquad \text{for } g \in G.$$

It is obvious that ε_A and Δ_A are then $*$-algebra homomorphisms and that $S_A \circ * \circ S_A \circ * = \mathrm{id}$ holds, so that $\mathrm{Fun}\, G$ is even a $*$-Hopf algebra. To turn $\mathrm{Fun}\, G$ and $\mathbb{K}G$ into a dual pair of $*$-Hopf algebras, the involution $* : \mathbb{K}G \to \mathbb{K}G$ has to satisfy

$$\langle \mathbb{1}_g, x^* \rangle = \langle \mathbb{1}_g^*, x^* \rangle = \overline{\langle \mathbb{1}_g, S(x^*)^* \rangle} = \overline{\langle \mathbb{1}_g, S^{-1}(x) \rangle},$$

for all $g \in G$, $x \in \mathbb{K}G$, so that we have to set

$$x^* = \sum_g \overline{x_{g^{-1}}} b_g, \qquad \text{for } x = \sum_g x_g b_g \in \mathbb{K}G.$$

3.2.2 *The bialgebras $l^\infty(\mathbb{Z})$ and $l^1(\mathbb{Z})$*

If we want to turn the space of functions on an infinite (semi-) group into a bialgebra, we have to be more careful. Consider for example the group $(\mathbb{Z}, +)$. The space $\mathbb{K}^{\mathbb{Z}}$ of all functions on $\mathbb{Z}$ is too large, we have $\mathbb{K}^{\mathbb{Z}} \otimes \mathbb{K}^{\mathbb{Z}} \subset \mathbb{K}^{\mathbb{Z}\times\mathbb{Z}}$, but not equality, because the algebraic tensor product contains only finite sums. On the other hand, if we restrict ourselves to the functions that are zero everywhere except on a finite number of elements (as a vector space this is exactly $\mathbb{K}\mathbb{Z}$, the free vector space generated by $\mathbb{Z}$)

$$\mathbb{K}\mathbb{Z} = \left\{ \sum_{k\in\mathbb{Z}} \lambda_k \mathbb{1}_k; \text{ only finitely many } \lambda_k \neq 0 \right\},$$

then this is too small. The constant function $e = \sum_{k\in\mathbb{Z}} \mathbb{1}_k$, our candidate for the unit element, is not contained in $\mathbb{K}\mathbb{Z}$.

But we can still define a (topological) Hopf algebra of functions on $\mathbb{Z}$, if we introduce an appropriate norm on $\mathbb{K}\mathbb{Z}$, complete $\mathbb{K}\mathbb{Z}$ with respect to this norm, and replace the algebraic tensor product by some appropriate topological tensor product. In the case of $\mathbb{Z}$ it turns out that we can take the norm $||\cdot||_\infty$ defined by

$$||\lambda||_\infty = \sup_{k\in\mathbb{Z}} |\lambda_k| \qquad \text{for } \lambda = \sum_{k\in\mathbb{Z}} \lambda_k b_k.$$

Completing $\mathbb{K}\mathbb{Z}$ with respect to the norm $||\cdot||_\infty$ we get the Banach space $l^\infty(\mathbb{Z})$ of bounded functions on $\mathbb{Z}$. The algebraic operations are given by

$$\begin{aligned} m_\infty(\lambda,\mu) &= \sum_{k\in\mathbb{Z}} \lambda_k \mu_k \mathbb{1}_k \\ e_\infty &= \sum_{k\in\mathbb{Z}} \mathbb{1}_k \\ \Delta_\infty(\lambda) &= \sum_{k,k'\in\mathbb{Z}} \lambda_{k+k'} \mathbb{1}_k \otimes \mathbb{1}_{k'} \\ \epsilon_\infty(\lambda) &= \lambda_0 \end{aligned}$$

for $\lambda = \sum \lambda_k \mathbb{1}_k$, $\mu = \sum m_k \mathbb{1}_k \in l^\infty(\mathbb{Z})$. We can also define an antipode $S : l^\infty(\mathbb{Z}) \to l^\infty(\mathbb{Z})$. It is determined by its action on the basis

$$S(\mathbb{1}_k) = \mathbb{1}_{-k}.$$

The operations defined above are continuous with respect to the norm $||\cdot||_\infty$:

$$\begin{aligned} ||m_\infty(\lambda,\mu)||_\infty &\leq ||\lambda||_\infty \cdot ||\mu||_\infty \\ ||e_\infty||_\infty &= 1 \\ ||\Delta_\infty(\lambda)||_\infty^\otimes &= ||\lambda||_\infty \\ |\epsilon_\infty(\lambda)| &\leq ||\lambda||_\infty \\ ||S(\lambda)||_\infty &= ||\lambda||_\infty \end{aligned}$$

(where $||\sum_{k,k'} \lambda_{k,k'} \mathbb{1}_k \otimes \mathbb{1}_{k'}||_\infty^\otimes = \sup_{k,k'} |\lambda_{k,k'}|$).

The Hopf algebra $(l^\infty(\mathbb{Z}), m_\infty, e_\infty, \Delta_\infty, \epsilon_\infty, S)$ can be dually paired with the Hopf algebra $(l^1(\mathbb{Z}), m_1, e_1, \Delta_1, \epsilon_1, S)$, where

$$\begin{aligned} ||x||_1 &= ||\sum_{k\in\mathbb{Z}} x_k b_k||_1 = \sum_{k\in\mathbb{Z}} |x_k| \\ m_1(x,y) &= \sum_{k\in\mathbb{Z}} \left(\sum_{l\in\mathbb{Z}} x_{k-l} y_l \right) b_k \\ e_1 &= b_0 \\ \Delta_1(x) &= \sum_{k\in\mathbb{Z}} x_k b_k \otimes b_k \\ \epsilon_1(x) &= \sum_{k\in\mathbb{Z}} x_k. \end{aligned}$$

and where the pairing is determined by

$$\langle b_k, \mathbb{1}_{k'} \rangle = \delta_{k,k'} \qquad \text{(Kronecker delta).}$$

3.2.3 *Polynomials on the real line*

Let us now take the algebra of polynomials

$$\mathbb{R}[x] = \mathrm{Pol}(\mathbb{R}) = \left\{ \sum_{n\in\mathbb{N}} a_n x^n;\ \text{only finitely many } a_n \neq 0 \right\}.$$

We will use the identification $\mathbb{R}[x]\otimes\mathbb{R}[x] = \mathbb{R}[x', x'']$ which can be obtained by replacing $x^n \otimes x^m$ by $x'^n x''^m$. We can define a multiplication, unit, comultiplication, counit and antipode on this space:

$$\begin{aligned}
m(p(x), q(x)) &= m\left(\sum p_n x^n, \sum q_n x^n\right) = \sum p_n q_n x^{n+m} \\
&= p(x)q(x) \qquad \text{(usual multiplication of polynomials)} \\
e &= 1 \\
\Delta(p(x)) &= \Delta\left(\sum p_n x^n\right) = \sum_n \sum_{\nu=0}^{n} p_n \begin{pmatrix} n \\ \nu \end{pmatrix} x^\nu \otimes x^{n-\nu} \\
&= p(x' + x'') \\
\epsilon(p(x)) &= p(0) = p_0 \\
S(p(x)) &= \sum (-1)^n p_n x^n = p(-x).
\end{aligned}$$

This Hopf algebra can be dually paired with itself. We set

$$\langle x^n, x^m \rangle = n!\delta_{n,m}.$$

This can also be written as

$$\langle p(x), q(x) \rangle = p\left(\frac{\mathrm{d}}{\mathrm{d}x}\right) q(x)\Big|_{x=0}.$$

3.2.4 *The continuous functions on the circle S^1*

Let S^1 be the unit circle $S^1 = \{z \in \mathbb{C}; |z| = 1\} = \{\exp(i\theta); \theta \in [-\pi, \pi[\}$. The restriction of the complex multiplication to the unit circle turns it into a compact topological group (using the trace topology). $S^1 \times S^1$, equipped with the product topology and product group structure, is also a compact topological group, and we have $C(S^1) \otimes C(S^1) \cong C(S^1 \times S^1)$, where $C(S^1)$ ($C(S^1 \times S^1)$) is the space of continuous functions on S^1 (resp. $S^1 \times S^1$), and $\otimes$ denotes the topological tensor product. The definitions

$$\begin{aligned}
(f \cdot g)(z) &= f(z)g(z) \\
e(z) &= 1 \\
(\Delta f)(z', z'') &= f(z'z'') \\
\epsilon(f) &= f(1) \\
(Sf) &= f(z^{-1})
\end{aligned}$$

turn $C(S^1)$ into a Hopf algebra. The dual space of $C(S^1)$, the space of signed Radon measures $M(S^1)$, can be turned into a Hopf algebra, too. We get

$$\begin{aligned} m(\mu,\nu) &= \mu\star\nu && \text{(convolution of measures)} \\ e &= \delta_1 && \text{(the Dirac measure at the unit element)} \\ \Delta(\mu) &= \mu\otimes\mu \\ \epsilon(\mu) &= \mu(S^1) && \text{(the total mass)} \\ S(\mu) &= i(\mu) = \mu\circ i \end{aligned}$$

where $i: S^1 \to S^1$, $i(z) = z^{-1}$.

3.2.5 *Universal enveloping algebras*

Another very important example for a Hopf algebra is the universal enveloping algebra $U(\mathfrak{g})$ of a Lie algebra $\mathfrak{g}$. We defined $U(\mathfrak{g})$ as an associative algebra in section 2.1. Define the mappings $\Delta : U(\mathfrak{g}) \to U(\mathfrak{g}) \otimes U(\mathfrak{g})$, $\epsilon : U(\mathfrak{g}) \to \mathbb{K}$ by setting

$$\begin{aligned} \Delta(X) &= X\otimes\mathbb{1} + \mathbb{1}\otimes X, \\ \epsilon(X) &= 0, \end{aligned}$$

for $X \in \mathfrak{g}$ and then extending them as algebra homomorphism to $U(\mathfrak{g})$. As antipode we take the map $S : U(\mathfrak{g}) \to U(\mathfrak{g})$ defined on page 7. $(U(\mathfrak{g}), m, \mathbb{1}, \Delta, \epsilon, S)$ is a Hopf algebra with these definitions.

3.2.6 *q-Numbers*

Before we can present examples of Hopf algebras that are neither commutative nor cocommutative, we have to introduce some notations.

To formulate the commutation relations and the formulas for the coproducts of quantum groups, we need the so-called q-numbers and several derived functions, such as the q-factorial, the q-binomial (Gauss polynomials), and the q-exponential. Readers already familiar with these functions may safely skip this paragraph.

We can consider the *q-numbers*

$$q_n = \sum_{\nu=0}^{n-1} q^\nu, \qquad \text{for } n \in \mathbb{N}, q \in \mathbb{C},$$

as a generalisation of the sequence of positive integers. For $q \neq 1$ this gives $q_n = \frac{1-q^n}{1-q}$, and for $q = 1$ we have $q_n = n$. This allows to define the q-numbers also for non-integer subscripts,

$$q_x = \begin{cases} x & \text{if } q = 1 \\ \frac{1-q^x}{1-q} & \text{else.} \end{cases}$$

For real and positive q they can be characterised as the solutions of a functional equation. It is well-known (see e. g. [AD89]) that the linear functions $f : \mathbb{R} \to \mathbb{R}$,

$f(x) = cx$ can be characterised by the (additive) Cauchy equation

$$f(x+y) = f(x) + f(y) \qquad \text{for all } x, y \in \mathbb{R}$$

plus certain mild regularity conditions such as measurability or boundedness (below or above on a set of positive inner measure). The q-numbers can be characterised by the generalised Cauchy equation

$$f(x+y) = f(x) + g(x)f(y) \qquad \text{for all } x, y \in \mathbb{R},$$

see e.g. [Fra94].

We will assume that q is a complex number, but not a root of unity, i. e. $q^n \neq 1$ for all $n \in \mathbb{N}$. We define the *q-factorial*

$$q_n! = \prod_{\nu=1}^{n} q_n = \prod_{\nu=1}^{n} \frac{1-q^\nu}{1-q},$$

and the *q-binomial*

$$\left[\begin{matrix} m \\ n \end{matrix}\right]_q = \frac{q_m \cdot q_{m-1} \cdots q_{m-n+1}}{q_1 \cdot q_2 \cdots q_n} = \frac{q_m!}{q_n! \cdot q_{m-n}!}$$

for $m, n \in N$, $m > n$. The q-binomial coefficients satisfy

$$\left[\begin{matrix} m \\ n \end{matrix}\right]_q = \left[\begin{matrix} m \\ m-n \end{matrix}\right]_q$$

and

$$\left[\begin{matrix} m+1 \\ n \end{matrix}\right]_q = \left[\begin{matrix} m \\ n \end{matrix}\right]_q + q^{m-n+1} \left[\begin{matrix} m \\ n-1 \end{matrix}\right]_q = q^n \left[\begin{matrix} m \\ n \end{matrix}\right]_q + \left[\begin{matrix} m \\ n-1 \end{matrix}\right]_q.$$

A symmetric version of the q-numbers is defined by $[n]_q = \frac{q^n - q^{-n}}{q - q^{-1}}$. Then $[n]_q! = \prod_{\nu=1}^{n} [n]_q$, and

$$\left[\left[\begin{matrix} m \\ n \end{matrix}\right]\right]_q = \frac{[m]_q [m-1]_q \cdots [m-n+1]_q}{[1]_q [2]_q \cdots [n]_q} = \frac{[m]_q!}{[n]_q! [m-n]_q!}.$$

These q-numbers are related to the ones defined before by $[n]_q = q^{-n+1} (q^2)_n$, $[n]_q! = q^{-\frac{n(n-1)}{2}} (q^2)_n!$, and

$$\left[\left[\begin{matrix} n \\ \nu \end{matrix}\right]\right]_q = q^{-\nu(n-\nu)} \left[\begin{matrix} n \\ \nu \end{matrix}\right]_{q^2} = q^{\nu(n-\nu)} \left[\begin{matrix} n \\ \nu \end{matrix}\right]_{q^{-2}}.$$

We will also need the *q-exponential*

$$e_q^x = \sum_{n=0}^{\infty} \frac{x^n}{q_n!}.$$

3.2.7 *The quantum algebra $U_q(aff(1))$*

Now we come to the first example that is neither commutative nor cocommutative.

The q-deformed universal enveloping algebra or quantum algebra that we present in this example was introduced by Drinfeld (see [Dri87] Example 6.1).

$U_q(aff(1))$ is generated by X_1 and X_2, with the relations

$$[X_1, X_2] = \alpha_0 X_2,$$

and

$$\begin{aligned}\Delta(X_1) &= X_1 \otimes 1 + 1 \otimes X_1, \\ \Delta(X_2) &= X_2 \otimes \exp(\frac{1}{2}h\beta_0 X_1) + \exp(-\frac{1}{2}h\beta_0 X_1) \otimes X_2.\end{aligned}$$

This is a deformation of the enveloping algebra of the affine group in one dimension $Aff(1)$ and can therefore be called the ***q-affine algebra.*** We use the equivalent generators $X = \beta X_1$, $Y = \exp(\frac{1}{2}h\beta_0 X_1)X_2$, and $\alpha = \alpha_0\beta_0$, because they will be more convenient for many calculations. Then

$$\begin{aligned}[X, Y] &= \alpha Y, \\ \Delta(X) &= X \otimes 1 + 1 \otimes X, \\ \Delta(Y) &= Y \otimes \exp(hX) + 1 \otimes Y.\end{aligned}$$

It is also possible to define an antipode for $U_q(Aff(1))$:

$$\begin{aligned}S(X) &= -X, \\ S(Y) &= -Y\exp(-hX).\end{aligned}$$

3.2.8 *The quantum algebra $U_q(sl(2))$*

Let h be such that $q = e^h$. The q-deformed enveloping algebra of $sl(2)$ can be defined as the algebra generated by H, X, Y with the relations

$$\begin{aligned}[H, X] &= 2X, \\ [H, Y] &= -2Y, \\ [X, Y] &= [H]_q.\end{aligned}$$

This becomes a Hopf algebra, if one defines Δ, S, ϵ by

$$\begin{aligned}\Delta(X) &= X \otimes 1 + \exp(hH) \otimes X, \\ \Delta(Y) &= Y \otimes \exp(-hH) + 1 \otimes Y, \\ \Delta(H) &= H \otimes 1 + 1 \otimes H, \\ S(H) &= -H, \\ S(X) &= -\exp(-hH)X, \\ S(Y) &= -Y\exp(hH), \\ \epsilon(H) &= \epsilon(X) = \epsilon(Y) = 0.\end{aligned}$$

The generators H, X, Y are related to the generators in [DHL91] by $H = H$, $X = \exp(\frac{h}{2}H)X_+$, and $Y = X_-\exp(-\frac{h}{2}H)$.

3.2.9 *The quantum algebra* $U(sl(2) \oplus_q sl(2))$

Another interesting example is the deformation of $SL(2) \otimes SL(2)$ where the two $SL(2)$ remain classical, and the deformation introduces a Quantum Link. The quantum group $Fun(SL(2) \otimes_q SL(2))$, the related quantum algebra and a quantum space have been constructed and studied by P. Truini, V.S. Varadarajan, and E. De Vito [TV91, TV93].

Truini et al. defined $U(sl(2) \oplus_q sl(2))$ as the Hopf algebra generated by H_1, H_2, $X_1^{\pm}, X_2^{\pm}$ with the relations

$$\begin{aligned}
[X_i^{\pm}, X_j^{\pm}] &= 0, \qquad\qquad (i \neq j) \\
[H_i, X_j^{\pm}] &= \pm\delta_{ij} 2X_i^{\pm}, \\
[X_i^{+}, X_j^{-}] &= \delta_{ij} H_i, \\
\Delta(X_1^{\pm}) &= X_1^{\pm} \otimes 1 + \exp(\pm hH_2) \otimes X_1^{\pm}, \\
\Delta(X_2^{\pm}) &= X_2^{\pm} \otimes \exp(\pm hH_1) + 1 \otimes X_2^{\pm}, \\
\Delta(H_i) &= H_i \otimes 1 + 1 \otimes H_i, \\
S(X_i^{\pm}) &= -\exp(\mp hH_j) X_i^{\pm}, \qquad\qquad (i \neq j) \\
S(H_i) &= -H_i, \\
\epsilon(H_i) &= \epsilon(X_i^{\pm}) = 0.
\end{aligned}$$

3.2.10 *The quantum algebra* $h_q(1)$

We consider the *q-Heisenberg algebra* $h_q(1)$ generated by $A^{\dagger}$, H, and A with the relations (see e. g. [CGST91])

$$\begin{aligned}
[H, A] &= [H, A^{\dagger}] = 0 \\
[A, A^{\dagger}] &= [H]_q \\
\Delta(A^{\dagger}) &= A^{\dagger} \otimes 1 + q^{-H} \otimes A^{\dagger} \\
\Delta(A) &= A \otimes q^{H} + 1 \otimes A \\
\Delta(H) &= H \otimes 1 + 1 \otimes H \\
S(A^{\dagger}) &= -A^{\dagger} q^{-H} \\
S(A) &= -Aq^{H} \\
S(H) &= -H \\
\varepsilon(A^{\dagger}) &= \varepsilon(A) = \varepsilon(H) = 0.
\end{aligned}$$

Because H is central, we get the classical Heisenberg-Weyl algebra relations if we change the basis to $A^{\dagger}$, $H' = [H]_q$, and A, but not with the standard coalgebra relations.

3.3 DUAL REPRESENTATIONS FOR QUANTUM GROUPS

In this section we will study the dual representations using bialgebra terminology.

Let $\mathfrak{g}$ be a Lie algebra and $U(\mathfrak{g})$ its universal enveloping algebra. Then actions $\rho_L, \rho_R : \mathfrak{g} \times U(\mathfrak{g}) \to U(\mathfrak{g})$ are defined by

$$\begin{aligned} \rho_L(X)u &= X \cdot u, \\ \rho_R(X)u &= u \cdot X, \end{aligned}$$

($X \in \mathfrak{g}$, $u \in U(\mathfrak{g})$), where $\cdot$ denotes the multiplication in $U(\mathfrak{g})$, and X has been identified with its image under the canonical injection $i : \mathfrak{g} \to U(\mathfrak{g})$. Choosing an associative algebra A and a non-degenerate pairing $\langle \cdot, \cdot \rangle : U(\mathfrak{g}) \times A \to \mathbb{C}$, we define the adjoint operators of $\rho_L(X)$ and $\rho_R(X)$ by requiring

$$\langle \rho_\bullet(X)u, a \rangle = \langle u, \rho_\bullet^*(X)a \rangle$$

for all $u \in U(\mathfrak{g})$, $a \in A$, where $\bullet$ stands for either L or R. One easily verifies

$$\rho_L^*(X) \circ \rho_L^*(Y) = \rho_L^*(YX), \tag{3.4}$$

$$\rho_R^*(X) \circ \rho_R^*(Y) = \rho_R^*(XY). \tag{3.5}$$

We want the $\rho_\bullet^*(X)$ to act as derivations on A. A sufficient condition for this can be formulated using the Hopf algebra structure of $U(\mathfrak{g})$.

Proposition 3.3.1 *Let $U(\mathfrak{g})$ be the enveloping algebra of $\mathfrak{g}$, A an associative algebra, and $\langle \cdot, \cdot \rangle : U(\mathfrak{g}) \times A \to \mathbb{C}$ a non-degenerate bilinear form. If for all $u \in U(\mathfrak{g})$, $a_1, a_2 \in A$,*

$$\langle \Delta(u), a_1 \otimes a_2 \rangle_\otimes = \langle u, a_1 a_2 \rangle; \tag{3.6}$$

is satisfied, then the operators $\rho_\bullet^*(X)$, $X \in \mathfrak{g}$, are derivations, i. e.

$$\rho_\bullet^*(X)(a_1 a_2) = (\rho_\bullet^*(X)a_1)a_2 + a_1(\rho_\bullet^*(X)a_2). \tag{3.7}$$

holds for all $X \in \mathfrak{g}$, $a_1, a_2 \in A$, where $\bullet$ stands for either L or R.

Proof: Let $\Delta(u) = \sum u_i^{(1)} \otimes u_i^{(2)}$, and notice

$$\begin{aligned} \Delta(\rho_R(X)u) &= \Delta(X \cdot u) = (X \otimes 1 + 1 \otimes X) \sum u_i^{(1)} \otimes u_i^{(2)} \\ &= \sum \Big((X \cdot u_i^{(1)}) \otimes u_i^{(2)} + u_i^{(1)} \otimes (X \cdot u_i^{(2)}) \Big) \\ &= \sum \Big((\rho_R(X) u_i^{(1)}) \otimes u_i^{(2)} + u_i^{(1)} \otimes (\rho_R(X) u_i^{(2)}) \Big) \\ &= (\rho_R(X) \otimes 1 + 1 \otimes \rho_R(X))\, \Delta(u), \end{aligned}$$

and similarly $\Delta(\rho_L(X)u) = (\rho_L(X) \otimes 1 + 1 \otimes \rho_L(X))\, \Delta(u)$. Then

$$\begin{aligned} \langle u, \rho_\bullet^*(X)(a_1 a_2) \rangle &= \langle \rho_\bullet(X)u, a_1 a_2 \rangle \\ &= \langle \Delta(\rho_\bullet(X)u), a_1 \otimes a_2 \rangle_\otimes \\ &= \langle (\rho_\bullet(X) \otimes 1 + 1 \otimes \rho_\bullet(X))\, \Delta(u), a_1 \otimes a_2 \rangle_\otimes \\ &= \langle \Delta(u), \rho_\bullet^*(X)a_1 \otimes a_2 + a_1 \otimes \rho_\bullet^*(X)a_2 \rangle_\otimes \\ &= \langle u, (\rho_\bullet^*(a_1))a_2 + a_1(\rho_\bullet^*(X)a_2) \rangle \end{aligned}$$

for all $X \in \mathfrak{g}$, $u \in U(\mathfrak{g})$, $a_1, a_2 \in A$. The non-degenerateness of $\langle\cdot,\cdot\rangle$ implies now the result. ∎

We see that the fact that the $\rho_\bullet^*(X)$ act as derivations on A is a direct consequence of the fact that the elements of $\mathfrak{g}$ are primitive in the Hopf algebra $U(\mathfrak{g})$.

The generators of q-deformed universal enveloping algebras (quantum algebras) are generally not primitive, but satisfy a slightly weakened condition. They are $(q^{\alpha H}, q^{\beta H})$-primitive with appropriate constants α, β, and appropriate H, i.e.

$$\Delta(X) = q^{\alpha H} \otimes X + X \otimes q^{\beta H}.$$

This means that if we can introduce a grading on A, such that $\rho_\bullet^*(H)a = \lambda deg(a)a$ for all homogeneous $a \in A$ with some constant λ, then the operators $\rho_\bullet^*(X)$ will be $(q^{\lambda\alpha}, q^{\lambda\beta})$-derivations, or what we one may call q-vector fields, i. e.

$$\rho_\bullet^*(X)(a_1a_2) = q^{\lambda\alpha \deg(a_1)}a_1(\rho_\bullet^*(X)a_2) + q^{\lambda\beta\deg(a_2)}(\rho_\bullet^*(X)a_1)a_2$$

for all homogeneous $a_1, a_2 \in A$.

Equation (3.6) suggests that A and $U(\mathfrak{g})$ should be dually paired. Indeed, if we suppose that A is a bialgebra, and that A and $U(\mathfrak{g})$ are in duality, in particular that

$$\langle u_1u_2, a\rangle = \langle u_1 \otimes u_2, \Delta(a)\rangle_\otimes$$

is satisfied, then we can give a very simple form for the dual representations (cf. [MMN+90a, MMN+90b, VS88]), i.e. the dual representation introduced in Chapter 2 for Lie algebras is nothing else than the dual representation introduced in Section 3.1 for dually paired bialgebras.

Proposition 3.3.2 *Let* $\Delta(a) = \sum a_i^{(1)} \otimes a_i^{(2)}$. *Then*

$$\begin{aligned}\rho_L^*(X)a &= \sum\langle X, a_i^{(1)}\rangle a_i^{(2)},\\ \rho_R^*(X)a &= \sum\langle X, a_i^{(2)}\rangle a_i^{(1)}.\end{aligned}$$

Remark: If we identify X with the linear functional defined by $X(a) = \langle X, a\rangle$, then

$$\begin{aligned}\rho_L^*(X)a &= (X \otimes \mathrm{id}) \circ \Delta(a),\\ \rho_R^*(X)a &= (\mathrm{id} \otimes X) \circ \Delta(a).\end{aligned}$$

Proof: We show the first equation, the second equation can be proved analogously. We have for all $u \in U(\mathfrak{g})$

$$\begin{aligned}\langle u, \rho_L^*(X)a\rangle &= \langle X \cdot u, a\rangle\\ &= \langle X \otimes u, \Delta(a)\rangle_\otimes\\ &= \sum\langle X, a_i^{(1)}\rangle\langle u, a_i^{(2)}\rangle\\ &= \langle u, \sum\langle X, a_i^{(1)}\rangle a_i^{(2)}\rangle.\end{aligned}$$

The pairing $\langle\cdot,\cdot\rangle$ was required to be non-degenerate, therefore this implies $\rho_L^*(X)a = \sum\langle X, a_i^{(1)}\rangle a_i^{(2)}$. ■

We will illustrate the above discussion by several examples. There are several methods to actually calculate the dual pairing and the dual actions, see for example [Dob92, DP93a, DP93b] or [FG93, BCG+94]. We shall use here dual bases and *formal pairings*, as in Section 2.3. Let $A = Fun_q(G)$ and $U = U_q(\mathfrak{g})$ be two dually paired Hopf algebras, and let $\{c_n; n \in \mathbb{N}^d\}$ and $\{\psi_n; n \in \mathbb{N}^d\}$ be a pair of dual bases, i. e.

$$\langle c_n, \psi_m\rangle = \delta_{nm}.$$

Then we write the pairing $\langle\cdot,\cdot\rangle : A \times U \to \mathbb{C}$ formally as

$$g = \sum c_n\psi_n.$$

Note that this is not unique. The key property of this formal pairing that we shall exploit continuously is

$$\begin{aligned}\sum c_n\rho_L(X)\psi_n &= \sum\left(\rho_L^*(X)c_n\right)\psi_n\\ \sum c_n\rho_R(X)\psi_n &= \sum\left(\rho_R^*(X)c_n\right)\psi_n,\end{aligned}$$

for all $X \in U$. We write the above identities often simply as

$$Xg = \rho_L^*(X)g \tag{3.8}$$

$$gX = \rho_R^*(X)g. \tag{3.9}$$

3.3.1 *Example: the quantum algebra $U_q(aff(1))$*

The generators of $U_q(aff(1))$ satisfy

$$\begin{aligned}\Delta(X^n) &= \sum_{\nu=0}^{n}\binom{n}{\nu} X^\nu \otimes X^{n-\nu},\\ \Delta(Y^m) &= \sum_{\mu=0}^{m}\begin{bmatrix} m\\ \mu\end{bmatrix}_{e^{\alpha h}} Y^\mu \otimes Y^{m-\mu}\exp(\mu hX),\end{aligned}$$

as can be proved by induction from the relations given in paragraph 3.2.7. It follows

$$\begin{aligned}\Delta(Y^mX^n) &= \sum_{\nu=0}^{n}\sum_{\mu=0}^{m}\binom{n}{\nu}\begin{bmatrix} m\\ \mu\end{bmatrix}_q Y^\mu X^\nu \otimes Y^{m-\mu}\exp(\mu hX)X^{n-\nu}\\ &= \sum_{\nu=0}^{n}\sum_{\mu=0}^{m}\sum_{\rho=0}^{\infty}\binom{n}{\nu}\begin{bmatrix} m\\ \mu\end{bmatrix}_q \frac{(\mu h)^\rho}{\rho!} Y^\mu X^\nu \otimes Y^{m-\mu}X^{n-\nu+\rho}\end{aligned}$$

(with $q = e^{\alpha h}$) and therefore the coefficients of Δ with respect to the basis $\{Y^m X^n; n, m \in \mathbb{N}\}$ are

$$\Delta_{(m,n)}^{(\mu,\nu),(\mu',\nu')} = \begin{cases} 0 & \text{if } \mu+\mu' \neq m \text{ or } \nu+\nu' < n, \\ \begin{pmatrix} n \\ \nu \end{pmatrix} \delta_{n,\nu+\nu'} & \text{if } \mu = 0 = m-\mu' \text{ and } \nu+\nu' \geq n, \\ \begin{pmatrix} n \\ \nu \end{pmatrix} \begin{bmatrix} m \\ \mu \end{bmatrix}_q \frac{(\mu h)^{\nu+\nu'-n}}{(\nu+\nu'-n)!} & \text{else.} \end{cases}$$

We want to define an algebra (A, m) and a mapping $\langle \cdot,\cdot \rangle : U_q(g) \times A \to \mathbb{C}$, such that equation (3.6) is satisfied. We let $A = \operatorname{span}\{A_{ij}; i,j \in N\}$ and $\langle Y^m X^n, A_{ij} \rangle = \delta_{im}\delta_{jn}$ and use (3.6) to define a multiplication on A. If the coefficients of $m : A \otimes A \to A$ are introduced by letting

$$A_{kl} \cdot A_{k'l'} = m(A_{kl} \otimes A_{k'l'}) = \sum_{\kappa\lambda} m_{(kl),(k'l')}^{\kappa\lambda} A_{\kappa\lambda},$$

then $m_{(kl),(k'l')}^{\kappa\lambda} = \Delta_{\kappa\lambda}^{(kl),(k'l')}$. Therefore

$$A_{kl} \cdot A_{k'l'} = \sum_{\kappa\lambda} \Delta_{\kappa\lambda}^{(kl),(k'l')} A_{\kappa\lambda},$$

in particular

$$\begin{aligned} A_{0\nu} \cdot A_{0\nu'} &= \begin{pmatrix} \nu+\nu' \\ \nu \end{pmatrix} A_{0,\nu+\nu'}, \\ A_{\mu 0} \cdot A_{\mu' 0} &= \begin{bmatrix} \mu+\mu' \\ \mu \end{bmatrix}_q A_{\mu+\mu',0}, \\ A_{0\nu} \cdot A_{\mu' 0} &= A_{\mu'\nu}, \\ A_{10} \cdot A_{01} &= hA_{10} + A_{11}. \end{aligned}$$

We set $a = A_{01}$ and $b = A_{10}$, and notice that A is the associative algebra generated by a and b with the relation

$$b \cdot a = a \cdot b + hb.$$

We have $A_{\mu\nu} = \frac{a^\nu b^\mu}{\nu! q_\mu!}$, and $< Y^m X^n, a^{n'} b^{m'} >= \delta_{mm'}\delta_{nn'} n! q_m!$. We write the pairing formally as

$$g = \sum_{n,m=0}^{\infty} \frac{Y^m X^n a^n b^m}{n! q_m!},$$

and calculate the actions of Y and X on g:

$$\begin{aligned} Yg &= \sum_{n,m=0}^{\infty} \frac{Y^{m+1} X^n a^n b^m}{n! q_m!} \\ &= \sum_{n,m=0}^{\infty} \frac{Y^m X^n a^n (\delta_b b^m)}{n! q_m!}, \end{aligned} \quad (3.10)$$

and therefore

$$\rho_L^*(Y) \quad = \quad \delta_b. \tag{3.11}$$

In the same manner we obtain $\rho_L^*(X)$, $\rho_R^*(Y)$ and $\rho_R^*(X)$.

Proposition 3.3.3 *The dual representations are given by*

$$\begin{aligned}
\rho_L^*(Y) &= \delta_b, \\
\rho_L^*(X) &= \partial_a + \alpha N_b, \\
\rho_R^*(Y) &= \exp(\alpha a)\delta_b, \\
\rho_R^*(X) &= \partial_a.
\end{aligned}$$

To obtain q-vector fields, we have to ensure $\rho_\bullet^*(X)a = \lambda\deg(a)a$. We can find one subalgebra for ρ_R^*, and one for ρ_L^*, on which this condition can be satisfied. Let A_1 be the subalgebra of A generated by $z = \exp(-\alpha a)b$, then

$$\begin{aligned}
\rho_R^*(X)z^n &= -\alpha n z^n, \\
\rho_R^*(X)(z_1 z_2) &= (\rho_R^*(X)z_1)z_2 + z_1\rho_R^*(X)z_2, \qquad z_1, z_2 \in A_1.
\end{aligned}$$

We define $\deg(z^n) = n$ and set $\lambda = -\alpha$. Since Y is $(1, \exp(hX))$-primitive, $\rho_R^*(y)$ is a $(1, q^{-1})$-derivation on A_1. One calculates

$$\begin{aligned}
\rho_R^*(Y)z^n &= \left(q^{-1}\right)_n z^{n-1}, \\
\rho_R^*(Y)(z^n z^m) &= q^{-m}(\rho_R^*(Y)z^n)z^m + z^n\rho_R^*(Y)z^m.
\end{aligned}$$

Note that because of the commutativity of A_1, $\rho_R^*(Y)$ is also a $(q^{-1}, 1)$-derivation.

We obtain similar results, if we take ρ_L^* and the subalgebra A_2 generated by b.

$$\begin{aligned}
\rho_L^*(X)b^n &= \alpha n b^n, \\
\rho_L^*(X)(b_1 b_2) &= (\rho_L^*(X)b_1)b_2 + b_1\rho_L^*(X)b_2, \qquad b_1, b_2 \in A_2, \\
\rho_L^*(Y)b^n &= q_n b^{n-1}, \\
\rho_L^*(Y)(b^n b^m) &= q^m(\rho_L^*(Y)b^n)b^m + b^n\rho_L^*(Y)b^m.
\end{aligned}$$

With the definition $\deg(b^n) = n$, $\rho_L^*(Y)$ acts as a $(1, q)$-derivation on A_2.

A more direct way to obtain the same realisations would have been to start with a representation V, in which $\rho_\bullet(X)$ is already diagonal. To follow the same procedure as before, we need one more structure on V, a coproduct $\Delta_V : V \to V \otimes V$. We can obtain such a V as the quotient space of $U_q(Aff(1))$ with respect to a subspace which is a coideal and also a left (right) ideal of $U_q(Aff(1))$. Such a space is the left (right) ideal $I_L = U_q(Aff(1))X$ (resp. $I_R = XU_q(Aff(1))$) generated by X.

Let now $V = U_q(Aff(1))/I_L$. Then $\{Y^n; n \in N\}$ is a basis of V, and the action of X, Y on V is determined by

$$\begin{aligned}
\rho_{L,V}(X)Y^n &= \alpha n Y^n, \\
\rho_{L,V}(Y)Y^n &= Y^{n+1}.
\end{aligned}$$

The coproduct Δ_V is given by

$$\Delta_V(Y^n) = \sum_{\nu=0}^{n} \begin{bmatrix} n \\ \nu \end{bmatrix}_q Y^\nu \otimes Y^{n-\nu},$$

therefore

$$(\Delta_V)_n^{\nu,\nu'} = \delta_{\nu+\nu',n} \begin{bmatrix} n \\ \nu \end{bmatrix}_q.$$

If we set $B_2 = \text{span}\{A_k; k \in N\}$ and $< Y^n, A_k >= \delta_{nk}$, and define a multiplication on B_2 as before, we find that B_2 is generated by $a = A_1$, with $A_k = \frac{a^k}{q_k!}$. We write formally

$$g = \sum_{n=0}^{\infty} \frac{Y^n a^n}{q_n!},$$

and compute $\rho^*_{L,V}$ in the same way as before:

$$\begin{aligned} \rho^*_{L,V}(X) &= aN_a, \\ \rho^*_{L,V}(Y) &= \delta_a, \end{aligned}$$

$\rho^*_{L,V}(X)$ is a derivation, and $\rho^*_{L,V}(Y)$ is a $(1,q)$-derivation. As we noticed in Equation (3.4), $\rho^*_{L,V}$ is an anti-homomorphism. If we compose $\rho^*_{L,V}$ with the antipode S, we get a homomorphism $\rho^\dagger_{L,V} = \rho^*_{L,V} \circ S$. This yields

$$\begin{aligned} \rho^\dagger_{L,V}(X) &= -aN_a, \\ \rho^\dagger_{L,V}(Y) &= -\delta_a q^{-N_a} = -{}_{q^{-1}}\delta_a. \end{aligned}$$

With respect to the grading introduced before, $\rho^\dagger_{L,V}(Y)$ is a $(1,q^{-1})$-derivation.

If we start with the right ideal generated by X, and set $V' = U_q(Aff(1))/I_R$, then $\{Y^n; n \in N\}$ is again a basis, and we get

$$\begin{aligned} \rho_{R,V'}(X)Y^n &= -anY^n, \\ \rho_{R,V'}(Y)Y^n &= Y^{n+1}, \\ \rho^*_{R,V'}(X) &= -aN_a, \\ \rho^*_{R,V'}(Y) &= {}_{q^{-1}}\delta_a, \end{aligned}$$

where $\rho^*_{R,V'}(X)$ and $\rho^*_{R,V'}(Y)$ act on the algebra B_1 generated by b, and the pairing is

$$g' = \sum_{n=0}^{\infty} \frac{Y^n b^n}{(q^{-1})_n!}.$$

Note that B_1 (resp. B_2) is isomorphic to the subalgebra A_1 (resp. A_2) of A that we used before.

3.3.2 *Example: the quantum algebra* $U_q(sl(2))$

From the defining commutation relations follows

$$\begin{aligned} f(H)X^n &= X^n f(H+2n), \\ f(H)Y^n &= Y^n f(H-2n), \\ YX^n &= X^nY - X^{n-1}[n]_q[H+n-1]_q, \\ Y^nX &= XY^n - [n]_q[H+n-1]_qY^{n-1}. \end{aligned}$$

Let $I_L = U_q(sl(2))H$ be the left ideal generated by H, and $V = U_q(sl(2))/I_L$. Then $\{X^nY^m; n, m \in N\}$ is a basis of V. The coproduct on V is given by

$$\Delta_V(X^nY^m) = \sum_{\nu=0}^{n}\sum_{\mu=0}^{m} \begin{bmatrix} n \\ \nu \end{bmatrix}_{q^2} \begin{bmatrix} m \\ \mu \end{bmatrix}_{q^2} q^{-2\nu\mu} X^{n-\nu}Y^{\mu} \otimes X^{\nu}Y^{m-\mu}.$$

Note that the coalgebra (V, Δ_V) is not cocommutative, although X and Y are cocommutative elements ($\Delta_V(X) = X \otimes 1 + 1 \otimes X$, $\Delta_V(Y) = Y \otimes 1 + 1 \otimes Y$). If we set $A = \mathrm{span}\{A_{kl}; k, l \in N\}$ and $\langle X^nY^m, A_{kl}\rangle = \delta_{nk}\delta_{ml}$, we find that A is isomorphic to the algebra generated by a, b with the relation $a \cdot b = q^2 b \cdot a$, which is $\mathbb{C}^2_{q^2}$ (cf. [RTF90]). The formal pairing is given by

$$g = \sum_{n,m=0}^{\infty} \frac{X^nY^m a^n b^m}{(q^2)_n!\,(q^2)_m!}.$$

The realisation of $U_q(sl(2))$ on $\mathbb{C}^2_{q^2}$ is given by

$$\begin{aligned} \rho^*_{L,V}(H) &= 2N_a - 2N_b, \\ \rho^*_{L,V}(X) &= {}_{q^2}\delta_a, \\ \rho^*_{L,V}(Y) &= {}_{q^2}\delta_b - aq^{-N_a}[N_a - 2N_b]_q, \end{aligned}$$

cf. [GP89]. In the limit $q \to 1$ this becomes

$$\begin{aligned} \rho^*_{L,V}(H) &\to 2a\partial_a - 2b\partial_b, \\ \rho^*_{L,V}(X) &\to \partial_a, \\ \rho^*_{L,V}(Y) &\to \partial_b - a^2\partial_a + 2ab\partial_b. \end{aligned}$$

If we take the right ideal $I_R = HU_q(sl(2))$ generated by H instead, set $W = U_q(sl(2))/I_R$, and apply the same procedure, then we get again a realisation of $U_q(sl(2))$ on $A = \mathbb{C}^2_{q^2}$. This time

$$\begin{aligned} \rho^*_{R,W}(H) &= 2N_b - 2N_a, \\ \rho^*_{R,W}(X) &= {}_{q^{-2}}\delta_a - \sigma_b q^{N_b}[N_b - 2N_a]_q, \\ \rho^*_{R,W}(Y) &= {}_{q^{-2}}\delta_b. \end{aligned}$$

In the limit $q \to 1$ this becomes

$$\begin{aligned} \rho^*_{R,W}(H) &\to 2b\partial_b - 2a\partial_a, \\ \rho^*_{R,W}(X) &\to \partial_a - b^2\partial_b + 2ab\partial_a, \\ \rho^*_{R,W}(Y) &\to \partial_b. \end{aligned}$$

3.3.3 *Example: the quantum algebra* $U(sl(2) \oplus_q sl(2))$

Let $I_R = H_1 U(sl(2) \oplus_q sl(2)) + H_2 U(sl(2) \oplus_q sl(2))$ be the right ideal generated by H_1 and H_2. We use $V = U(sl(2) \oplus_q sl(2))/I_R$ and construct an associative algebra A and a dual pairing $\langle \cdot, \cdot \rangle : U(sl(2) \oplus_q sl(2)) \times A \to \mathbb{C}$. We find that A is generated by x_1, x_2, y_1, y_2 with the relations

$$\begin{aligned} x_1 x_2 &= q^2 x_2 x_1, & x_1 y_1 &= y_1 x_1, \\ x_1 y_2 &= q^{-2} y_2 x_1, & x_2 y_1 &= q^2 y_1 x_2, \\ x_2 y_2 &= y_2 x_2, & y_1 y_2 &= q^2 y_2 y_1. \end{aligned}$$

The dual pairing can be written as

$$g = \sum \frac{(X_1^+)^{n_1} (X_2^+)^{n_2} (X_1^-)^{m_1} (X_2^-)^{m_2} x_1^{n_1} y_1^{m_1} x_2^{n_2} y_2^{m_2}}{n_1! m_1! n_2! m_2!}.$$

We obtain the following realisation:

$$\begin{aligned} \rho_R^*(X_1^+) &= \partial_{x_1} + 2x_1 y_1 \partial_{y_1} - y_1^2 \partial_{y_1}, \\ \rho_R^*(X_1^-) &= \partial_{y_1}, \\ \rho_R^*(H_1) &= 2N_{y_1} - 2N_{x_1}, \\ \rho_R^*(X_2^+) &= \partial_{x_2} + 2\sigma_{x_2} \sigma_{y_2} \partial_{y_2} - \sigma_{y_2}^2 \partial_{y_2}, \\ \rho_R^*(X_2^-) &= \partial_{y_2}, \\ \rho_R^*(H_2) &= 2N_{y_2} - 2N_{x_2}. \end{aligned}$$

3.3.4 *Example: the quantum algebra* $h_q(1)$

We proceed as before. By induction $\Delta(A^{\dagger^n} H^m A^r) =$

$$\sum \binom{n}{\nu} \binom{m}{\mu} \binom{r}{\rho} A^{\dagger^\nu} H^\mu q^{(\nu - n)H} A^\rho \otimes A^{\dagger^{n-\nu}} H^{m-\mu} q^{\rho H} A^{r-\rho}.$$

If we expand this we can read off the coefficients $\Delta^{(n,m,r)}_{(n_1,m_1,r_2),(n_2,m_2,r_2)}$. We define functionals A_{nmr} by

$$\langle A_{nmr}, A^{\dagger^{n'}} H^{m'} A^{r'} \rangle = \delta_{nn'} \delta_{mm'} \delta_{rr'},$$

and turn $H_q(1) = \text{span}\{A_{nmr}; n, m, r \in \mathbb{N}\}$ into an algebra by setting

$$m(A_{n_1 m_1 r_1}, A_{n_1 m_2 r_2}) = \sum \Delta^{(n,m,r)}_{(n_1,m_1,r_2),(n_2,m_2,r_2)} A_{nmr}.$$

This leads to

$$\begin{aligned} A_{n_1,0,r_1} \cdot A_{n_2,0,r_2} &= \frac{(n_1+n_2)!(r_1+r_2)!}{n_1! n_2! r_1! r_2!} A_{n_1+n_2,0,r_1+r_2} \\ A_{0,m_1,0} \cdot A_{0,m_2,0} &= \frac{(m_1+m_2)!}{m_1! m_2!} A_{0,m_1+m_2,0} \\ A_{n,0,0} \cdot A_{0,m,0} \cdot A_{0,0,r} &= A_{n,m,r} \end{aligned}$$

and

$$\begin{aligned} A_{100} \cdot A_{010} &= A_{010} \cdot A_{100} + hA_{100} \\ A_{001} \cdot A_{010} &= A_{010} \cdot A_{001} + hA_{001}. \end{aligned}$$

$H_q(1)$ is the associative algebra generated by three elements $a = A_{100}$, $b = A_{010}$, and $c = A_{001}$, $A_{nmr} = \frac{a^n b^m c^r}{n!m!r!}$, with the relations

$$\begin{aligned} a \cdot b &= b \cdot a + ha \\ c \cdot b &= b \cdot c + hc \\ a \cdot c &= c \cdot a \end{aligned}$$

The pairing between $h_q(1)$ and $H_q(1)$ can be formally written as

$$g = e^{aA^\dagger} e^{bH} e^{cA}.$$

With this the left and right dual representations can be calculated.

Proposition 3.3.4 *The dual representations are given by*

$$\begin{aligned} \rho_L^*(A^\dagger) &= \partial_a \\ \rho_L^*(H) &= \partial_b \\ \rho_L^*(A) &= \partial_c + a[\partial_b]_q \\ \rho_R^*(A^\dagger) &= \partial_a + \sigma_c[\partial_b]_q \\ \rho_L^*(H) &= \partial_b \\ \rho_L^*(A) &= \partial_c, \end{aligned}$$

where $[\partial]_q$ *acts as*

$$[\partial]_q f(x) = \frac{q^\partial - q^{-\partial}}{q - q^{-1}} f(x) = \frac{f(x+h) - f(x-h)}{q - q^{-1}}.$$

Note that $[\partial]_q \to \partial$ for $q \to 1$ on differentiable functions. On the monomials $(q \neq 1)$

$$[\partial]_q x^n = \frac{2}{q - q^{-1}} \sum_{\nu=0}^{[\frac{n-1}{2}]} \binom{n}{2\nu+1} x^{n-2\nu-1} h^{2\nu+1}.$$

That the sum runs only over a finite number of terms shows in particular that no convergence problems arise as long as we act on polynomials in the variables a, b and c. If we switch to a different set of generators for $H_q(1)$, namely a, $B = e^b$, and c, then the formal pairing becomes

$$g(a, B, c; A^\dagger, H, A) = e^{aA^\dagger} B^H e^{cA},$$

and the algebraic relations between a, B, and c are

$$aB = qBa, \qquad cB = qBc, \qquad ac = ca.$$

The dual representations change to

$$\begin{aligned} \rho_L^*(A^\dagger) &= \partial_a, & \rho_R^*(A^\dagger) &= \partial_a + \sigma_c[N_B]_q, \\ \rho_L^*(H) &= N_B, & \rho_R^*(H) &= N_B, \\ \rho_L^*(A) &= \partial_c + a[N_B]_q, & \rho_R^*(A) &= \partial_c, \end{aligned}$$

where $[N_B]_q = [B\partial_B]_q$ is the q-number operator defined by

$$\begin{aligned} [N]_q f(x) &= \frac{f(qx) - f(q^{-1}x)}{q - q^{-1}}, \\ [N]_q x^n &= [n]_q x^n. \end{aligned}$$

3.4 A COMPOSITION LAW FOR QUANTUM GROUPS

In Chapter 1 we used a composition law for formal pairing to construct stochastic on Lie groups. We will again need such a composition law, but this time we have no group multiplication that could produce such a relation. We will now show how to combine two formal series $g(a'; X)$ and $g(a''; X)$ to a series $g(a; X)$, and it will be this composition law that will allow us in the following chapter to construct stochastic processes on quantum groups.

Consider the dual pairing for the q-affine group

$$g = \sum_{n,m=0}^{\infty} \frac{Y^m X^n a^n b^m}{n! q_m!}.$$

We can rewrite this in the form

$$\begin{aligned} g &= \sum_{n,m=0}^{\infty} \frac{(X - m\alpha)^n Y^m a^n b^m}{n! q_m!} \\ &= \sum_{m=0}^{\infty} e^{a(X - m\alpha)} \frac{Y^m b^m}{q_m!} \\ &= e^{aX} e_q^{e^{-\alpha a} bY} \\ &= e^{AX} e_q^{BY}, \end{aligned}$$

where $e_q^x = \sum \frac{x^n}{q_n!}$, $A = a$, and $B = e^{-\alpha a} b$. If we take two commuting copies of $q\text{-}aff(1)$, i. e.

$$\begin{aligned} A_1 &= A \otimes 1, & B_1 &= B \otimes 1, \\ A_2 &= 1 \otimes A, & B_2 &= 1 \otimes B, \end{aligned}$$

then we find

$$g(A_1, B_1; X, Y) g(A_2, B_2; X, Y) = e^{A_1 X} e_q^{B_1 Y} e^{A_2 X} e_q^{B_2 Y}$$

$$\begin{aligned} &= e^{A_1 X} \sum e^{A_2(X-\alpha m)} \frac{B_1^m Y^m}{q_m!} e_q^{B_2 Y} \\ &= e^{(A_1+A_2)X} e_q^{e^{-\alpha A_2} B_1 Y} e_q^{B_2 Y} \\ &= e^{(A_1+A_2)X} e_q^{(e^{-\alpha A_2} B_1 + B_2) Y}, \end{aligned}$$

since $B_2 e^{-\alpha A_2} = e^{-\alpha(A_2-h)} B_2 = q e^{-\alpha A_2} B_2$, and $e_q^x e_q^y = e_q^{x+y}$ for $xy = qyx$ (see e. g. [Sch53]). The dual of the multiplication on $U_q(aff(1))$ defines a coproduct on the algebra generated by A and B,

$$\begin{aligned} \Delta(A) &= A \otimes 1 + 1 \otimes A, \\ \Delta(B) &= B \otimes e^{-\alpha A} + 1 \otimes B, \end{aligned}$$

(see (3.21), (3.23)) and with this we can write

$$g(A \otimes 1, B \otimes 1; X, Y) g(1 \otimes A, 1 \otimes B; X, Y) = g(\Delta(A), \Delta(B); X, Y). \tag{3.12}$$

That the formal pairing ($g \in Fun_q(G) \otimes U_g(\mathfrak{g})$) always has this property has been realised before (see [FG93, BCG+94]). In the literature the names transfer matrix or universal T-matrix are also used. Bonechi et al. remarked that g can generally be expressed as a product of exponentials and q-exponentials. We list several examples (see also [MV94].

3.4.1 *Example:* $SL_q(2)$

The dual pairing for $U_q(sl(2))$ can be written as

$$g = e_{q'}^{\phi F} e^{\gamma J_3} e_{1/q'}^{\eta E},$$

where $q' = e^{h'} = q^2$, $h' = 2h$, and ϕ, η, γ are generators of $Fun_q(SL(2))$ with

$$\begin{aligned} [\phi, \eta] &= 0 \\ [\gamma, \phi] &= h'\phi \\ [\gamma, \eta] &= h'\eta \\ m^*(\phi) &= \phi \otimes 1 + (e^{-\gamma/2} \otimes \phi)(1 \otimes 1 + \eta \otimes \phi)^{-1}(e^{-\gamma/2} \otimes 1) \\ m^*(\eta) &= 1 \otimes \eta + (1 \otimes e^{-\gamma/2})(1 \otimes 1 + \eta \otimes \phi)^{-1}(\eta \otimes e^{-\gamma/2}) \\ m^*(\gamma) &= 1 \otimes \gamma + \gamma \otimes 1 - 2h' \sum \frac{(-\eta \otimes \phi)^n}{1 - q^{-1}} \end{aligned}$$

and $E = \alpha Y$, $J_3 = -H/2$, $F = \alpha X$ is a basis of $U_g(sl(2))$ with $\alpha = 1/\sqrt{q}$. The generators ϕ, η, γ do not satisfy the standard $SL_q(2)$-relations, but one can choose functions of ϕ, η, γ that do. We show this in detail in the next paragraph for $GL_{q,q'}(2)$.

3.4.2 *Example:* $GL_{q,q'}(2)$

For $U_{q,q'}(gl(2))$ we have

$$g = e_{qq'}^{zp_-} e^{xp_0} e^{\hat{x}p_1} e_{1/qq'}^{yp_+},$$

where $q = e^h$, $q' = e^{h'}$, and x, $\hat{x}$, y, z are generators of $Fun_{q,q'}(GL(2))$ with

$$\begin{aligned}[x,y] &= h'y \\ [x,z] &= hz \\ [y,z] &= 0 \\ [\hat{x},y] &= hy \\ [\hat{x},z] &= h'z \\ [x,\hat{x}] &= 0\end{aligned}$$

and p_0, p_1, p_-, p_+ is a basis of $U_{q,q'}(gl(2))$ with

$$\begin{aligned}[p_0,p_\pm] &= [p_1,p_\pm] = \pm p_\pm \\ [p_0,p_1] &= 0 \\ [p_+,p_-] &= (q-q'^{-1})(e^{hp_0}e^{h'p_1} - e^{h'p_0}e^{hp_1}) \\ \Delta(p_0) &= p_0 \otimes 1 + 1 \otimes p_0 \\ \Delta(p_1) &= p_1 \otimes 1 + 1 \otimes p_1 \\ \Delta(p_+) &= p_+ \otimes e^{-h'p_0 - hp_1} + 1 \otimes p_+ \\ \Delta(p_-) &= p_- \otimes 1 + e^{hp_0 + h'p_1} \otimes p_-.\end{aligned}$$

If we set

$$a = e^x, \qquad b = e^x y, \qquad c = e^{-\hat{x}} z, \qquad d = e^{-\hat{x}} + ze^x y,$$

then we find generators of $GL_{q,q'}(2)$ that satisfy the standard relations

$$\begin{aligned}ab &= q'ba & \qquad ac &= qca \\ bd &= qdb & cd &= q'dc \\ q'bc &= qcb & ad - da &= (q' - q^{-1})bc.\end{aligned}$$

The coproduct of these generators is

$$\begin{aligned}\Delta(a) &= a \otimes a + b \otimes c & \qquad \Delta(b) &= a \otimes b + b \otimes c \\ \Delta(c) &= c \otimes a + d \otimes c & \Delta(d) &= c \otimes b + d \otimes d.\end{aligned}$$

It is also interesting to note that the quantum algebra $U_{q,q'}(gl(2))$ splits as an algebra into the direct product $U_{\bar{q}}(sl(2)) \otimes U(k)$, where $\bar{q} = \sqrt{qq'}$ is the geometric mean of q and q' and $U(k)$ is the enveloping algebra of the one-dimensional Lie algebra, i. e. $U(k)$ is the free commutative algebra with one generator k. The generators of $U_{\bar{q}}(sl(2))$ are

$$\begin{aligned}\hat{p} &= \frac{1}{\hbar}(h'p_0 + hp_1) \\ \hat{p}_\pm &= e^{-\Delta hk/2} p_\pm\end{aligned}$$

where $\Delta h = h - h'$, $\hbar = \frac{h+h'}{2}$, and $k = p_0 - p_1$ is the generator to $U(k)$. The coalgebra structure does not respect this decomposition [Dob92].

3.4.3 *Example: q-Euclidean group $E_q(2)$*

For the q-Euclidean algebra $e_q(2)$ we have

$$g = e_q^{\pi_- b_-} e^{\pi J} e_{1/q}^{\pi_+ b_+}$$

where $q = e^h$ and π_-, π_+, π are generators of $E_q(2)$ with

$$\begin{aligned}
[\pi_-, \pi_+] &= 0 \\
[\pi, \pi_-] &= -h\pi_- \\
[\pi, \pi_+] &= -h\pi_+ \\
m^*(\pi_-) &= \pi_- \otimes 1 + e^{-\pi} \otimes \pi_- \\
m^*(\pi) &= \pi \otimes 1 + 1 \otimes \pi \\
m^*(\pi_+) &= 1 \otimes \pi_+ + \pi_+ \otimes e^{-\pi}.
\end{aligned}$$

Generators that satisfy the standard relations have to be constructed from π_+, π_-, π in the same manner as in the previous paragraph. For

$$v = e^{-\pi}, \qquad \bar{v} = e^{\pi}, \qquad n = \pi_-, \qquad \bar{n} = \pi_+,$$

we get the algebraic relations

$$\begin{aligned}
vn &= qnv, & v\bar{n} &= q\bar{n}v, \\
n\bar{v} &= q\bar{v}n, & \bar{n}\bar{v} &= q\bar{v}\bar{n}, \\
v\bar{v} &= \bar{v}v = 1, & n\bar{n} &= \bar{n}n,
\end{aligned}$$

and the coalgebraic relations

$$\begin{aligned}
\Delta(v) &= v \otimes v, & \Delta(n) &= n \otimes 1 + v \otimes v, \\
\Delta(\bar{v}) &= \bar{v} \otimes \bar{v}, & \Delta(\bar{n}) &= \bar{n} \otimes v + 1 \otimes \bar{n},
\end{aligned}$$

(cf. [Wor91, SWZ92]).

3.4.4 *Example: q-Galilean group $\Gamma_q(1)$*

For the q-Galilean group $U_q(\Gamma(1))$ [BCG+92] we have

$$g = e^{\mu m} e^{xP} e^{tT} e^{vb}$$

where $m = e^{-ihP}M$, $b = e^{-ihP}B$, and μ, x, t, v can be considered as generators of $\Gamma_q(1)$ with

$$\begin{aligned}
[v, x] &= 2ihv \\
[x, \mu] &= -2ih\mu \\
[v, \mu] &= -hv^2 \\
[t, \mu] &= [t, x] = [t, v] = 0 \\
m^*(v) &= 1 \otimes v + v \otimes 1 \\
m^*(t) &= 1 \otimes t + t \otimes 1 \\
m^*(\mu) &= 1 \otimes \mu + \mu \otimes 1 + iv \otimes x - \frac{1}{2}v^2 \otimes t \\
m^*(x) &= 1 \otimes x + x \otimes 1 + iv \otimes t,
\end{aligned}$$

and B, P, M, T are generators of $U(\Gamma_q(1))$ with

$$\begin{aligned}
[B,P] &= iM \\
[B,T] &= \frac{i\sin(hP)}{h} \\
[P,T] &= 0 \\
[M,B] &= [M,P] = [M,T] = 0 \\
\Delta(B) &= e^{-ihP}\otimes B + B\otimes e^{ihP} \\
\Delta(M) &= e^{-ihP}\otimes M + M\otimes e^{ihP} \\
\Delta(P) &= 1\otimes P + P\otimes 1 \\
\Delta(T) &= 1\otimes T + T\otimes 1.
\end{aligned}$$

For $q\to 1$ this converges to the (centrally extended) Galilean algebra, which is isomorphic to the nilpotent algebra introduced in 2.2.2 (for $N=3$)

$$B\to X_0, \qquad T\to X_1, \qquad iP\to X_2, \qquad -M\to X_3.$$

We can use the formal pairing introduced above to calculate the representations (as in section 2.3). We get for the left dual representation of b for example

$$\begin{aligned}
be^{\mu m}e^{xP}e^{tT}e^{vb} &= e^{\mu m}(b-h\mu m^2)e^{xP}e^{tT}e^{vb} = \cdots \\
&= e^{\mu m}e^{xP}e^{tT}be^{vb} + e^{\mu m}e^{xP}e^{tT}t\frac{e^{-2ihP}-1}{2h}e^{vb} \\
&\quad + ie^{\mu m}xme^{xP}e^{tT}e^{vb} - h\mu m^2 e^{\mu m}e^{xP}e^{tT}e^{vb}
\end{aligned}$$

From this we can read the action of $\rho^*_L(b)$ on $\Gamma_q(1)$.

Proposition 3.4.1 *We have for the left dual representation*

$$\begin{aligned}
\rho^*_L(b) &= \partial_v - h\sigma_\mu\partial_\mu^2 + i\sigma_x\partial_\mu + \sigma_t\frac{e^{-2ih\partial_x}-1}{2h}, \\
\rho^*_L(T) &= \partial_t, \\
\rho^*_L(P) &= \partial_x, \\
\rho^*_L(m) &= \partial_\mu,
\end{aligned}$$

and for the right dual representation

$$\begin{aligned}
\rho^*_R(b) &= \partial_v, \\
\rho^*_R(T) &= \partial_t + \frac{\sigma_v}{2h}\left(1 - e^{-2ih\partial_x}\sum_{n=0}^{\infty} h^n\sigma_v^n\partial_\mu^n\right), \\
\rho^*_R(P) &= \partial_x + \frac{i}{h}\sum_{n=1}^{\infty}\frac{h^n}{n}\sigma_v^n\partial_\mu^n, \\
\rho^*_R(m) &= \partial_\mu\sum_{n=0}^{\infty} h^n\sigma_v^n\partial_\mu^n.
\end{aligned}$$

Of interest is also the dual of the quotient representation induced by $\Omega M = \Omega\lambda_1$ and $\Omega B = \Omega\lambda_2$,

$$
\begin{aligned}
\rho^*_{R,\lambda_1,\lambda_2}(B) &= \lambda_2 - ix\lambda_1 - t\frac{e^{ih\partial_x} - e^{-ih\partial_x}}{2h},\\
\rho^*_{R,\lambda_1,\lambda_2}(T) &= \partial_t,\\
\rho^*_{R,\lambda_1,\lambda_2}(P) &= \partial_x,\\
\rho^*_{R,\lambda_1,\lambda_2}(M) &= \lambda_1,
\end{aligned}
$$

acting on the polynomials in the two commuting variables x and t.

3.5 q-EXPONENTIALS

In this section we consider a class of Hopf algebras characterised by the relations of their generators (see below), and calculate the dual algebras and dual pairings. The pairing between such an algebra and its dual can formally be written as a product of q-exponentials, a result that is helpful later when we study multiplicative processes on these algebras.

Let $\mathcal{U}$ be a Hopf algebra with generators $X_1, \ldots, X_{d_X}, H_1, \ldots, H_{d_H}, Y_1, \ldots, Y_{d_Y}$. We add generators $K_1, \ldots, K_{d_H}$ that commute with $H_1, \ldots, H_{d_H}$. We will later see that they play the rôle of $e^{h_1 H_1}, \ldots, e^{h_{d_H} H_{d_H}}$, $h_i \in \mathbb{C}$, and suppose the following conditions are satisfied:

$$
\mathbf{H}: \left\{
\begin{array}{rcl}
\multicolumn{3}{l}{\{\psi_{klmn} = Y^k H^l K^m X^n; k \in \mathbb{N}^{d_Y}, l \in \mathbb{N}^{d_H}, m \in \mathbb{Z}^{d_H}, n \in \mathbb{N}^{d_X}\} \text{ spans } \mathcal{U}}\\
\Delta H_i &=& H_i \otimes 1 + 1 \otimes H_i,\\
\Delta X_k &=& \prod K_i^{s_{ik}} \otimes X_k + X_k \otimes \prod K_i^{t_{ik}},\\
\Delta Y_l &=& \prod K_i^{p_{il}} \otimes Y_l + Y_l \otimes \prod K_i^{q_{il}},\\
{[H_i, H_j]} &=& 0, \quad [H_i, K_j] = 0, \quad [K_i, K_j] = 0\\
{[H_i, X_k]} &=& \chi_{ik} X_k, \quad K_i X_k = e^{\chi_{ik} h_i} X_k K_i\\
{[H_i, Y_l]} &=& \eta_{il} Y_l, \quad K_i Y_l = e^{\eta_{il} h_i} Y_l K_i
\end{array}
\right.
$$

for some constants $s_{ik}, t_{ik}, p_{il}, q_{il} \in \mathbb{Z}$, $\chi_{ik}, \eta_{il} \in \mathbb{C}$.

Remark: Note that we impose no conditions on the relations between the X_k and the Y_l; they are not needed for the calculations in this section. But the condition that the ψ_{klmn} span $\mathcal{U}$ implies that such relations exist, namely that their commutator is a linear combination of the ψ_{klmn}. Furthermore, they are restricted by the condition that $\mathcal{U}$ is a Hopf algebra, but this leaves many possibilities. Just looking at Lie algebras we see among the possibilities that a commutator of X's and/or Y's may be zero, or an element of the Cartan subalgebra (i. e., a linear combination of the H's), or a linear combination of X's and Y's.

The conditions **H** are satisfied for most quantum algebras, in particular for the standard semi-simple quantum groups introduced by Drinfeld and Jimbo[Dri87, Jim85, Jim86], as well as many others, e. g. , that have been considered by physicists.

We also introduce the subspace $\mathcal{U}_0 = \text{span}\{\psi_{kl0r}\}$.

Then the dual $\mathcal{U}^*$ of $\mathcal{U}$ is an algebra with the multiplication defined by

$$(f_1 \cdot f_2)(u) = (f_1 \otimes f_2)(\Delta u). \tag{3.13}$$

We define functionals $A_{kln} \in \mathcal{U}^*$ by

$$A_{k'l'n'}(\psi_{klmn}) = \begin{cases} \delta_{kk'}\delta_{nn'} \prod \frac{(h_i m_i)^{m'_i}}{(l'_i - l_i)!} & \text{if } l'_i \geq l_i \text{ for all } i \\ 0 & \text{otherwise,} \end{cases} \tag{3.14}$$

and set $\mathcal{A}_0 = \text{span}\{A_{kln}; k \in \mathbb{N}^{d_Y}, l \in \mathbb{N}^{d_H}, n \in \mathbb{N}^{d_X}\}$. It turns out that $\mathcal{A}_0$ is a subalgebra of $\mathcal{U}^*$. This definition guarantees that $\sum_{n=0}^{N} \frac{(h_i H_i)^n}{n!}$ will tend to K_i in the weak topology, i. e., that K_i can be considered as $e^{h_i H_i}$, if $\mathcal{U}$ is interpreted as a subspace of $\mathcal{A}_0^*$.

Lemma 3.5.1 *Let $X \in \mathcal{U}$ be (A, B)-primitive, i. e., $\Delta X = A \otimes X + X \otimes B$ and suppose $XA = \alpha AX$, $XB = \beta BX$, $\alpha, \beta \in \mathbb{C}$. Then*

$$\begin{aligned} \Delta X^n &= \sum_{\nu=0}^{n} \begin{bmatrix} n \\ \nu \end{bmatrix}_{\alpha/\beta} \beta^{(n-\nu)\nu} A^{n-\nu} X^{\nu} \otimes B^{\nu} X^{n-\nu} \\ &= \sum_{\nu=0}^{n} \begin{bmatrix} n \\ \nu \end{bmatrix}_{\alpha/\beta} \alpha^{-(n-\nu)\nu} X^{\nu} A^{n-\nu} \otimes X^{n-\nu} B^{\nu}. \end{aligned}$$

Proof: We set $\Delta X^n = \sum_{\nu=0}^{n} C_\nu^n A^{n-\nu} X^\nu \otimes B^\nu X^{n-\nu}$, consider $\Delta X^{n+1} = \Delta X \cdot \Delta X^n$, and find the following recursion relation

$$C_\nu^{n+1} = \beta^\nu C_\nu^n + \alpha^{n-\nu+1} C_{\nu-1}^n.$$

Solving this relation now completes the proof. ∎

We introduce the following constants:

$$\begin{aligned} \alpha_k &= e^{-\sum_i s_{ik}\chi_{ik}h_i}, & \beta_k &= e^{-\sum_i t_{ik}\chi_{ik}h_i}, \\ \gamma_l &= e^{-\sum_i p_{il}\eta_{il}h_i}, & \delta_l &= e^{-\sum_i q_{il}\eta_{il}h_i}, \\ \alpha_{kk'} &= e^{-\sum_i s_{ik}\chi_{ik'}h_i}, & \beta_{kk'} &= e^{-\sum_i t_{ik}\chi_{ik'}h_i}, \\ \gamma_{ll'} &= e^{-\sum_i p_{il}\eta_{il'}h_i}, & \delta_{ll'} &= e^{-\sum_i q_{il}\eta_{il'}h_i}. \end{aligned}$$

WARNING: The double-indexed δ used here is not a Kronecker delta.

Lemma 3.5.2

$$\Delta\psi_{klmn} = \sum_{(\kappa,\lambda,\nu) \leq (kln)} C_{(\kappa,\lambda,\nu)}^{(kln)} L_{(\kappa,\lambda,\nu)}^{(kln)} \otimes R_{(\kappa,\lambda,\nu)}^{(kln)}$$

where

$$C_{(\kappa,\lambda,\nu)}^{(kln)} = \begin{bmatrix} k_1 \\ \kappa_1 \end{bmatrix}_{\gamma_1/\delta_1} \cdots \begin{bmatrix} k_{d_Y} \\ \kappa_{d_Y} \end{bmatrix}_{\gamma_{d_Y}/\delta_{d_Y}} \binom{l_1}{\lambda_1} \cdots \binom{l_{d_H}}{\lambda_{d_H}}$$

$$
\begin{aligned}
&\cdot \begin{bmatrix} n_1 \\ \nu_1 \end{bmatrix}_{\alpha_1/\beta_1} \cdots \begin{bmatrix} n_{d_X} \\ \nu_{d_X} \end{bmatrix}_{\alpha_{d_X}/\beta_{d_X}} \prod_l \gamma^{-(k_l-\kappa_l)\kappa_l} \prod \beta_r^{(n_r-\nu_r)\nu_r} \\
&\cdot \prod_{l'>l} \gamma_{ll'}^{-(k_l-\kappa_l)\kappa_{l'}} \prod_{l'>l} \delta_{ll'}^{-\kappa_l(k_{l'}-\kappa_{l'})} \prod_{k'<k} \alpha_{kk'}^{(n_k-\nu_k)\nu_{k'}} \prod_{k'<k} \beta_{kk'}^{\nu_k(n_{k'}-\nu_{k'})} \\
L^{(kln)}_{(\kappa,\lambda,\nu)} &= Y_1^{\kappa_1} \cdots Y_{d_Y}^{\kappa_{d_Y}} H_1^{\lambda_1} \cdots H_{d_H}^{\lambda_{d_H}} \prod K_i^{s_{ik}(n_k-\nu_k)+p_{il}(k_l-\kappa_l)+m_i} X_1^{\nu_1} \cdots X_{d_X}^{\nu_{d_X}} \\
R^{(kln)}_{(\kappa,\lambda,\nu)} &= Y_1^{k_1-\kappa_1} \cdots Y_{d_Y}^{k_{d_Y}-\kappa_{d_Y}} H_1^{l_1-\lambda_1} \cdots H_{d_H}^{l_{d_X}-\lambda_{d_X}} \prod K_i^{t_{ik}\nu_k+q_{il}\kappa_l+m_i} \\
&\cdot X_1^{n_1-\nu_1} \cdots X_{d_X}^{n_{d_X}-\nu_{d_X}}.
\end{aligned}
$$

Proof: Apply Lemma 3.5.1 and reorder the terms. ∎

We set $a_l = A_{e_l,0,0}$, $b_i = A_{0,e_i,0}$, $c_k = A_{0,0,e_k}$. The formula of the previous lemma allows us to calculate the algebraic relations on a_l, b_i, c_k.

Lemma 3.5.3

$$
\begin{aligned}
b_i \cdot A_{0,m,0} &= A_{0,m,0} \cdot b_i = (m_i+1) A_{0,m+e_i,0}, \\
a_l \cdot A_{nmr} &= \sum_{m'\le m} \binom{\gamma_l}{\delta_l}_{n_l+1} \gamma_l^{-n_l} \prod_{l'<l} \gamma_{l'l}^{-n_{l'}} \prod_{l'>l} \delta_{ll'}^{-n_l} \prod_{i=1}^{d_H} \frac{q_{il}^{m_i-m'_i}}{(m_i-m'_i)!} A_{n+e_l,m',r} \\
A_{nmr} \cdot a_l &= \sum_{m'\le m} \binom{\gamma_l}{\delta_l}_{n_l+1} \gamma_l^{-n_l} \prod_{l'>l} \gamma_{ll'}^{-n_{l'}} \prod_{l>l'} \delta_{l'l}^{-n_{l'}} \prod_{i=1}^{d} \frac{p_{il}^{m_i-m'_i}}{(m_i-m'_i)!} A_{n+e_l,m',r} \\
c_k \cdot A_{nmr} &= \sum_{m'\le m} \binom{\alpha_k}{\beta_k}_{r_k+1} \beta_k^{r_k} \prod_{k<k'} \alpha_{k'k}^{r_{k'}} \prod_{k'<k} \beta_{kk'}^{r_{k'}} \prod_{i=1}^{d} \frac{t_{ik}^{m_i-m'_i}}{(m_i-m'_i)!} A_{n,m',r+e_k}, \\
A_{nmr} \cdot c_k &= \sum_{m'\le m} \binom{\alpha_k}{\beta_k}_{r_k+1} \beta_k^{r_k} \prod_{k'<k} \alpha_{kk'}^{r_{k'}} \prod_{k<k'} \beta_{k'k}^{r_k} \prod_{i=1}^{d} \frac{s_{ik}^{m_i-m'_i}}{(m_i-m'_i)!} A_{n,m',r+e_k},
\end{aligned}
$$

where $\binom{\gamma_l}{\delta_l}_{n_l+1}$ *and* $\binom{\alpha_k}{\beta_k}_{r_k+1}$ *are the q-numbers introduced in the previous section.*

Proof: Follows from the definition of the multiplication in $\mathcal{U}^*$ in Equation (3.13) and Lemma 3.5.2. ∎

These relations show that $\mathcal{A}_0$ is in fact a subalgebra of $\mathcal{U}^*$.

Proposition 3.5.4 *If γ_k/δ_k and α_k/β_k are not roots of unity, then the algebra $\mathcal{A}_0$ is generated by $a_1, \ldots, a_{d_Y}, b_1, \ldots, b_{d_H}, c_1, \ldots, c_{d_X}$ with the relations*

$$
\begin{aligned}
a_l \cdot c_k &= c_k \cdot a_l, & b_i \cdot b_j &= b_j \cdot b_i, \\
\delta_{ll'}\, a_l \cdot a_{l'} &= \gamma_{ll'}\, a_{l'} \cdot a_l, & \beta_{k'k}\, c_k \cdot c_{k'} &= \alpha_{k'k}\, c_{k'} \cdot c_k, \\
[b_i, a_l] &= (p_{il} - q_{il}) a_l, & [b_i, c_k] &= (s_{ik} - t_{ik}) c_k.
\end{aligned}
$$

Proof: Follows directly from Lemma 3.5.3. ∎

If some of the γ_k/δ_k, α_k/β_k are roots of unity, then the algebra generated by $a_1,\ldots,a_{d_Y}, b_1,\ldots,b_{d_H}, c_1,\ldots,c_{d_X}$ is a subalgebra of $\mathcal{A}_0$, which we shall denote by $\hat{\mathcal{A}}_0$. For this case we introduce the algebra $\hat{\mathcal{U}} = \mathcal{U}/\mathcal{I}$, where $\mathcal{I} = \{u \in \mathcal{U}; \forall a \in \hat{\mathcal{A}}_0 :< u, a >= 0\}$.

For the following calculations we will assume that

$$q_{il} = 0, \qquad s_{ik} = 0.$$

This can always be achieved by an appropriate choice of the generators, e. g., set $\tilde{Y}_l = \prod K_i^{-q_{il}} Y_l$, $\tilde{X}_k = \prod K_i^{-s_{ik}} X_k$. Then $\alpha_k = 1$, $\alpha_{kk'} = 1$, $\delta_l = 1$, $\delta_{ll'} = 1$.

We find in this case

$$\begin{aligned} a_l \cdot A_{nmr} &= \left(\frac{1}{\gamma_l}\right)_{n_l+1} A_{n+e_l,m,r} && \text{if } n_{l'} = 0 \text{ for } l' < l, \\ A_{nmr} \cdot c_k &= (\beta_k)_{r_k+1} A_{n,m,r+e_k} && \text{if } r_{k'} = 0 \text{ for } k' > k. \end{aligned}$$

If we assume also that $1/\gamma_i$ and β_i are not roots of unity, then we have

$$A_{nmr} = \frac{a_1^{n_1} \cdots a_{d_Y}^{n_{d_H}} b_1^{m_1} \cdots b_{d_H}^{n_{d_H}} c_1^{r_1} \cdots c_{d_X}^{r_{D_X}}}{\left(\frac{1}{\gamma_1}\right)_{n_1}! \cdots \left(\frac{1}{\gamma_{d_Y}}\right)_{n_{d_Y}}! m_1! \cdots m_{d_H}! (\beta_1)_{r_1}! \cdots (\beta_{r_{d_X}})_{r_{d_X}}!}.$$

We consider the sequence

$$g^{(N)}(A, \psi) = \sum_{nmr}^{N} A_{nmr} \otimes \psi_{nmr}.$$

The sequence $(g^{(N)})_{N \in \mathbb{N}} \subset \mathcal{A} \otimes \mathcal{U}_0 \subset \mathcal{U}_0^* \otimes \mathcal{U}_0 \subset \mathrm{End}(\mathcal{U}_0)$ converges weakly towards the identity $\mathrm{id}_{\mathcal{U}_0}$. If we omit the tensor product we can also formally write for the sequence $g = (g^{(N)})_{N \in \mathbb{N}}$, $g(\{a\}, \{b\}, \{c\}; \{Y\}, \{H\}, \{X\}) =$

$$e_{1/\gamma_1}^{a_1 Y_1} \cdots e_{1/\gamma_{d_Y}}^{a_{d_Y} Y_{d_Y}} e^{b_1 H_1} \cdots e^{b_{d_H} H_{d_H}} e_{b_1}^{c_1 X_1} \cdots e_{\beta_{d_X}}^{c_{d_X} X_{d_X}},$$

i. e., we have found that we can write the formal pairing as a product of q-exponentials if hypothesis **H** is satisfied and the generators are chosen appropriately. The duality between $\Delta_{\mathcal{A}}$ and $m_{\mathcal{U}}$ implies that we can formally write (see e. g. [BCG+94, FG93])

$$g(\Delta_{\mathcal{A}}(a); \phi) = g(a'; \psi) g(a''; \psi), \tag{3.15}$$

where $a' = a \otimes 1$, $a'' = 1 \otimes a$.

With some modifications this result remains valid also for the case where some of the parameters γ_k/δ_k, α_k/β_k are roots of unity. The sum has to be restricted to the terms where all q-factorials are different from zero, and we get a dual pairing between $\hat{\mathcal{U}}_0$ and $\hat{\mathcal{A}}_0$ in this case.

Example: The q-affine algebra is the quantum algebra $\mathcal{U}$ with two generators X, Y and relations

$$\begin{aligned} XY - YX &= \alpha Y, \\ \Delta(X) &= X \otimes 1 + 1 \otimes X, \\ \Delta(Y) &= Y \otimes \exp(\beta X) + 1 \otimes Y, \\ \varepsilon(X) &= \varepsilon(Y) = 0, \end{aligned}$$

with $\alpha, \beta \in \mathbb{C}$, $q = e^{\alpha\beta}$ not a root of unity. Then we get $\mathcal{A} = \text{span}\{a^n b^m; n, m \in \mathbb{N}\}$, $ba - ab = \beta b$, and

$$g(a, b; X, Y) = \sum_{n,m=0}^{\infty} \frac{a^n b^m}{n! q_m!} X^n Y^m = e^{aX} e_q^{bY}$$

We will also call $\mathcal{A}$ the *q-affine group.*

If the dual pairing can be written as a product of q-exponentials, then we can really compute the dual representations in the same way as in Section 2.3 for Lie groups. The key property of the formal pairing $g = \sum A_n \otimes \psi_n = \prod e_{q_i}^{a_i X_i}$ is that multiplying the ψ_n in each term from the right (left) by X_i leads to the same result as applying the right (left) dual representation $\rho_R^*(X_i)$ ($\rho_L^*(X_i)$) to A_n in each term, i. e.,

$$\begin{aligned} \rho_R^*(X_i) g(a; X) &= g(a; X) X_i \\ \rho_L^*(X_i) g(a; X) &= X_i g(a; X). \end{aligned}$$

To compute the dual representations, commute X_i past the factors $e_{q_j}^{a_j X_j}$ in $g(a; X)$ until it is next to $e_{q_i}^{a_i X_i}$, and then replace it by the operator δ_i,

$$\delta_i f(a_i) = \begin{cases} \frac{f(q_i a_i) - f(a_i)}{(q_i - 1) a_i} & \text{if } q_i \neq 1, \\ \frac{\partial f(a_i)}{\partial a_i} & \text{if } q_i = 1, \end{cases}$$

since on the individual factors we have $X_i e_{q_i}^{a_i X_i} = \delta_i e_{q_i}^{a_i X_i}$. For factors where q_j is equal to 1, we can apply the relation $e^{a_j X_j} X_i e^{-a_j X_j} = e^{a_j \, \text{ad} X_j} X_i$ (where $\text{ad}X_j\,(X_i) = [X_j, X_i]$).

Example: We get for the q-affine algebra (cf. Proposition 3.3.3)

$$\begin{aligned} \rho_L^*(X) &= \partial_a + \alpha b \partial_b, & \rho_L^*(Y) &= \delta_b, \\ \rho_R^*(X) &= \partial_a, & \rho_R^*(Y) &= \exp(\alpha a) \delta_b. \end{aligned}$$

Since the dual of these representations is the coproduct of $\mathcal{A}$ (interpreted as right or left corepresentation of $\mathcal{A}$ on itself), we can use their matrix elements to get the coproduct of $\mathcal{A}$:

$$\begin{aligned} \Delta(a) &= a \otimes 1 + 1 \otimes a, \\ \Delta(b) &= b \otimes 1 + \exp(\alpha a) \otimes b. \end{aligned}$$

We shall use the quotient representation that arises from the relation $X = r$:

$$\rho_r(X) = r - \alpha b \partial_b, \qquad \rho_r(Y) = \delta_b. \tag{3.16}$$

3.6 MATRIX ELEMENTS

We introduce *matrix elements* for quantum groups. We define left and right matrix elements corresponding to the left and right multiplication

$$\begin{aligned} g(a;X)\psi_n &= \sum_{n+N\geq 0} M_n^N(a;L)\psi_{n+N} \\ \psi_n g(a;X) &= \sum_{n+N\geq 0} M_n^N(a;R)\psi_{n+N}. \end{aligned}$$

Note that in Section 2.6 we introduced only the left matrix elements. We will again switch freely between this notation and the subscript notation defined by $M_{nm}(a;\bullet) = M_n^{m-n}(a;\bullet)$ and we will write shorter $\psi(X)^*$ (resp. $\psi(X)^\dagger$) for the right (left) dual representation $\rho_R^*(\psi(X))$ (resp. $\rho_L^*(\psi(X))$).

Proposition 3.6.1 *We have the* principal formulas

$$\begin{aligned} M_n^N(a;L) &= \psi_n(X)^* c_{n+N}(a) \\ M_n^N(a;R) &= \psi_n(X)^\dagger c_{n+N}(a). \end{aligned}$$

Remark: With the identities $X^\dagger = (\mathrm{id}\otimes X)\circ\Delta$ and $X^* = (X\otimes\mathrm{id})\circ\Delta$ we can write the principal formulas also as

$$\begin{aligned} M_n^N(a;L) &= (\mathrm{id}\otimes\psi_n)\circ\Delta c_{n+N}(a) \\ M_n^N(a;R) &= (\psi_n\otimes\mathrm{id})\circ\Delta c_{n+N}(a). \end{aligned}$$

Proof: We have (with equations (3.8), (3.9))

$$\begin{aligned} \sum M_n^N(a;L)\psi_{n+N}(X) &= g(a;X)\psi_n(X) \\ &= \psi_n(X)^* g(a;X) \\ &= \sum \psi_n(X)^* c_m(a)\psi_m(a) \end{aligned}$$

and, respectively,

$$\begin{aligned} \sum M_n^N(a;R) &= \psi_n(X)g(a;X) \\ &= \psi_n(X)^\dagger g(a;X) \\ &= \sum \psi_n(X)^\dagger c_m(a)\psi_m(X). \end{aligned}$$

∎

Just as in the Section 2.6, summation theorems and recurrence relations follow readily from the definitions. We get for example[1]

$$\Delta c_n(a) = \sum M_k^{n-k}(a;L)\otimes c_k(a) \tag{3.17}$$

[1] To see e. g. equation (3.17), note that $\Delta c_n(a)$ can be written as $\Delta c_n(a) = \sum L_{nl}\otimes c_l(a)$ since $\{c_l(a)\}$ is a basis. With $<\psi_k(X), c_l(a)>=\delta_{kl}$ we get

$$M_{kn}(a;L) = M_k^{n-k}(a;L) = (\mathrm{id}\otimes\psi_k(X))\circ\Delta c_n(a) = \sum L_{nl}\delta_{kl} = L_{nk}.$$

Equations (3.18) and (3.19) can be derived by applying the coproduct Δ twice to $c_n(a)$ and using the coassociativity (i. e. $(\mathrm{id}\otimes\Delta)\circ\Delta = (\Delta\otimes\mathrm{id})\circ\Delta$).

$$= \sum c_k(a) \otimes M_k^{n-k}(a;R),$$

and

$$\Delta M_{nm}(a;L) = \sum_l M_{lm}(a;L) \otimes M_{nl}(a;L) \tag{3.18}$$

$$\Delta M_{nm}(a;R) = \sum_l M_{nl}(a;R) \otimes M_{lm}(a;R) \tag{3.19}$$

Equation (3.17) shows immediately

$$\begin{aligned} g(\Delta(a);X) &= \sum \Delta c_n(a)\psi_n = \sum M_k^{n-k}(a;L) \otimes c_k(a)\psi_n \\ &= \sum \psi_k^* c_n(a) \otimes c_k(a)\psi_n = \sum (c_n(a)\otimes 1)(1 \otimes c_k(a))\psi_n \cdot \psi_k \\ &= g(a\otimes 1;X)g(1\otimes a;X). \end{aligned} \tag{3.20}$$

i. e. the composition law that was mentioned in Section 3.4.

Recurrence relations follow from

$$\begin{aligned} X_i^* M_n^N(a;L) &= X_i^*\psi_n(X)^* c_{n+N}(a) = (X_i \cdot \psi_n(X))^* c_{n+N}(a) \\ &= \sum m_n^k(X_i;L)\psi_{n+k}(X)^* c_{n+N}(a) \\ &= \sum m_n^k(X_i;L) M_{n+k}^{n+N-k}(a;L) \\ X_i^\dagger M_n^N(a;R) &= X_i^\dagger\psi_n(X)^\dagger c_{n+N}(a) = (\psi_n(X)\cdot X_i)^\dagger c_{n+N}(a) \\ &= \sum m_n^k(X_i;R)\psi_{n+k}(X)^\dagger c_{n+N}(a) \\ &= \sum m_n^k(X_i;R) M_{n+k}^{n+N-k}(a;R), \end{aligned}$$

where the matrix elements $m_n^N(X_i;L)$ and $m_n^N(X_i;R)$ are defined by

$$\begin{aligned} X_i\psi_n(X) &= \sum m_n^N(X_i;L)\psi_{n+N}(X) \\ \psi_n(X)X_i &= \sum m_n^N(X_i;R)\psi_{n+N}(X). \end{aligned}$$

3.6.1 *Example: the quantum algebra $U_q(aff(1))$*

With the principal formulas we calculate

$$\begin{aligned} M_{nm}^{NM}(a,b;L) &= (e^{\alpha a}\delta_b)^m \partial_a^n \frac{a^{n+N}b^{m+M}}{(n+N)!q_{m+M}!} \\ &= e^{m\alpha a}\frac{a^N b^M}{N!q_M!} \\ M_{nm}^{NM}(a,b;R) &= (\partial_a + \alpha N_b)^n \delta_b^m \frac{a^{n+N}b^{m+M}}{(n+N)!q_{m+M}!} \\ &= \sum_{\nu=0}^{n} \binom{n}{\nu} (\alpha M)^\nu \frac{a^{N+\nu}}{(N+\nu)!}\frac{b^M}{q_M!} \\ &= {}_1F_1(-n;N+1;\alpha M a)\frac{a^N b^M}{N!q_M!}. \end{aligned}$$

In particular

$$M_{nm,01}(a,b;L) = \begin{cases} b & (n,m)=(0,0) \\ e^{\alpha a} & (n,m)=(0,1) \\ 0 & \text{else} \end{cases}$$

$$M_{nm,10}(a,b;L) = \begin{cases} a & (n,m)=(0,0) \\ 1 & (n,m)=(1,0) \\ 0 & \text{else.} \end{cases}$$

With equation (3.17) this implies

$$\Delta(a) = a\otimes 1 + 1\otimes a \tag{3.21}$$

$$\Delta(b) = b\otimes 1 + e^{\alpha a}\otimes b. \tag{3.22}$$

We also calculate the coproduct of $B = e^{-\alpha a}b$. We get

$$\Delta(B) = B\otimes e^{-\alpha a} + 1\otimes B. \tag{3.23}$$

3.6.2 *Example: the quantum algebra $h_q(1)$*

With the principal formula we get

$$\begin{aligned} M_{nmr}^{NMR}(a,b,c;L) &= (\partial_a + \sigma_c[\partial_b]_q)^n \partial_b^m \partial_c^r \frac{a^{n+N} b^{m+M} c^{r+R}}{(n+N)!(m+M)!(r+R)!} \\ &= \sum \binom{n}{\nu} \frac{a^{N+\nu}}{(N+\nu)!} \frac{([\partial_b]_q)^\nu\, b^M}{M!} \frac{c^{R+\nu}}{R!}. \end{aligned}$$

In particular,

$$M_{nmr,100}(a,b,c;L) = \begin{cases} a & (nmr)=(000) \\ 1 & (nmr)=(100) \\ 0 & \text{else} \end{cases}$$

$$M_{nmr,010}(a,b,c;L) = \begin{cases} b & (nmr)=(000) \\ 1 & (nmr)=(010) \\ \frac{2hc}{q-q^{-1}} & (nmr)=(100) \\ 0 & \text{else} \end{cases}$$

$$M_{nmr,001}(a,b,c;L) = \begin{cases} c & (nmr)=(000) \\ 1 & (nmr)=(001) \\ 0 & \text{else.} \end{cases}$$

With formula (3.17) these matrix elements allow us to calculate the defining relations for the coproduct $\Delta = m^*$ of $H_q(1)$ that turns $h_q(1)$ and $H_q(1)$ into a pair of dual bialgebras,

$$\begin{aligned} \Delta(a) &= a\otimes 1 + 1\otimes a \\ \Delta(b) &= b\otimes 1 + 1\otimes b + \frac{2h}{q-q^{-1}} c\otimes a \\ \Delta(c) &= c\otimes 1 + 1\otimes c. \end{aligned}$$

3.7 BRAIDED TENSOR CATEGORIES

Braided categories were introduced by André Joyal and Ross Street (Macquarie Mathematics Reports 850067 (Dec. 1985) and 86081 (Nov. 1986), see [JS91a, JS91b, JS93, Kas95, Maj95b]).

Definition 3.7.1 *[Mac71] A* tensor category *(also called* monoidal category*) is a category $\mathcal{C}$ with a tensor product $\otimes : \mathcal{C}\times\mathcal{C} \to \mathcal{C}$, a unit I, an associativity constraint a, a left unit constraint l w.r.t. I, and a right unit constraint r w.r.t. I, such that the* Pentagon Axiom

$$\begin{array}{ccc} & \Big((U\otimes V)\otimes W\Big)\otimes X & \\ a_{U,V,W}\otimes \mathrm{id}_X \swarrow & & \searrow a_{U\otimes V,W,X} \\ \Big(U\otimes (V\otimes W)\otimes X\Big) & & \\ a_{U,V\otimes W,X}\downarrow & & (U\otimes V)\otimes (W\otimes X) \\ U\otimes \Big((V\otimes W)\otimes X\Big) & & \\ \mathrm{id}_U\otimes a_{V,W,X}\searrow & & \swarrow a_{U,V,X\otimes X} \\ & U\otimes\Big(V\otimes (W\otimes X)\Big) & \end{array}$$

and the Triangle Axiom

$$\begin{array}{ccc} (V\otimes I)\otimes W & \overset{a_{V,I,W}}{\longrightarrow} & V\otimes (I\otimes W) \\ r_V\otimes \mathrm{id}_W\searrow & & \swarrow \mathrm{id}_V\otimes l_W \\ & V\otimes W & \end{array}$$

are satisfied for all objects U, V, W, X of $\mathcal{C}$.

Example: The most fundamental example is the category Vec($\mathbb{K}$) of vector spaces over a field $\mathbb{K}$. It comes with a tensor product $\otimes$: Vec($\mathbb{K}$) $\times$ Vec($\mathbb{K}$) $\ni U\times V \mapsto U\otimes V \in$ Vec($\mathbb{K}$), the unit object is the ground field $\mathbb{K}$, and the associativity and unit constraints are the natural isomorphisms $a_{U,V,W} : (U\otimes V)\otimes W \to U\otimes (V\otimes W)$, $l_v : \mathbb{K}\otimes V\to V$, $r_V : V\otimes\mathbb{K}\to V$,

$$a_{U,V,W}\Big((u\otimes v)\otimes w\Big) = u\otimes (v\otimes w), \qquad l_V(\lambda\otimes v) = \lambda v = r_V(v\otimes\lambda)$$

for $u\in U$, $v\in V$, $w\in W$, $U, V, W \in$ Vec($\mathbb{K}$), $\lambda\in\mathbb{K}$. The pentagon and triangle axioms are clearly satisfied.

Example: Let $\mathcal{A}$ be a bialgebra. The category of left (right) $\mathcal{A}$-modules $\mathcal{A}$-Mod (Mod-$\mathcal{A}$, resp.) and the category of right (left) $\mathcal{A}$-comodules Comod-$\mathcal{A}$ ($\mathcal{A}$-Comod, resp.) can be equipped with tensor product.

Definition 3.7.2 *(a) A* left (right) $\mathcal{A}$-module *over an algebra $\mathcal{A}$ is a pair (M, μ_M) of a vector space M and a linear map $\mu_M : \mathcal{A}\otimes M\to M$ ($\mu_M : M\otimes\mathcal{A}\to M$, resp.) such that $\mu_M(a\otimes\mu_M(b\otimes u)) = \mu_M(m(a\otimes b)\otimes b)$ and $\mu_M(e\otimes u) = u$ (or $\mu_M(\mu_M(u\otimes a)\otimes b) = \mu_M(u\otimes m(a\otimes b))$ and $\mu_M(u\otimes 1) = u$, resp.) for all $a, b\in\mathcal{A}$, $u\in M$.*

(b) *A* right (left) $\mathcal{C}$-comodule *over a coalgebra* $\mathcal{C}$ *is a pair* (N, δ_N) *of a vector space* N *and a linear map* $\delta_N : N \to N \otimes \mathcal{C}$ $(\delta_N : N \to \mathcal{C} \otimes N$, *resp.)* *such that* $(\delta_N \otimes \mathrm{id}_\mathcal{C}) \circ \delta_N = (\mathrm{id}_N \otimes \Delta) \circ \delta_N$ *and* $(\mathrm{id}_N \otimes \epsilon) \circ \delta_N = \mathrm{id}_N$ *(or* $(\mathrm{id}_\mathcal{C} \otimes \delta_N) \circ \delta_N = (\Delta \otimes \mathrm{id}_N) \circ \delta_N$ *and* $(\varepsilon \otimes \mathrm{id}_N) \circ \delta_N = \mathrm{id}_N$, *resp.)*.

If M, M' are left (right) $\mathcal{A}$-modules, then $M \otimes M'$ has a natural left (right) $\mathcal{A} \otimes \mathcal{A}$-module structure, just take $(\mu_M \otimes \mu_{M'}) \circ (\mathrm{id} \otimes \tau \otimes \mathrm{id})$, and check the axioms. If $\mathcal{A}$ is a bialgebra, then we get back an $\mathcal{A}$-module, if we set $\mu_{M \otimes M'} = (\mu_M \otimes \mu_{M'}) \circ (\mathrm{id}_\mathcal{A} \otimes \tau \otimes \mathrm{id}'_M) \circ (\Delta \otimes \mathrm{id}_{M \otimes M'})$ (or $\mu_{M \otimes M'} = (\mu_M \otimes \mu_{M'}) \circ (\mathrm{id}_M \otimes \tau \otimes \mathrm{id}_\mathcal{A}) \circ (\mathrm{id}_{M \otimes M'} \otimes \Delta)$ for right modules).

Similarly, for two right (left) $\mathcal{A}$-comodules N, N' over a bialgebra $\mathcal{A}$ their tensor product has again a right (left) $\mathcal{A}$-comodule structure, take $\delta_{N \otimes N'} = (\mathrm{id}_{N \otimes N'} \otimes m) \circ (\mathrm{id}_N \otimes \tau \otimes \mathrm{id}_\mathcal{A}) \circ (\delta_N \otimes \delta_{N'})$ (or $\delta_{N \otimes N'} = (m \otimes \mathrm{id}_{N \otimes N'}) \circ (\mathrm{id}_\mathcal{A} \otimes \tau \mathrm{id}_{N'}) \circ (\delta_N \otimes \delta_{N'})$, resp.).

Definition 3.7.3 *Let* $(\mathcal{C}, \otimes, I, a, l, r)$ *be a tensor category. A* braiding Ψ *in* $\mathcal{C}$ *is a commutativity constraint satisfying the* Hexagon Axiom, *i.e. the diagrams*

$$\begin{array}{ccccc} & U \otimes (V \otimes W) & \overset{\Psi_{U,V\otimes W}}{\longrightarrow} & (V \otimes W) \otimes U & \\ a_{U,V,W} \nearrow & & & & \searrow a_{V,W,U} \\ (U \otimes V) \otimes W & & & & V \otimes (W \otimes U) \\ \Psi_{U,V} \otimes \mathrm{id}_W \searrow & & & & \nearrow \mathrm{id}_V \otimes \Psi_{U,W} \\ & (V \otimes U) \otimes W & \overset{a_{V,U,W}}{\longrightarrow} & V \otimes (U \otimes W) & \end{array}$$

and

$$\begin{array}{ccccc} & (U \otimes V) \otimes W & \overset{\Psi_{U\otimes V,W}}{\longrightarrow} & W \otimes (U \otimes V) & \\ a^{-1}_{U,V,W} \nearrow & & & & \searrow a^{-1}_{W,U,V} \\ U \otimes (V \otimes W) & & & & (W \otimes U) \otimes V \\ \mathrm{id}_U \otimes \Psi_{V,W} \searrow & & & & \nearrow \Psi_{U,W} \otimes \mathrm{id}_V \\ & U \otimes (W \otimes V) & \overset{a^{-1}_{U,W,V}}{\longrightarrow} & (U \otimes W) \otimes V & \end{array}$$

commute for all objects U, V, W, X *of* $\mathcal{C}$.

Example: A trivial example for a braiding is the flip automorphism $\tau_{U,V} : U \otimes V \to V \otimes U$, $\tau(u \otimes v) = v \otimes u$, in the category of left (or right) modules of a cocommutative bialgebra. That this works on the level of vector spaces, is obvious. But we also have $\mu_{M' \otimes M} = \tau^{-1}_{M' \otimes M} \circ \mu_{M \otimes M'} \circ (\mathrm{id}_\mathcal{A} \otimes \tau_{M' \otimes M})$. In general, the category of modules of a bialgebra $\mathcal{A}$ has a braiding if and only if $\mathcal{A}$ is *quasi-triangular* (or braided), i.e. if there exists an invertible element $R \in \mathcal{A} \otimes \mathcal{A}$, called *universal R-matrix* that 'controls' the non-cocommutativity of $\mathcal{A}$ in the sense that

$$\tau \circ \Delta(a) = R \Delta(a) R^{-1}, \qquad \text{for all } a \in \mathcal{A}$$

and that satisfies $(\Delta\otimes\mathrm{id})(R) = R_{13}R_{23}$ and $(\mathrm{id}\otimes\Delta)(R) = R_{13}R_{12}$, see e.g. [Kas95, Proposition XIII.1.4]. Here $R_{12} = R\otimes 1$, $R_{23} = 1\otimes R$, $R_{13} = (\mathrm{id}\otimes\tau)(R\otimes 1)$ are elements of the algebra $\mathcal{A}\otimes\mathcal{A}\otimes\mathcal{A}$.

Definition 3.7.4 *A* braided tensor category $(\mathcal{C},\Psi)$ *consists of a tensor category $\mathcal{C}$ and a braiding Ψ of $\mathcal{C}$.*

3.8 BRAIDED BIALGEBRAS AND BRAIDED HOPF ALGEBRAS

Consider two algebras A and B who 'live' in a braided tensor category $(\mathcal{C},\Psi)$, i.e. A and B are objects of $\mathcal{C}$, and their multiplications $m_A : A\otimes A \to A$, $m_B : B\otimes B \to B$ and unit maps $e_A : I \to A$, $e_B : I \to B$ are morphisms of the category (here the unit object I is the underlying field $\mathbb{K}$). Then there exists a *braided tensor product algebra* $A\underline{\otimes}B$, build on the object $A\otimes B$, with multiplication $m_{A\underline{\otimes}B} = (m_A\otimes m_B)\circ(\mathrm{id}\otimes\Psi\otimes\mathrm{id})$ and unit map $e_{A\underline{\otimes}B} = e_A\otimes e_B$, cf. [Maj93a, Lemma 4.1]. Thus it makes sense to look for *braided bialgebras*, i.e. for algebras that are objects of a braided category, and for which there exist morphisms $\underline{\Delta}_A : A \to A\underline{\otimes}A$, $\underline{\varepsilon} : A \to \mathbb{K}$, that satisfy the conditions corresponding to those in Definition 3.1.2, i.e.

$$(\underline{\Delta}\otimes\mathrm{id})\circ\underline{\Delta} = (\mathrm{id}\otimes\underline{\Delta})\circ\underline{\Delta}, \qquad (\underline{\varepsilon}\otimes\mathrm{id})\circ\underline{\Delta} = \mathrm{id} = (\mathrm{id}\otimes\underline{\varepsilon})\circ\underline{\Delta},$$

and $\underline{\Delta}$ and $\underline{\varepsilon}$ are algebra homomorphisms.

A *braided antipode* $\underline{S} : A \to A$ is defined by the condition

$$m_A\circ(\underline{S}\otimes\mathrm{id})\circ\underline{\Delta} = e\circ\underline{\varepsilon} = m_A\circ(\mathrm{id}\otimes\underline{S})\circ\underline{\Delta},$$

it is again unique, but no longer an algebra anti-homomorphism. Instead we have $\underline{S}\circ m = m\circ\Psi\circ(\underline{S}\otimes\underline{S})$ (but note that we still have $\underline{S}\circ e = e$). With respect to the coproduct and counit it satisfies $\underline{\Delta}\circ\underline{S} = (S\circ S)\circ\Psi\circ\underline{\Delta}$ and $\underline{\varepsilon}\circ\underline{S} = \underline{S}$.

Most notions related to algebras and coalgebras as e.g. modules, comodules, dual pairing or dual representations can immediately be defined in braided categories as well. One should note though that for the extension of the dual pairing to tensor products often a slightly different convention is adopted, and, related to this, the factors are numbered in opposite direction.. When calculating $\langle,\rangle_{\underline{\otimes}n} : A^{\underline{\otimes}n}\times B^{\underline{\otimes}n} \to \mathbb{K}$, often one evaluates the last factor of A against the first of B, etc., i.e. $\langle a_1\otimes a_2\otimes\cdots\otimes a_n, b_n\otimes b_{n-1}\otimes\cdots\otimes b_1\rangle_{\underline{\otimes}n} = \prod_{\nu=1}^{n}\langle a_\nu, b_\nu\rangle$.

If we use the same dual pairing as in the symmetric case, i.e.

$$\langle a_1\otimes b_1, a_2\otimes b_2\rangle_{\otimes} = \langle a_1, a_2\rangle\langle b_1, b_2\rangle, \qquad \text{for all } a_i, b_i \in \mathcal{B}_i$$

then we get the following Leibniz formula for the right dual representations.

Lemma 3.8.1 *a) The right dual representation is an homomorphism, i.e.*

$$\rho(xy) = \rho(x)\rho(y)$$

for all $x, y \in \mathrm{Hom}(\mathcal{B},\mathbb{C})$.

b) *If x is in a bialgebra dually paired with $\mathcal{B}$, then $\rho(x)$ satisfies a generalised* Leibniz formula, *which has the form*

$$\rho(x)(ab) = \sum_i (m \otimes x_i^{(1)}) \circ (\mathrm{id} \otimes \Psi)(\Delta(a) \otimes \rho(x_i^{(2)})b)$$

where $x \circ m = \sum_i x_i^{(1)} \otimes x_i^{(2)}$ is the coproduct of x. In the symmetric case we can also write

$$\rho(x)(ab) = \sum_i \left(\rho(x_i^{(1)})a\right)\left(\rho(x_i^{(2)})b\right).$$

Proof:

a) Using the definition of the right dual representation and the coassociativity of Δ, we get $\rho(xy)a = (\mathrm{id} \otimes xy) \circ \Delta(a) = (\mathrm{id} \otimes x \otimes y) \circ (\mathrm{id} \otimes \Delta) \circ \Delta(a) = (\mathrm{id} \otimes x \otimes y) \circ (\Delta \otimes \mathrm{id}) \circ \Delta(a) = \rho(x)\rho(y)a$.

b) We have to compute $\rho(x)(ab) = (\mathrm{id} \otimes x) \circ \Delta \circ m(a \otimes b)$. The braiding now shows up because Δ is a homomorphism between the algebras $(\mathcal{B}, m)$ and $(\mathcal{B} \otimes \mathcal{B}, m^{\otimes} = (m \otimes m) \circ (\mathrm{id} \otimes \Psi \otimes \mathrm{id}))$, i.e. we get

$$\begin{aligned} \rho(x)(ab) &= (\mathrm{id} \otimes x) \circ \Delta \circ m(a \otimes b) \\ &= (\mathrm{id} \otimes x)(m \otimes m) \circ (\mathrm{id} \otimes \Psi \otimes \mathrm{id}) \circ (\Delta \otimes \Delta)(a \otimes b) \end{aligned}$$

We get the desired expression, if we substitute $x \circ m = \sum_i x_i^{(1)} \otimes x_i^{(2)}$. With Sweedlers notation $\Delta a = \sum_j a_j^{(1)} \otimes a_j^{(2)}$ and $\Delta b = \sum_k b_k^{(1)} \otimes b_k^{(2)}$, we can also write this as

$$\rho(x)(ab) = \sum_{i,j,k} (m \otimes x_i^{(1)})(a_j^{(1)} \otimes \Psi(a_j^{(2)} \otimes b_k^{(1)}))\langle x_i^{(2)}, b_k^{(2)}\rangle.$$

In the symmetric case we have $\Psi(a_j^{(2)} \otimes b_k^{(1)}) = b_k^{(1)} \otimes a_j^{(2)}$ and everything simplifies to

$$\rho(x)(ab) = \sum_{i,j,k} a_j^{(1)} b_k^{(1)} \langle x_i^{(1)}, a_j^{(2)}\rangle \langle x_i^{(2)}, b_k^{(2)}\rangle = \sum_i \left(\rho(x_i^{(1)})a\right)\left(\rho(x_i^{(2)})b\right).$$

■

It is not clear what the correct definition of $*$-structures on braided Hopf algebras is, see [Maj94, Maj95a, Maj97] for one set of axioms. But for the study of Lévy processes these axioms are not adequate since the braided coproduct is not a $*$-algebra homomorphism.

We propose an alternative definition.

Definition 3.8.2 *A* braided $*$-bialgebra *is a braided bialgebra $(\mathcal{A}, \Delta, \varepsilon, m, 1, \Psi)$ over $\mathbb{C}$ with an anti-linear map $* : \mathcal{A} \to \mathcal{A}$ that satisfies*

- $(\mathcal{A}, m, 1, *)$ *is a* $*$*-algebra,*
- $\mathcal{A} \otimes \mathcal{A}$ *admits a* $*$*-structure such that the canonical inclusions are* $*$*-algebra homomorphisms and* $\Delta : \mathcal{A} \to \mathcal{A} \otimes \mathcal{A}$ *is a* $*$*-homomorphism for this* $*$*-structure.*

If there exists antipode S *(i.e. a linear map* $S : \mathcal{A} \to \mathcal{A}$ *s.t.* $(\mathcal{A}, \Delta, \varepsilon, m, 1, S, \Psi)$ *is a Hopf algebra), then we will call* $(\mathcal{A}, \Delta, \varepsilon, m, 1, S, \Psi, *)$ *a* braided $*$-Hopf algebra.

Remarks:

- From the condition that the canonical inclusions $\mathcal{A} \xrightarrow{i_1} \mathcal{A}\otimes\mathcal{A} \xleftarrow{i_2} \mathcal{A}$ are $*$-homomorphisms follows $(a\otimes 1)^* = a^*\otimes 1$, $(1\otimes a)^* = 1\otimes a^*$ for all $a \in \mathcal{A}$. Now the condition that this map is an anti-homomorphism uniquely determines it, $(a\otimes b)^* = \Big((a\otimes 1)(1\otimes b)\Big)^* = (1\otimes b^*)(a^*\otimes 1) = \Psi(b^*\otimes a^*)$ for all $a, b \in \mathcal{A}$. This defines a $*$-structure on $\mathcal{A} \otimes \mathcal{A}$ if and only if it is also an involution, i.e. if $(\Psi \circ (* \otimes *) \circ \tau)^2 = \mathrm{id}$. To see this, let $\{x_i\}$ be a basis of $\mathcal{A}$ consisting of self-adjoint elements, i.e. such that $x_i^* = x_i$. Define the coefficients of the braiding Ψ in this basis by $\Psi(x_i \otimes x_j) = \sum_{kl} \Psi_{ij}^{kl} x_l \otimes x_k$ (note that this sum is finite, even though the index set can be infinite). $\Psi \circ (* \otimes *) \circ \tau$ is an involution, if and only if $\sum_{nm} \overline{\Psi_{ji}^{nm}} \Psi_{nm}^{kl} = \delta_i^l \delta_j^k$. With this relation we can show that $\Psi \circ (* \otimes *) \circ \tau$ is an anti-homomorphism. It is sufficient to check this on the basis elements, we get

$$
\begin{aligned}
&\Big((x_i \otimes x_j) \cdot (x_k \otimes x_l)\Big)^* \\
=\;& \sum_{nm} (\Psi_{jk}^{nm} x_i x_m \otimes x_n x_l)^* = \sum_{nm} \overline{\Psi_{jk}^{nm}} \Psi\Big((x_n x_l)^* \otimes (x_i x_m)^*\Big) \\
=\;& \sum_{nm} \overline{\Psi_{jk}^{nm}}\, m_\Psi^\otimes \circ (\Psi \otimes \Psi) \circ (\mathrm{id} \otimes \Psi \otimes \mathrm{id})(x_l \otimes x_n \otimes x_m \otimes x_i) \\
=\;& \sum_{nmpq} \overline{\Psi_{jk}^{nm}} \Psi_{nm}^{pq}\, m_\Psi^\otimes \circ (\Psi \otimes \Psi)(x_l \otimes x_q \otimes x_p \otimes x_i) \\
=\;& m_\Psi^\otimes \circ (\Psi \otimes \Psi)(x_l \otimes x_k \otimes x_j \otimes x_i) \\
=\;& m_\Psi^\otimes \Big(\Psi(x_l \otimes x_k) \otimes \Psi(x_j \otimes x_i)\Big) \\
=\;& (x_k \otimes x_l)^* \cdot (x_i \otimes x_j)^*,
\end{aligned}
$$

  where $m_\Psi^\otimes = (m \otimes m) \circ (\mathrm{id} \otimes \Psi \otimes \mathrm{id})$ is the multiplication of $\mathcal{A} \otimes \mathcal{A}$. If $\Psi \circ (* \otimes *) \circ \tau$ is an involution, then we also have $\Psi \circ * \circ \Psi \circ * = \mathrm{id}_{H\otimes H}$.

- If the coproduct is fixed, then the counit and the antipode (if it exists) are unique. Therefore one should expect their axioms to be a consequence of those given in the definition.

3.9 EXAMPLES OF BRAIDED BIALGEBRAS AND BRAIDED HOPF ALGEBRAS

3.9.1 *The braided line* $\mathbb{R}_q$

Let $q \in \mathbb{C}$, q not a root of unity. The braiding $\Psi(x^n \otimes x^m) = q^{nm} x^m x^n$ turns the algebra of polynomials in one variable into a braided Hopf algebra (with $\Delta x = x + x'$). This Hopf algebra, denoted by $\mathbb{R}_q$ can be dually paired with itself, set $\langle p^n, x^m \rangle = q_n! \delta_{nm}$. The dual copy acts on $\mathbb{R}_q$ via $\rho(u)a = \sum \langle u, a^{(1)} \rangle a^{(2)}$, where $\Delta a = \sum a^{(1)} \otimes a^{(2)}$. One finds that $\rho(p)$ is the q-difference operator $\delta_q : f(x) \mapsto \frac{f(qx)-f(x)}{x(q-1)}$.

3.9.2 *The braided plane* $\mathbb{C}_q^{2|0}$

The ***braided plane*** is the braided Hopf algebra with two q-commuting primitive generators x, y and braid relations $x'x = q^2 xx'$, $x'y = qyx'$, $y'x = qxy' + (q^2-1)yx'$, $y'y = q^2 yy'$.

$\mathbb{C}_q^{2|0}$ is dually paired with $\mathbb{C}_{q^{-1}}^{2|0}$ and thus there are partial derivatives,

$$\partial_1 x^n y^m = (q^2)_n x^{n-1} y^m, \qquad \partial_2 x^n y^m = (q^2)_m q^n x^n y^{m-1},$$

3.9.3 *The braided Heisenberg-Weyl group* HW_q

In this Subsection we introduce a braided Hopf algebra structure for the q-oscillator algebra HW_q defined by

$$ac - qca = (1-q)b, \qquad ab = ba, \qquad cb = bc, \tag{3.24}$$

where $q \in \mathbb{C}\backslash\{0\}$, q not a root of unity. For other quantum or braided groups related to the Heisenberg group see [BM93] and the references therein.

We can also regard this algebra as generated by two generators a, c with the cubic relations

$$aac + qcaa = (1+q)aca \tag{3.25}$$
$$acc + qcca = (1+q)cac \tag{3.26}$$

with $q \in \mathbb{C}$, q not a root of unity. If we define the braiding by

$$a'a = qaa' \qquad a'c = ca' \qquad c'a = \frac{1}{q}ac' \qquad c'c = qcc' \tag{3.27}$$

then

$$\Delta a = a + a', \qquad \Delta c = c + c'$$

defines an algebra homomorphism from HW_q to $\mathrm{HW}_q \tilde{\otimes} \mathrm{HW}_q$. The counit is defined by $\varepsilon(a) = \varepsilon(c) = 0$ (extend as a unital algebra homomorphism). The braided bialgebra defined in this way also admits an antipode. Define S by

$$S(a) = -a, \qquad S(c) = -c,$$

and extend by $S \circ m = m \circ \Psi \circ (S \otimes S)$.

The advantage of introducing a third generator, e.g. by $ac - qca = (1-q)b$ or $d = ac - ca$, is that $\{a^n b^m c^r; n, m, r \in \mathbb{N}\}$ and $\{a^n d^m c^r; n, m, r \in \mathbb{N}\}$ are bases (Poincaré-Birkhoff-Witt bases) of HW_q.

One can calculate the coproduct on a general basis element

$$\Delta(a^n b^m c^r) = \sum_{\nu=0}^{n}\sum_{\mu=0}^{m}\sum_{\sigma=0}^{\mu}\sum_{\rho=0}^{r}\begin{bmatrix} n \\ \nu \end{bmatrix}_q \begin{bmatrix} m \\ \mu \end{bmatrix}_q \begin{bmatrix} \mu \\ \sigma \end{bmatrix}_q \begin{bmatrix} r \\ \rho \end{bmatrix}_q q^{\rho(m-\mu)} q^{\sigma(n-\nu)} a^\nu b^\sigma c^{\rho+\mu-\sigma} a'^{n-\nu+\mu-\sigma} b'^{m-\mu} c'^{r-\rho}. \tag{3.28}$$

We will construct the dual of HW_q. The coproduct of HW_q defines an algebra structure in its dual. Let $X, Y : \mathrm{HW}_q \to \mathbb{C}$ be two functionals on HW_q, then their product is defined by

$$(X \star Y)(u) = (X \otimes Y)\Delta(u) = \sum X(u_{(2)})Y(u_{(1)}) \qquad \forall u \in \mathrm{HW}_q,$$

where we used Sweedler's notation, $\Delta(u) = \sum u_{(1)} \otimes u_{(2)}$. For the functionals defined by

$$X_{ijk}(a^n b^m c^r) = \delta_{in}\delta_{jm}\delta_{kr}$$

we see that they span an algebra, denoted by $\mathrm{HW}_q^\vee$, generated by the three elements $x = X_{100}$, $y = X_{010}$, $z = X_{001}$ such that $X_{ijk} = \frac{z^k y^j x^i}{q_k! q_j! q_i!}$ and

$$xy = qyx, \quad yz = qzy, \quad xz - zx = y.$$

We see that this algebra is isomorphic to HW_q, the isomorphism being $x \mapsto a$, $z \mapsto c$.

The following formula is useful to compute products of general basis elements of the HW_q algebra, it can be derived by induction.

$$c^r a^n = \sum_{k=0}^{r\wedge n} \frac{(q^{-1})_r!(q^{-1})_n!(-1)^k}{(q^{-1})_{r-k}!(q^{-1})_{n-k}!(q^{-1})_k!} a^{n-k} d^k c^{r-k}. \tag{3.29}$$

If q is real, then the involution $* : \mathrm{HW}_q \to \mathrm{HW}_q$ defined by

$$a^* = c, \qquad b^* = b, \qquad c^* = a.$$

satisfies

$$\begin{aligned} * \circ * &= \mathrm{id} \qquad (\text{i.e. } (a^*)^* = a \quad \forall a \in \mathcal{A}), \\ * \circ m &= m \circ \tau \circ (* \otimes *) \qquad \text{i.e. } ^* \text{ is an anti-homomorphism}, \\ (* \otimes *) \circ \Delta &= \tau \circ \Delta \circ *, \\ \varepsilon \circ * &= \bar{} \circ \varepsilon, \\ S \circ * &= * \circ S, \end{aligned}$$

i.e. turns HW_q into a *braided *-Hopf algebra in the sense of Majid* (see [Maj95a] for the origin of the last three conditions).

An action of $\mathrm{HW}_q^\vee$ of HW_q can be defined by defining $\rho(x)$ and $\rho(z)$ as

$$\rho(x)a = 1, \quad \rho(x)c = 0,$$
$$\rho(z)a = 0, \quad \rho(z)c = 1,$$

on the generators a and c, and extending with the *Leibniz formulas*

$$\begin{array}{rclrcl} \rho(x)(au) & = & u + qa\rho(x)u, & \rho(x)(cu) & = & \frac{1}{q}c\rho(x)u, \\ \rho(z)(au) & = & a\rho(z)u, & \rho(z)(cu) & = & u + qc\rho(z)u, \end{array}$$

3.10 BRAIDED SPACES

Braided spaces are braided Hopf algebras with a particular form of the product and coproduct, see below. They are a generalisation of super-spaces, there are braided lines, braided planes, braided matrices, etc., all in analogy with super-lines, super-planes, etc. Braided spaces are the elementary building blocks of the 'braided world' or braided linear algebra.

For a pair of invertible matrices (R, R') that satisfy

(i) $R_{12}R_{13}R_{23} = R_{23}R_{13}R_{12}$ (i.e. the Quantum-Yang-Baxter equation),

(ii) $R_{12}R_{13}R'_{23} = R'_{23}R_{13}R_{12}$, $R_{23}R_{13}R'_{12} = R'_{12}R_{13}R_{23}$,

(iii) $(PR + 1)(PR' - 1) = 0$,

(iv) $R_{21}R'_{12} = R'_{21}R_{12}$,

where P is the *permutation matrix* ($\mathbf{x}_2 \otimes \mathbf{x}_1 P = \mathbf{x}_1 \otimes \mathbf{x}_2$), and $R_{12} = R \otimes \mathrm{id}$, $R_{23} = \mathrm{id} \otimes R$, etc., there exist two braided Hopf algebras $V(R, R')$ and $V\check{}(R, R')$ with generators $v_1, \ldots, v_n$ and $x_1, \ldots, x_n$ and relations

- for $V(R, R')$:

$$\begin{array}{rcl} \underline{\Delta} v^i & = & v^i \otimes 1 + 1 \otimes v^i, \\ \Psi(v^i \otimes v^j) & = & R^i{}_k{}^j{}_l v^l \otimes v^k, \\ v^i v^j & = & R'^i{}_k{}^j{}_l v^l v^k, \\ \underline{S}(v^i) & = & -v_i, \\ \epsilon(v^i) & = & 0, \end{array}$$

- for $V\check{}(R, R')$:

$$\begin{array}{rcl} \underline{\Delta} x_i & = & x_i \otimes 1 + 1 \otimes x_i, \\ \Psi(x_i \otimes x_j) & = & x_l \otimes x_k R^k{}_i{}^l{}_j, \\ x_i x_j & = & x_l x_k R'^k{}_i{}^l{}_j, \\ \underline{S}(x_i) & = & -x_i, \\ \epsilon(x_i) & = & 0, \end{array}$$

see [Maj93b, Theorem 1]. Here the first two indices (i,k) correspond to the first factor in the tensor product, and the last two (j,l) to the second factor. We use the summation convention, i.e. indices that appear twice (once as subscript and once as superscript) are summed over. In a shorter notation this becomes

- for $V(R,R')$: $\underline{\Delta}\mathbf{v}=\mathbf{v}\otimes 1+1\otimes\mathbf{v}$, $\Psi(\mathbf{v}_1\otimes\mathbf{v}_2)=R_{12}\mathbf{v}_2\otimes\mathbf{v}_1$, $R'_{12}\mathbf{v}_2\mathbf{v}_1=\mathbf{v}_1\mathbf{v}_2$, $\underline{S}(\mathbf{v})=-\mathbf{v}$, $\epsilon(\mathbf{v})=0$,
- for $V\check{}(R,R')$: $\underline{\Delta}\mathbf{x}=\mathbf{x}\otimes 1+1\otimes\mathbf{x}$, $\Psi(\mathbf{x}_1\otimes\mathbf{x}_2)=\mathbf{x}_2\otimes\mathbf{x}_1R_{12}$, $\mathbf{x}_2\mathbf{x}_1R'_{12}=\mathbf{x}_1\mathbf{x}_2$, $\underline{S}(\mathbf{x})=-\mathbf{x}$, $\epsilon(\mathbf{x})=0$.

The braided Hopf algebras $V(R,R')$ and $V\check{}(R,R')$ are called the ***braided vector space*** and the ***braided covector space*** associated to (R,R'). For a detailed development of their properties see also [Maj93c, Maj93a, Maj96]. We will summarise here only some facts that are used in this presentation.

Let $\mathcal{S}\{1,\ldots,n\}$ be the set of all finite sequences composed of the elements $1,\ldots,n$, i.e.

$$\mathcal{S}\{1,\ldots,n\}=\{\emptyset;1,\ldots,n;11,12,\ldots,nn;111,\ldots\}.$$

Set $\mathbf{v}^a=v^{a_k}\cdots v^{a_1}$ if $a=(a_1\cdots a_k)\in\mathcal{S}\{1,\ldots,n\}\backslash\{\emptyset\}$, and $\mathbf{v}^\emptyset=1$, and analogously, but in reversed order, for $\mathbf{x}_a=x_{a_1}\cdots x_{a_k}$. Then $\{\mathbf{v}^a;a\in\mathcal{S}\{1,\ldots,n\}\}$, $\{\mathbf{x}_a;a\in\mathcal{S}\{1,\ldots,n\}\}$ span $V(R,R')$ and $V\check{}(R,R')$, respectively. Notice that in general these sets do not form a basis. But a basis can easily be extracted from them, if one orders them, and eliminates all elements that are linearly dependent on those that precede them.

The algebras $V(R,R')$ and $V\check{}(R,R')$ are $\mathbb{N}$-graded with

$$\begin{aligned}V(R,R')^{(r)} &= \text{span}\,\{\mathbf{v}^a;a\in\mathcal{S}^r\{1,\ldots,n\}\}\\ V\check{}(R,R')^{(r)} &= \text{span}\,\{\mathbf{x}_a;a\in\mathcal{S}^r\{1,\ldots,n\}\},\end{aligned}$$

where $\mathcal{S}^r\{1,\ldots,n\}\subseteq\mathcal{S}\{1,\ldots,n\}$ contains only the sequences of length r. This allows us to define a scaling $s(\lambda):V\check{}(R,R')\to V\check{}(R,R')$ by $s(\lambda)a=\lambda^{\deg(a)}a$ for homogeneous elements $a\in V\check{}(R,R')$ or $V(R,R')$.

$V(R,R')$ and $V\check{}(R,R')$ are mutually dual with the definition

$$\langle\mathbf{v}^a,\mathbf{x}_a\rangle=\delta_{|a|,|b|}[|a|,R]!^a_b$$

where $|a|$, $|b|$ is the length of $a,b\in\mathcal{S}\{1,\ldots,n\}$ and

$$\begin{aligned}[m;R] &= 1+(PR)_{12}+(PR)_{12}(PR)_{23}+\cdots+(PR)_{12}\cdots(PR)_{m-1,m} &(3.30)\\ [m;R]! &= [2;R]_{m-1,m}[3;R]_{m-2,m-1,m}\cdots[m;R]_{1,\ldots,m} &(3.31)\end{aligned}$$

(cf. [Maj96, Section 5]) are the *braided-integers* and the *braided-factorials*. This defines also an action of $V(R,R')$ on $V\check{}(R,R')$ (and vice versa) by

$$\begin{aligned}\rho(v^i) &= \partial^i=(\langle v^i,\cdot\rangle\otimes\text{id})\circ\underline{\Delta},\\ \rho(x_i) &= \hat{\partial}_i=(\langle\cdot,x_i\rangle\otimes\text{id})\circ\underline{\Delta}.\end{aligned}$$

Braided-exponentials $\exp(\mathbf{x}|\mathbf{v})$ are defined as solutions of

$$\partial^i \exp(\mathbf{x}|\mathbf{v}) = \exp(\mathbf{x}|\mathbf{v})v^i, \qquad (\epsilon\otimes \mathrm{id})\exp(\mathbf{x}|\mathbf{v}) = 1$$
$$\exp(\mathbf{x}|\mathbf{v})\hat{\partial}_i = x_i \exp(\mathbf{x}|\mathbf{v}), \qquad (\mathrm{id}\otimes\epsilon)\exp(\mathbf{x}|\mathbf{v}) = 1.$$

If the braided integers $[m;R]$ are all invertible, the $\exp(\mathbf{x}|\mathbf{v})$ is given by

$$\exp(\mathbf{x}|\mathbf{v}) = \sum \mathbf{x}_a([m;R]!^{-1})^a_b \mathbf{v}^b.$$

We will assume that this is the case in our examples and applications, or rather that there exists a braided-exponential $\exp(\mathbf{x}|\mathbf{v}) = \sum \mathbf{x}_a F(m;R)^a_b \mathbf{v}^b$, see [Maj96, Section 5.4].

Again, as in Section 3.8 it is not clear, what the correct axioms for a $*$-structure are, let us give the following tentative definition.

Definition 3.10.1 *A* braided $*$-space *is a braided space (cf. [Maj96]) equipped with an involution that turns it into a braided $*$-Hopf algebra (in the sense of Definition 3.8.2).*

3.10.1 *Examples*

We have already seen two examples, the braided line $\mathbb{R}_q$ and the braided plane $\mathbb{C}_q^{2|0}$, see Subsections 3.9.1 and 3.9.2. There are defined by the R-matrices $R = (q)$ and $R' = (1)$ and

$$R = \begin{pmatrix} q^2 & 0 & 0 & 0 \\ 0 & q & q^2-1 & 0 \\ 0 & 0 & q & 0 \\ 0 & 0 & 0 & q^2 \end{pmatrix}, \qquad R' = q^{-2}R,$$

respectively.

Even though the second braided-integer for the R-matrix of the braided plane $\mathbb{C}_q^{2|0}$

$$[2;R] = \begin{pmatrix} 1+q^2 & 0 & 0 & 0 \\ 0 & 1 & q & 0 \\ 0 & q & q^2 & 0 \\ 0 & 0 & 0 & 1+q^2 \end{pmatrix},$$

is not invertible, there exists nonetheless a braided-exponential [Maj96, Example 5.4]

$$\exp(\mathbf{x}|\mathbf{v}) = \sum_{a\in S\{1,\ldots,n\}} \frac{\mathbf{x}_a\mathbf{v}^a}{[|a|;q^2]!} = \sum_{p=0}^{\infty} \frac{(\mathbf{x}\cdot\mathbf{v})^p}{[p;q^{-2}]!},$$

where $[m;q^2] = \frac{1-q^{2m}}{1-q^2}$ and $[m;q^2]! = \prod_{\mu=1}^m [\mu;q^2]$.

The free braided-space

Here $R' = P$, there are no relations in the algebras $V(R, R')$ and $V^\check{}(R, R')$ since the ideal generated by $R'_{12}\mathbf{v}_2\mathbf{v}_1 = \mathbf{v}_1\mathbf{v}_2$ (respectively $\mathbf{x}_1\mathbf{x}_2R'_{12} = \mathbf{x}_2\mathbf{x}_1$) is equal to $\{0\}$, i.e. we have the free algebra with n generators. $\{\mathbf{v}^a; a \in \mathcal{S}\{1, \ldots, n\}\}$ (respectively $\{\mathbf{x}_a; a \in \mathcal{S}\{1, \ldots, n\}\}$) is thus a basis of $V(R, R')$ (respectively $V^\check{}(R, R')$).

We will assume that R is also equal to P, i.e. that the braiding is given by $\Psi(v^i \otimes v^j) = (v^i \otimes v^j)$ (respectively $\Psi(x_i \otimes x_j) = (x_i \otimes x_j)$). In this case the braided-integers $[m; P]^a_b = m\delta_{ab}$, $m = |a| = |b|$, are invertible and thus

$$\exp(\mathbf{x}|\mathbf{v}) = \sum_{k=0}^{\infty} \frac{1}{k!} \sum_{a \in \mathcal{S}^k\{1,\ldots,n\}} \mathbf{x}_a \mathbf{v}^a,$$

The dual action is

$$\rho(v^i)\mathbf{x}_a = \begin{cases} |a|\, \mathbf{x}_{a_2\cdots a_{|a|}} & \text{if } a_1 = i, \\ 0 & \text{else} \end{cases}$$

on the basis elements.

3.11 COMPACT QUANTUM GROUPS

Definition 3.11.1 *[Wor98, Definition 2.1] A pair (A, Δ) where A is a separable unital C^*-algebra and $\Delta : A \to A\hat{\otimes}A$ is a unital $*$-algebra homomorphism from A into its minimal tensor product $A\hat{\otimes}A$ is called a* compact quantum group, *if*

1. *Δ is coassociative, i.e. the diagram*

$$\begin{array}{ccc} A & \xrightarrow{\Delta} & A\hat{\otimes}A \\ \Delta\downarrow & & \downarrow \Delta\otimes\mathrm{id} \\ A\hat{\otimes}A & \xrightarrow{\mathrm{id}\otimes\Delta} & A\hat{\otimes}A\hat{\otimes}A \end{array}$$

is commutative, and

2. *the sets*

$$\{(b \otimes 1)\Delta(c); b, c \in A\}$$
$$\{(1 \otimes b)\Delta(c); b, c \in A\}$$

are linearly dense subsets of $A\hat{\otimes}A$.

If A is commutative, then there exists a compact topological group Λ such that $A = C(\Lambda)$, see the third remark following [Wor98, Definition 2.1].

We recall two fundamental results of the theory of compact quantum groups. The first states that a compact quantum group contains (densely) a Hopf $*$-algebra.

Theorem 3.11.2 *[Wor98, Theorem 2.2] Let (A, Δ) be a compact quantum group and $\mathcal{A}$ be the set of all linear combinations of matrix elements of all finite-dimensional unitary representations of (A, Δ). Then $\mathcal{A}$ is a dense $*$-subalgebra of A and $\Delta\mathcal{A} \subseteq \mathcal{A} \otimes \mathcal{A}$. Moreover, $(\mathcal{A}, \Delta|_{\mathcal{A}})$ is a Hopf $*$-algebra.*

The second result is the existence of a *Haar state.* Let $\mathcal{B}$ be a $\mathbb{K}$-bialgebra. A functional $\int : \mathcal{B} \to \mathbb{K}$ is called a *left (right) Haar functional* or a *left (right) invariant integral,* if it satisfies

$$(\mathrm{id} \otimes \int) \circ \Delta = e \circ \int, \qquad \left((\int \otimes \mathrm{id}) \circ \Delta = e \circ \int \quad \text{resp.} \right).$$

If $\mathcal{B}$ is a Hopf algebra and if $\int$ is a right Haar functional with $\int 1 \neq 0$, then it is even biinvariant and unique, if we normalise so that $\int 1 = 1$. Woronowicz has shown that on compact quantum groups such a normalised, biinvariant Haar functional always exists and that is is positive.

Theorem 3.11.3 *[Wor98, Theorem 2.3] Let (A, δ) be a compact quantum group. Then there exists a unique positive normalised linear functional $h : A \to \mathbb{C}$ such that*

$$(\mathrm{id} \otimes h) \circ \Delta(a) = (h \otimes \mathrm{id}) \circ \Delta(a) = h(a)1$$

for all $a \in A$.

REFERENCES

In addition to the two classic textbooks on bialgebras and Hopf algebras [Swe69, Abe80], a number of books on this topic have appeared during the last 10 years [DHL91, Fuc92, FK93, Lus93, Mon93, SS93, CP95, Gui95, Kas95, Maj95b].

Chapter 4

Stochastic processes on quantum groups

In this chapter we define Lévy processes on involutive bialgebras. These processes appeared first in the work of von Waldenfels[Wal73], the algebraic framework for the general theory of these processes was formulated in [ASW88]. For the development of the theory until 1993 see also [Sch93].

In the first section of this chapter we recall the basic terminology of non-commutative or quantum probability. In Section 4.2 we introduce the notion of independence for quantum random variables. The next section (Section 4.3) contains the definition of quantum stochastic processes with independent and stationary increments, i.e. Lévy processes. In Sections 4.4 and 4.5 we describe two constructions of realisations of Lévy process. The first starts from the generator and gives a realisation on a Bose Fock space, whereas the second requires a convolution semi-group of normalised functionals as input and uses an inductive limit procedure. In Section 4.6, we construct convolution semi-groups on quantum groups from classical Lévy processes and in Section 4.7 we prove an analogue of the Feynman-Kac formula for these semi-groups. The last section (Section 4.8) deals with duality and time-reversal.

4.1 QUANTUM PROBABILITY

We summarise the basic definitions of quantum probability or non-commutative probability, for a detailed introduction see e.g. the books by Biane [Bia93], Meyer [Mey93], and Parthasarathy [Par92].

A *quantum probability space* is defined as a pair $(\mathcal{A}, \Phi)$ consisting of a $*$-algebra $\mathcal{A}$ and a *state* (i.e. a normalised positive linear functional) Φ on $\mathcal{A}$. A classical probability space $(\Omega, \mathcal{F}, P)$ gives rise to a quantum probability space by taking a $*$-algebra of complex-valued integrable functions on Ω, e.g. $L^\infty(\Omega, \mathcal{F}, P)$, and the functional defined by $\Phi : f \mapsto \int_\Omega f dP$.

A *quantum random variable* j over a quantum probability space $(\mathcal{A}, \Phi)$ on a

$*$-algebra $\mathcal{B}$ is a $*$-algebra homomorphism $j : \mathcal{B} \to \mathcal{A}$. A classical random variable $X : \Omega \to E$ over a probability space $(\Omega, \mathcal{F}, P)$ with values in a measurable space $(E, \mathcal{E})$ defines a quantum random variable via $j_X(f) = f \circ X$ for $f \in \mathcal{B}$ (where $\mathcal{B}$ is an appropriately chosen algebra of functions on E, e.g., $\mathcal{B} = L^\infty(E, \mathcal{E})$). The functional $\varphi_j = \Phi \circ j$ is called the *distribution* of j in the state Φ.

A *quantum stochastic process* is simply a family of quantum random variables $\{j_t; t \in I\}$ over the same quantum probability space, indexed by some set I, just as in the classical context. Its *one-dimensional* or *marginal distributions* are the functionals $\varphi_t = \Phi \circ j_t$.

Two quantum stochastic processes $\{j_t; t \in I\}$ and $\{k_t; t \in I\}$, indexed by the same set I, on the same $*$-algebra $\mathcal{B}$ over the quantum probability spaces $(\mathcal{A}_j, \Phi_j)$ and $(\mathcal{A}_k, \Phi_k)$ are called *equivalent*, if all their finite-dimensional distributions agree, i.e. if

$$\Phi_j\Big(j_{t_1}(b_1)\cdots j_{t_n}(b_n)\Big) = \Phi_k\Big(k_{t_1}(b_1)\cdots k_{t_n}(b_n)\Big),$$

for all $n \in \mathbb{N}$, $t_1, \ldots, t_n \in I$, $b_1, \ldots, b_n \in \mathcal{B}$.

Every element a of a quantum probability space $(\mathcal{A}, \Phi)$ defines a quantum random variable on $\mathbb{C}\langle z, z^*\rangle$ (= the free algebra over $\mathbb{C}$ with generators z, z^*, and the obvious $*$-structure), or on $\mathbb{C}[z]$ (= the algebra of complex-valued polynomials on $\mathbb{R}$), if a is self-adjoint. Simply set $j(z) = a$ and extend as a $*$-algebra homomorphism. In the same way a family $\{a_t; t \in I\}$ of elements of $\mathcal{A}$ gives rise to a quantum stochastic process indexed by I. We will call such a family $\{a_t; t \in I\}$ an *operator process*. Given a quantum stochastic process $\{j_t : \mathcal{B} \to \mathcal{A}; t \in I\}$ one gets an operator process of every element $x \in \mathcal{B}$ of $\mathcal{B}$ by simply setting $a_t = j_t(x)$, $t \in I$.

Let us associate densities on the real line to self-adjoint elements a of a quantum probability space. By a *density* of a (in the state Φ) we will mean a measure μ on $\mathbb{R}$ such that $\Phi(a^n) = \int_{\mathbb{R}} x^n \mathrm{d}\mu(x)$ for all $n \in \mathbb{N}$. Note that μ is determined by a moment problem and therefore not necessarily unique. Then we might call a random variable X on some probability space $(\Omega, \mathcal{F}, P)$ a classical version of a if its distribution P_X is a density of a. For a *classical version* $\{\tilde{X}_t; t \in I\}$ of a family of quantum random variables $\{a_t; t \in I\}$, i.e. of a process, indexed by an ordered set I, we only require that the time-ordered moments agree, i.e.

$$\Phi(a_{t_1}^{k_1} \cdots a_{t_n}^{k_n}) = \mathbb{E}(\tilde{X}_{t_1}^{k_1} \cdots \tilde{X}_{t_n}^{k_n})$$

for all $n, k_1, \ldots, k_n \in \mathbb{N}$, $t_1, \ldots, t_n \in I$, $t_1 \le t_2 \le \cdots \le t_n$. Again, a classical version, if it exists, may not be unique.

4.2 INDEPENDENCE

Several different definitions of independence in non-commutative probability theory have been proposed and studied. There is Voiculescu's[VDN92] free independence, related to the free product of algebras, the tensor independence (see, e.g., [Sch93, Section 1.3]) related to the tensor product, and finally the boolean independence (see e.g. [SW97]). Work by Schürmann and Ben Ghorbal[BGS99],

Schürmann[Sch95a], and R. Speicher[Spe97] indicates that, under certain natural conditions, these are the only ones, at least if one insists that it should be possible to reconstruct the joint distribution of two independent quantum random variables from their marginal distributions. But there do exist many other notions of independence not satisfying this condition, see e.g. [Küm90, ALV94], that are also used in non-commutative probability, for example to derive limit theorems or to develop a theory of stochastic integration[HKK98].

The appropriate notion of independence for this work is that of (braided) tensor independence. Schürmann has considered non-symmetric tensor products, where the braiding arises from an action α and a coaction γ of some group algebra $\mathbb{C}\mathbb{L}$ as

$$\Psi = (\alpha \otimes \mathrm{id}) \circ (\mathrm{id} \otimes \tau) \circ (\gamma \otimes \mathrm{id}), \qquad \text{or} \qquad \Psi = (\mathrm{id} \otimes \alpha) \circ (\tau \otimes \mathrm{id}) \circ (\mathrm{id} \otimes \gamma). \quad (4.1)$$

We will generalise this to arbitrary braidings (but we still assume that the associativity constraint is the trivial one).

Definition 4.2.1 *Let $(\mathcal{A}, \Phi)$ be a quantum probability space, and $\mathcal{B}$ a $*$-algebra in some braided category $(\mathcal{C}, \Psi)$. An n-tuple $(j_1, \ldots, j_n)$ of quantum random variables $j_i : \mathcal{B} \to \mathcal{A}$, $i = 1, \ldots, n$, over $(\mathcal{A}, \Phi)$ on $\mathcal{B}$ is* Ψ-independent *or* braided independent, *if*

(i) $\Phi\Big(j_{\sigma(1)}(b_1) \cdots j_{\sigma(n)}(b_n)\Big) = \Phi(j_{\sigma(1)}(b_1)) \cdots \Phi(j_{\sigma(n)}(b_n))$ *for all permutations* $\sigma \in \mathcal{S}(n)$ *and all* $b_1, \ldots, b_n \in \mathcal{B}$, *and*

(ii) $m_{\mathcal{A}} \circ (j_k \otimes j_l) = m_{\mathcal{A}} \circ (j_l \otimes j_k) \circ \Psi$ *for all* $1 \leq k < l \leq n$.

Remark:

1. Note that, in general, the order of $(j_1, \ldots, j_n)$ is important, unlike the classical case.

2. If $\mathcal{A}$ is commutative, then (i) is equivalent to the condition used in classical probability theory:

   (i') $\Phi\Big(j_1(b_1) \cdots j_n(b_n)\Big) = \Phi(j_1(b_1)) \cdots \Phi(j_n(b_n))$ for all $b_1, \ldots, b_n \in \mathcal{B}$.

   We will call $j_1, \ldots, j_n$ *pseudo-(Ψ-)independent*, if they satisfy only (i') and (ii). We will use the notion of pseudo-independence also, if there are no $*$-structures defined on $\mathcal{A}$ or $\mathcal{B}$, or if Φ is not positive.

3. More generally, we say that a family $\{j_\imath | \imath \in I\}$ of quantum random variables indexed by some partially ordered set I is independent, if the n-tuple $(j_{\imath_1}, \ldots, j_{\imath_n})$ is independent for all finite n-tuples $(\imath_1, \ldots, \imath_n)$ such that $\imath_1 < \imath_2 < \cdots < \imath_n$).

Both independence and the weaker pseudo-independence imply that $\Phi \circ m_{\mathcal{A}}^{(n-1)} \circ (j_1 \otimes \cdots \otimes j_n)$ can be factorised into a tensor product of the marginal distributions $\phi_i = \Phi \circ j_i$, $i = 1, \ldots, n$, and thus that it is uniquely determined by the marginal distributions. But the 'true' independence implies in addition an invariance or commutativity condition, namely,

$$\hat{\phi}_1 = \phi_1 \otimes \underbrace{\varepsilon \otimes \cdots \otimes \varepsilon}_{(n-1)\text{ times}}, \quad \hat{\phi}_2 = \varepsilon \otimes \phi_2 \otimes \underbrace{\varepsilon \otimes \cdots \otimes \varepsilon}_{(n-2)\text{ times}}, \quad \ldots \quad \hat{\phi}_n = \underbrace{\varepsilon \otimes \cdots \otimes \varepsilon}_{(n-1)\text{ times}} \otimes \phi_n$$

have to commute (in the convolution algebra of functionals on $\mathcal{B}^{\otimes n}$).

We call a functional ϕ on an algebra $\mathcal{A}$ in some braided category $(\mathcal{C}, \Psi)$ Ψ-*invariant*, if we have

$$(\phi \otimes \theta) \circ \Psi = \theta \otimes \psi.$$

for all functionals $\theta : \mathcal{A} \to \mathbb{K}$,

In general, the convolution of two positive functionals on a braided $*$-bialgebra is not again positive, but if a functional ϕ on a braided $*$-bialgebra $\mathcal{A}$ is positive and Ψ-invariant, then convolution $\phi \star \theta = (\phi \otimes \theta) \otimes \Delta$ of ϕ with another positive functional θ on $\mathcal{A}$ is also positive, see the lemma below.

Lemma 4.2.2 *Let ϕ, θ be two positive functionals, and ϕ Ψ-invariant or θ Ψ^{-1}-invariant. Then $\phi \star \theta = (\phi \otimes \theta) \circ \Delta$ is also positive.*

Proof: Let $a \in \mathcal{A}$, and $\Delta(a) = \sum_i a_i^{(1)} \otimes a_i^{(2)}$ (Sweedler's notation). Then $\Delta(a^*) = \sum_j \Psi\left(\left(a_j^{(2)}\right)^* \otimes \left(a_j^{(1)}\right)^*\right)$. If ϕ is Ψ-invariant, then $(\phi \otimes \theta) \circ \Delta(a^* a) = (\phi \otimes \theta) \circ (m \otimes m) \circ (\mathrm{id} \otimes \Psi \otimes \mathrm{id}) \circ (\Psi \otimes \mathrm{id} \otimes \mathrm{id}) \left(\sum_{i,j} \left(a_j^{(2)}\right)^* \otimes \left(a_j^{(1)}\right)^* \otimes a_i^{(1)} \otimes a_i^{(2)}\right) = \theta \circ (\phi \otimes m) \circ (\Psi \circ \mathrm{id}) \circ (\mathrm{id} \otimes m \otimes \mathrm{id}) \left(\sum_{i,j} \left(a_j^{(2)}\right)^* \otimes \left(a_j^{(1)}\right)^* \otimes a_i^{(1)} \otimes a_i^{(2)}\right) = \sum_{i,j} \theta\left(\left(a_j^{(2)}\right)^* a_i^{(2)}\right) \phi\left(\left(a_j^{(1)}\right)^* a_i^{(1)}\right)$ is positive, since it is the Schur product of two positive definite matrices.

If instead θ is Ψ^{-1}-invariant, then we can show in the same way that $(\phi \otimes \theta) \circ \Delta(aa^*) = \sum_{i,j} \phi\left(a_j^{(1)} \left(a_i^{(1)}\right)^*\right) \theta\left(a_j^{(2)} \left(a_i^{(2)}\right)^*\right)$ is positive. ∎

4.3 LÉVY PROCESSES ON BIALGEBRAS

Let us now introduce one of the central themes of this book, processes with independent and stationary increments (Lévy processes) on bialgebras. To define the notion of increments one needs an operation to compose them. In classical probability Lévy processes can therefore be defined, if the state space that the processes take values in has a semi-group structure. A semi-group structure on the state space can usually be used to define a bialgebra structure on an appropriate algebra of functions on this space, see, e.g., the first examples in Section 3.2. This is therefore the right structure to study Lévy processes in non-commutative probability (but it is not the only possibility, cf. [Sch95b]).

Definition 4.3.1 *[Sch93] Let $\mathcal{B}$ be a (braided) $*$-bialgebra. A quantum stochastic process $\{j_{st} | 0 \leq s \leq t \leq T\}$, $T \in \mathbb{R}_+ \cup \{\infty\}$ on $\mathcal{B}$ over some quantum probability space $(\mathcal{A}, \Phi)$ is called a* Lévy process *if the following conditions are satisfied.*

1. *(increment property)*

$$\begin{aligned} j_{rs} \star j_{st} &= j_{rt} \quad \textit{for all } 0 \leq r \leq s \leq t \leq T, \\ j_{tt} &= e \circ \varepsilon \quad \textit{for all } 0 \leq t \leq T, \end{aligned}$$

2. *(independence of increments) the family* $\{j_{st}|0 \leq s \leq t \leq T\}$ *is independent,*

3. *(stationarity of increments) the distribution* $\varphi_{st} = \Phi \circ j_{st}$ *of* j_{st} *depends only on the difference* $t - s$,

4. *(weak continuity)* j_{st} *converges to* j_{ss} $(= e \circ \varepsilon)$ *in distribution for* $t \searrow s$.

Remarks:

1. Recall that the convolution $j_1 \star j_2 : \mathcal{C} \to \mathcal{A}$ of two linear maps $j_1, j_2 : \mathcal{C} \to \mathcal{A}$ from a coalgebra $\mathcal{C}$ to an algebra $\mathcal{A}$ is defined by

$$j_1 \star j_2 = m_{\mathcal{A}} \circ (j_1 \otimes j_2) \circ \Delta_{\mathcal{C}}.$$

2. If $\{j_t|0 \leq t \leq T\}$ is a quantum stochastic process on a (braided) $*$-Hopf algebra, then we can define its increments by

$$j_{st} = (j_s \circ S) \star j_t,$$

this automatically satisfies the increment property. We call $\{j_t|0 \leq t \leq T\}$ a Lévy process, if $\{j_{st} = (j_s \circ S) \star j_t|0 \leq s \leq t \leq T\}$ is one.

3. If the increments are only pseudo-independent, or if Φ in not positive, but only a unital linear functional, or if no $*$-structures are specified on $\mathcal{B}$ and $\mathcal{A}$, then we will call (j_{st}) a pseudo-Lévy process.

We know (cf. [Sch93]) that the process is uniquely determined by its marginal laws $\varphi_t = \Phi \circ j_{0t}$ (up to equivalence). This can also be seen from Section 4.5, since all we need to recover the joint distributions are the marginal distributions. The marginal distributions form a convolution semi-group, since

$$\begin{aligned} \varphi_s \star \varphi_t &= \varphi_s \star \varphi_{s,s+t} = (\Phi \circ j_{0,s}) \otimes (\Phi \circ j_{s,s+t}) \circ \Delta \\ &= \Phi \circ m_{\mathcal{A}} \circ (j_{0,s} \otimes j_{s,s+t}) \circ \Delta = \Phi \circ j_{0,s+t} \\ &= \varphi_{s+t}, \end{aligned}$$

where we used first the stationarity, then the factorisation property from the definition of independence, and then the increment property.

Using the fundamental theorem of coalgebras (Theorem 3.1.1) one can show that there exists a unique, conditionally positive (i.e. positive on ker ε: $L(a^*a) \geq 0$ for all $a \in \mathcal{B}$ with $\varepsilon(a) = 0$), hermitian, linear functional $L : \mathcal{B} \to \mathbb{C}$ with $L(\mathbb{1}) = 0$, such that $\varphi_t = \exp_\star tL$, see [Sch93, Theorem 1.9.2]. This functional is called the *generator* of the process (j_{st}). Furthermore, the independence of the increments implies $(L \otimes L) \circ \Psi = L \otimes L$. If we restrict ourselves to Ψ-invariant generators (for $\Psi = \tau$, i.e. in the symmetric or 'unbraided' case all generators are invariant), then we can also go back. Using Schoenberg correspondence, we see that the convolution semi-group generated by L is again positive. For the symmetric case, this is [Sch93, Theorem 3.2.8], for the graded case it follows from [Sch93, Section 3.3]. Thanks to Lemma 4.2.2 this result remains true in the general braided case. Thus the construction in Section 4.5 or on Pages 38-40 in [Sch93] gives a kind of 'canonical representation' of the process on an inductive limit of the tensor product algebras of $\mathcal{B}$. It is interesting to note that this construction does not depend on the positivity, we can get these 'canonical representations' also for 'pseudo-Lévy processes' (i.e. 'processes' that have only pseudo-independent increments, or that are defined without the $*$-structure or with a functional Φ that is not positive, but that satisfy all the other conditions).

We summarise this in the following proposition, for details see [Sch93, Corollary 1.9.7, Theorem 3.2.8].

Proposition 4.3.2 *Let $\mathcal{B}$ be a $*$-bialgebra over $\mathbb{C}$ in a braided category $(\mathcal{C}, \Psi)$.*

(i) Let $\Psi = \tau$. Then there is a one-to-one correspondence between (equivalence classes of) Lévy processes $\{j_{st}\}$, normed positive convolution semi-groups of functionals $\varphi_t = \Phi \circ j_{0t} = \exp_\star tL$, and hermitian, conditionally positive functionals $L = \frac{\mathrm{d}}{\mathrm{d}t}\varphi_t\big|_{t=0}$ on $\mathcal{B}$.

(ii) The correspondence holds also for general braidings, if we restrict to Ψ-invariant generators, and their semi-groups and processes.

The essential ingredient of the generalisation from α-invariant generators in [Sch93] to Ψ-invariant generators is Lemma 4.2.2.

In the next two sections we will present two constructions of realisations of Lévy processes.

4.4 REALISATION OF LÉVY PROCESSES ON FOCK SPACES

In this section we will discuss Schürmann's representation theorem [Sch93, Theorem 2.5.3]. This theorem states that every Lévy process on a symmetric involutive bialgebra can be realized as the solution of a quantum stochastic differential equation on some Boson Fock space. Using the symmetrisation from [Sch93, Chapter 3] this result extends to all involutive bialgebras in braided tensor categories whose braiding comes from an action and coaction of a group as in Equation (4.1). For the detailed proof see Schürmann's monograph [Sch93], for a brief outline see also [Mey93, Section VII.2].

So let now $\mathcal{B}$ be a symmetric involutive bialgebra and let $L : \mathcal{B} \to \mathbb{C}$ be a conditionally positive, hermitian, linear functional with $L(\mathbb{1}) = 0$, i.e. a generator of a Lévy process. A bialgebra $\mathcal{B}$ can always be written as a direct sum $\mathcal{B} = \mathbb{K}\mathbb{1} \oplus \mathcal{B}_0$, where $\mathcal{B}_0 = \ker\varepsilon$. Note that because of the multiplicativity of the counit $\mathcal{B}_0$ is even an ideal. The decomposition of an element $a \in \mathcal{B}$ is given by $a = \varepsilon(a)\mathbb{1} + (a - \varepsilon(a)\mathbb{1})$. The generator L can be used to define a semi-positive inner product $\langle\cdot,\cdot\rangle_L : \mathcal{B}_0 \otimes \mathcal{B}_0 \to \mathbb{C}$ on $\mathcal{B}_0$,

$$\langle a_0, b_0\rangle_L = L(a_0^* b_0), \qquad \text{for } a_0, b_0 \in \mathcal{B}_0.$$

This inner product can be extended to all of $\mathcal{B}$ by setting

$$\langle a, b\rangle_L = \langle P_0 a, P_0 b\rangle_L = L\Big((a - \varepsilon(a)\mathbb{1})^*(b - \varepsilon(b)\mathbb{1})\Big), \qquad \text{for } a, b \in \mathcal{B},$$

where P_0 denotes the projection from $\mathcal{B}$ onto $\mathcal{B}_0$. Let

$$\mathcal{N}_L = \{a \in \mathcal{B} | \langle a, a\rangle_L = 0\}$$

be the null-space of $\langle\cdot,\cdot\rangle_L$, then $\mathcal{H}_L = \mathcal{B}/\mathcal{N}_L$ is a pre-Hilbert space with the inner product induced by $\langle\cdot,\cdot\rangle_L$ (which we will again denote by $\langle\cdot,\cdot\rangle_L$). Let us denote the canonical projection by $\eta_L : \mathcal{B} \to \mathcal{H}_L$.

$\mathcal{B}$ acts on $\mathcal{B}_0$ by left multiplication. The null-space $\mathcal{N}_L^0 = \mathcal{N}_L \cap \mathcal{B}_0 = P_0(\mathcal{N}_L)$ is invariant under this action, since

$$\begin{aligned} ||ab_0||_L^2 &= L\Big((ab_0 - \varepsilon(ab_0)\mathbb{1})^*(ab - \varepsilon(ab_0)\mathbb{1})\Big) = \langle b_0, a^*ab_0\rangle_L \\ &\leq ||b_0||_L ||a^*ab_0||_L = 0, \qquad \text{for all } a \in \mathcal{B}, b_0 \in \mathcal{N}_L^0, \end{aligned}$$

by the Cauchy-Schwarz inequality, so that it induces an action on $\mathcal{H}_L = \mathcal{B}/\mathcal{N}_L \cong \mathcal{B}_0/\mathcal{N}_L^0$. Let us denote this action by ρ_L. By the definition of ρ_L we have $\rho_L(a)\eta_L(b_0) = \eta_L(ab_0)$ for $a \in \mathcal{B}$, $b_0 \in \mathcal{B}_0$. For a general element $b \in \mathcal{B}$ we have $\eta_L(b) = \eta_L(P_0 b) = \eta_L(b - \varepsilon(b)\mathbb{1})$ and therefore

$$\rho_L(a)\eta_L(b) = \eta_L(ab) - \eta_L(a)\varepsilon(b), \qquad \text{for } a, b \in \mathcal{B},$$

i.e. η_L is a (ρ_L, ε)-1-cocycle.

Let $\tilde{\eta}_L = \eta_L \circ *$, then we have

$$\begin{aligned} L(ab) &= L\Big((a - \varepsilon(a)\mathbb{1})(b - \varepsilon(b)\mathbb{1})\Big) + L(a)\varepsilon(b) + \varepsilon(a)L(b) \\ &= \langle\tilde{\eta}_L(a), \eta_L(b)\rangle_L + L(a)\varepsilon(b) + \varepsilon(a)L(b), \end{aligned}$$

for all $a, b \in \mathcal{B}$. Bringing $\langle\tilde{\eta}_L(a), \eta_L(b)\rangle_L$ to the left, we see that $\langle\tilde{\eta}_L(\cdot), \eta_L(\cdot)\rangle_L$ is the $(\varepsilon, \varepsilon)$-2-coboundary of $-L$, i.e.

$$\langle\tilde{\eta}_L(a), \eta_L(b)\rangle_L = L(ab) - L(a)\varepsilon(b) - \varepsilon(a)L(b),$$

for all $a, b \in \mathcal{B}$.

Note that in this construction we did not use the coproduct, but only the counit.

Definition 4.4.1 *A* Schürmann triple *on a unital $*$-algebra with a one-dimensional $*$-representation $\varepsilon : \mathcal{B} \to \mathbb{C}$ is a triple (ρ, η, L) where ρ is a $*$-representation of $\mathcal{B}$ on some pre-Hilbert space $\mathcal{K}$, η is a (ρ, ε)-1-cocycle, and L is a hermitian linear functional that satisfies*

$$\langle\eta(a), \eta(b)\rangle = L(a^*b) - L(a^*)\varepsilon(b) - \overline{\varepsilon(a)}L(b),$$

for all $a, b \in \mathcal{B}$.

By the preceding discussion we see that we can always construct a Schürmann triple from a given generator. On the other hand the Schürmann triple uniquely characterises the generator (since it is part of the triple) and therefore the process. Furthermore, if (ρ, η, L) is a Schürmann triple on some unital $*$-algebra $\mathcal{B}$ with a one-dimensional representation ε, then it is not difficult to show that L is conditionally positive[1], i.e. that $L(a^*a) \geq 0$ for all $a \in \ker \varepsilon$, and that $L(\mathbb{1}) = 0$. So the classification of all Lévy processes on a given $*$-bialgebra can be done by first classifying all its $*$-representations, then determining all cocycles, and finally finding all solutions of the cohomological equation $\langle \tilde{\eta}(\cdot), \eta(\cdot) \rangle = -\partial L$ with $\tilde{\eta} = \eta \circ *$. For the case of $SU_q(2)$ this programme was carried out by M. Skeide, cf. [Ske94, SS98].

Schürmann's *representation theorem for Lévy processes* [Sch93, Theorem 2.5.3] says that if (ρ, η, L) is a Schürmann triple on a $*$-bialgebra $\mathcal{B}$, then the quantum stochastic differential equations

$$\mathrm{d}j_{st} = j_{st} \star (\mathrm{d}A_t^* \circ \eta + \mathrm{d}\Lambda_t \circ (\rho - \varepsilon) + \mathrm{d}A_t \circ \tilde{\eta} + L\mathrm{d}t) \tag{4.2}$$

with the initial conditions

$$j_{ss} = \varepsilon \mathrm{id}, \tag{4.3}$$

i.e. with Sweedler's notation $\Delta b = \sum b^{(1)} \otimes b^{(2)}$,

$$\begin{aligned} \mathrm{d}j_{st}(b) &= j_{st}(b^{(1)}) \Big(\mathrm{d}A_t^*(\eta(b^{(2)})) + \mathrm{d}\Lambda_t(\rho(b^{(2)}) - \varepsilon(b^{(2)})\mathrm{id}) + \mathrm{d}A_t(\eta(b^{(2)*})) \\ &\quad + L(b^{(2)})\mathrm{d}t \Big), \\ j_{ss}(b) &= \varepsilon(b)\mathrm{id}, \end{aligned}$$

for all $b \in \mathcal{B}$, have solutions (defined on some appropriate common domain in the Boson Fock space $\Gamma(L^2(\mathbb{R}_+, \mathcal{K})))$ and that the process (j_{st}) is a Lévy process on $\mathcal{B}$ with generator L in the vacuum state. Conversely, if (k_{st}) is a Lévy process with generator L on a symmetric involutive bialgebra, then it is equivalent to the process (j_{st}) defined by Equations (4.2) and (4.3) with $\rho = \rho_L$ and $\eta = \eta_L$.

For the definition of the creator, conservator, and annihilator integrals see, e.g., [Par92, Mey93]. We will only show here that

$$\mathrm{d}M_t = \mathrm{d}A_t^* \circ \eta + \mathrm{d}\Lambda_t \circ (\rho - \varepsilon) + \mathrm{d}A_t \circ \tilde{\eta} + L\mathrm{d}t$$

formally defines a $*$-homomorphism on $\ker \varepsilon = \mathcal{B}_0$, if we define the algebra of quantum stochastic differentials (or Itô algebra, cf. [Bel98] and the references therein) over some pre-Hilbert space $\mathcal{K}$ as follows. Let $L(\mathcal{K})$ denote the algebra of all linear operators on $\mathcal{K}$ and denote by $H(\mathcal{K})$ the subalgebra of $L(\mathcal{K})$ of all operators $F \in L(\mathcal{K})$ with the property that there exists a $G \in L(\mathcal{K})$ with

$$\langle u, Fv \rangle = \langle Gu, v \rangle, \qquad \text{for all } u, v \in \mathcal{K}.$$

[1] But note that it is necessary to impose the condition that L is hermitian in the definition of a Schürmann triple. Let $\mathcal{B} = \mathbb{C}\langle x, x^* \rangle$ be the free algebra with two generators x and x^* and the obvious $*$-structure. The words in x and x^* form a basis of $\mathcal{B}$. If we set $\rho = 0$, $\eta = 0$, and define L on the words in x and x^* by $L(x) = z_1$, $L(x^*) = z_2$, and $L(w) = 0$ else, then (ρ, η, L) satisfies all conditions of the definition of a Schürmann triple except the hermitianity of L for all $z_1, z_2 \in \mathbb{C}$. But the process defined by Equation (4.2) is an involutive homomorphism if and only if $\overline{z_1} = z_2$, i.e. only if L is hermitian.

If we set $F^* = G$, then this defines an involution $H(\mathcal{K})$. The algebra of quantum stochastic differentials $\mathcal{I}(\mathcal{K})$ over $\mathcal{K}$ is the $*$-algebra generated by

$$\{\mathrm{d}\Lambda(F)|F \in H(\mathcal{K})\} \cup \{\mathrm{d}A^*(u)|u \in \mathcal{K}\} \cup \{\mathrm{d}A(u)|u \in \mathcal{K}\} \cup \{\mathrm{d}t\},$$

if we identify

$$\begin{aligned}
\mathrm{d}\Lambda(\lambda F + \mu G) &\equiv \lambda \mathrm{d}\Lambda(F) + \mu \mathrm{d}\Lambda(G), \\
\mathrm{d}A^*(\lambda u + \mu v) &\equiv \lambda \mathrm{d}A^*(u) + \mu \mathrm{d}A^*(v), \\
\mathrm{d}A(\lambda u + \mu v) &\equiv \overline{\lambda} \mathrm{d}A(u) + \overline{\mu} \mathrm{d}A(v),
\end{aligned}$$

for all $F, G \in H(\mathcal{K})$, $u, v \in \mathcal{K}$, $\lambda, \mu \in \mathbb{C}$. The involution of $\mathcal{I}(\mathcal{K})$ is defined by

$$\begin{aligned}
\mathrm{d}\Lambda(F)^* &= \mathrm{d}\Lambda(F^*), \\
\Big(\mathrm{d}A^*(u)\Big)^* &= \mathrm{d}A(u), \\
\mathrm{d}A(u)^* &= \mathrm{d}A^*(u),
\end{aligned}$$

for $F \in H(\mathcal{K})$, $u \in \mathcal{K}$, and the multiplication by the Itô table

$\bullet$	$\mathrm{d}A^*(u)$	$\mathrm{d}\Lambda(F)$	$\mathrm{d}A(u)$	$\mathrm{d}t$
$\mathrm{d}A^*(v)$	0	0	0	0
$\mathrm{d}\Lambda(G)$	$\mathrm{d}A^*(Gu)$	$\mathrm{d}\Lambda(GF)$	0	0
$\mathrm{d}A(v)$	$\langle v, u\rangle \mathrm{d}t$	$\mathrm{d}A(F^*v)$	0	0
$\mathrm{d}t$	0	0	0	0

for all $F, G \in H(\mathcal{K})$, $u, v \in \mathcal{K}$, i.e. we have, for example,

$$\mathrm{d}A(v) \bullet \mathrm{d}A^*(u) = \langle v, u\rangle \mathrm{d}t, \qquad \text{and} \qquad \mathrm{d}A^*(u) \bullet \mathrm{d}A(v) = 0.$$

Proposition 4.4.2 *Let $\mathcal{B}$ be an involutive unital algebra with a one-dimensional $*$-representation ε, $\mathcal{B}_0 = \ker \varepsilon$, and $(\rho : \mathcal{B} \to H(\mathcal{K}), \eta : \mathcal{B} \to \mathcal{K}, L : \mathcal{B} \to \mathbb{C})$ a Schürmann triple. Then the map $\mathrm{d}M : \mathcal{B}_0 \to \mathcal{I}(\mathcal{K})$ defined by*

$$\mathrm{d}M = \mathrm{d}A^* \circ \eta + \mathrm{d}\Lambda \circ (\rho - \varepsilon) + \mathrm{d}A \circ \tilde{\eta} + L\mathrm{d}t$$

is a $*$-homomorphism.

Proof: Let $a, b \in \mathcal{B}_0$. Using the Itô table we get

$$\begin{aligned}
\mathrm{d}M(a) \bullet \mathrm{d}M(b) &= \mathrm{d}A^*\Big((\rho(a) - \varepsilon(a)\mathrm{id})\eta(b)\Big) + \mathrm{d}A\Big((\rho(b) - \varepsilon(b)\mathrm{id})\eta(a^*)\Big) \\
&\quad + \mathrm{d}\Lambda\Big((\rho(a) - \varepsilon(a)\mathrm{id})(\rho(b) - \varepsilon(b)\mathrm{id})\Big) + \langle \eta(a^*), \eta(b)\rangle \mathrm{d}t \\
&= \mathrm{d}A^*(\eta(ab)) + \mathrm{d}\Lambda(\rho(ab) - \varepsilon(ab)\mathrm{id}) + \mathrm{d}A(\tilde{\eta}(ab)) + L(ab)\mathrm{d}t.
\end{aligned}$$

And we also have

$$\begin{aligned}
\mathrm{d}M(a)^* &= \Big(\mathrm{d}A^*(\eta(a))\Big)^* + \mathrm{d}\Lambda(\rho(a)) - \varepsilon(a)\mathrm{id})^* + \mathrm{d}A(\tilde{\eta}(a))^* + \Big(L(a)\mathrm{d}t\Big)^* \\
&= \mathrm{d}A(\tilde{\eta}(a^*)) + \mathrm{d}\Lambda(\rho(a^*)) - \varepsilon(a^*)\mathrm{id}) + \mathrm{d}A^*(\eta(a^*))^* + L(a^*)\mathrm{d}t \\
&= \mathrm{d}M(a^*),
\end{aligned}$$

because L is hermitian. ∎

Remark: $\mathrm{d}M$ does not define a $*$-homomorphism on $\mathcal{B}$, because we always have $\mathrm{d}M(\mathbb{1}) = 0$, but it can be turned into a $*$-homomorphism, if we define $\widetilde{\mathcal{I}}(\mathcal{K})$ by adjoining a unit to the algebra $\mathcal{I}(\mathcal{K})$ and $\widetilde{\mathrm{d}M} : \mathcal{B} \to \widetilde{\mathcal{I}}(\mathcal{K})$ by $\widetilde{\mathrm{d}M}(a) = \varepsilon(a)\mathbb{1} + \mathrm{d}M(a - \varepsilon(a)\mathbb{1})$.

4.5 REALISATION OF LÉVY PROCESSES BY AN INDUCTIVE LIMIT

We will now present an alternative construction. It has the advantage to be applicable in a more general framework. We do not need the positivity of the state or the convolution semi-group for this construction. Also, it easily applies directly to the braided case also, without the need of bosonisation or symmetrisation. This construction was introduced in [ASW88, Sch93], but see also [FFS97, Section 8].

Let $\mathcal{B}$ be a braided Hopf algebra. We set $T_n = \{t = (t_1, \ldots, t_n); t_i \in \mathbb{R}_+, t_1 < t_2 < \cdots < t_n\} \subset \mathbb{R}_+^n$, $T_0 = \{\emptyset\}$, $T = \bigcup_{n=0}^{\infty} T_n$, i. e., T_n contains all subsets of $\mathbb{R}_+$ with n elements, and T all finite subsets of $\mathbb{R}_+$. The set T is partially ordered by inclusion. For each $t = (t_1, \ldots, t_n) \in T$ we set $\mathcal{A}_t = \mathcal{B}^{\otimes|t|}$ as a vector space, where $|t| = n$ is the length of t. Let $\psi_t : \mathcal{A}_t \to \mathcal{B}^{\otimes|t|}$ be defined by

$$\psi_t(a_1 \otimes \cdots \otimes a_{|t|}) = (a_1 \otimes 1 \otimes \cdots \otimes 1)(\Delta a_2 \otimes 1 \otimes \cdots \otimes 1) \cdots (\Delta^{|t|-1} a_{|t|}) \quad (4.4)$$

where the multiplication on the right hand side is the usual multiplication $m^{\otimes|t|}$ in the braided tensor product algebra $\mathcal{B}^{\otimes|t|}$.

Lemma 4.5.1 *These maps are vector space isomorphisms.*

Proof: Let $\psi_2 : \mathcal{B} \otimes \mathcal{B} \to \mathcal{B} \otimes \mathcal{B}$ be defined by $\psi_2(a \otimes b) = (a \otimes 1)(\Delta b)$, i.e., $\psi_2 = (m \otimes \mathrm{id}_\mathcal{B}) \circ (\mathrm{id}_\mathcal{B} \otimes \Delta)$. ψ_2 is an isomorphism, and its inverse is given by $\phi_2 = (m \otimes \mathrm{id}_\mathcal{B}) \circ (\mathrm{id}_\mathcal{B} \otimes S \otimes \mathrm{id}_\mathcal{B}) \circ (\mathrm{id}_\mathcal{B} \otimes \Delta)$. Here is a diagrammatic proof of the relation $\psi_2 \circ \phi_2 = \mathrm{id}_{\mathcal{B}\otimes\mathcal{B}}$, and $\phi_2 \circ \psi_2 = \mathrm{id}_{\mathcal{B}\otimes\mathcal{B}}$ can be shown similarly.

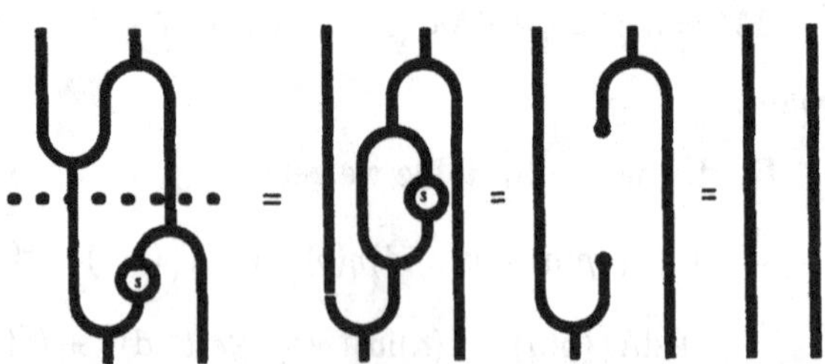

For the interpretation of this diagram see [Maj93a, Maj95b]. Not using the diagram technique we can write the proof as $\psi_2 \circ \phi_2 = (\mathrm{id}_\mathcal{B} \otimes \Delta) \circ (m \otimes \mathrm{id}_\mathcal{B}) \circ (\mathrm{id}_\mathcal{B} \otimes \Delta) \circ (\mathrm{id}_\mathcal{B} \otimes S \otimes \mathrm{id}_\mathcal{B}) \circ (m \otimes \mathrm{id}_\mathcal{B}) = (\mathrm{id}_\mathcal{B} \otimes \Delta) \circ (\mathrm{id}_\mathcal{B} \otimes \Delta \otimes \mathrm{id}_\mathcal{B}) \circ (\mathrm{id}_\mathcal{B} \otimes \mathrm{id}_\mathcal{B} \otimes S \otimes \mathrm{id}_\mathcal{B}) \circ (\mathrm{id}_\mathcal{B} \otimes m \otimes \mathrm{id}_\mathcal{B}) \circ (m \otimes \mathrm{id}_\mathcal{B}) = (\mathrm{id}_\mathcal{B} \otimes \Delta) \circ (\mathrm{id}_\mathcal{B} \otimes \varepsilon \otimes \mathrm{id}_\mathcal{B}) \circ (\mathrm{id}_\mathcal{B} \otimes e \otimes \mathrm{id}_\mathcal{B})(m \otimes \mathrm{id}_\mathcal{B}) = \mathrm{id}_{\mathcal{B}\otimes\mathcal{B}}$

Note now that ψ_t can be written as

$$\psi_t = (\psi_2 \otimes \mathrm{id}_\mathcal{B} \otimes \cdots \mathrm{id}_\mathcal{B}) \circ (\mathrm{id}_\mathcal{B} \otimes \psi_2 \otimes \mathrm{id}_\mathcal{B} \otimes \cdots \otimes \mathrm{id}_\mathcal{B}) \circ \cdots \circ (\mathrm{id}_\mathcal{B} \otimes \cdots \otimes \mathrm{id}_\mathcal{B} \otimes \psi_2). \quad (4.5)$$

This completes the proof. ∎

This lemma allows us to equip $\mathcal{A}_t$ with an algebra structure by the definition

$$m_t(a,b) = \psi_t^{-1}(m^{\otimes|t|}(\psi_t(a), \psi_t(b))). \tag{4.6}$$

Define $i_{t',t} : \mathcal{A}_{t'} \to \mathcal{A}_t$ for $t' \in t$ by inserting the unit element 1 in the factors corresponding to the indices of t that are missing in t', e. g. for $t = (t_1, \ldots, t_n)$ and $t' = (t_1, \ldots, t_{j-1}, t_{j+1}, \ldots, t_n)$ we have $i_{t',t}(a_1 \otimes \cdots \otimes a_{n-1}) = a_1 \otimes \cdots a_{j-1} \otimes 1 \otimes a_{j+1} \otimes \cdots a_{n-1}$. Note $i_{t',t} \circ i_{t'',t'} = i_{t'',t}$ and that these mappings are injective. Related to these are the maps $j_{t',t} = \psi_t \circ i_{t',t} \circ \psi_{t'}^{-1} : \mathcal{B}^{\otimes|t'|} \to \mathcal{B}^{\otimes|t|}$. One verifies that for these maps also $j_{t',t} \circ j_{t'',t'} = j_{t'',t}$, and that for $t = (t_1, \ldots, t_n)$ and $t' = (t_1, \ldots, t_{j-1}, t_{j+1}, \ldots, t_n)$ we have $j_{t',t}(a_1 \otimes \cdots \otimes a_{n-1}) = a_1 \otimes \cdots a_{j-1} \otimes \Delta a_j \otimes a_{j+1} \otimes \cdots a_{n-1}$, i. e., $j_{t',t}$ decomposes the increment from t_{j-1} to t_{j+1} into the increments from t_{j-1} to t_j and from t_j to t_{j+1} via the coproduct, whereas for $t = (t_1, \ldots, t_n)$ and $t' = (t_1, \ldots, t_{n-1})$ we have $j_{t',t}(a_1 \otimes \cdots \otimes a_{n-1}) = a_1 \otimes \cdots a_{n-1} \otimes 1$, i. e., for increments corresponding to times larger than the maximal time represented in t' the unit element 1 is adjoined.

Proposition 4.5.2 *The algebras $(\mathcal{A}_t, m_t)$ and the mappings $i_{t,t'}$ form an inductive system and thus define an algebra $\tilde{\mathcal{A}}$ with injective algebra homomorphisms $i_t : \mathcal{A}_t \to \tilde{\mathcal{A}}$ such that $i_t \circ i_{t',t} = i_{t'}$.*

Proof: To prove the proposition we have to show $m_t \circ (i_{t',t} \otimes i_{t',t}) = i_{t',t} \circ m_{t'}$, i. e., that the maps $i_{t',t}$ are algebra homomorphisms with respect to the multiplications m_t, $m_{t'}$. It is clear that $j_{t',t}$ is an algebra homomorphism, since the coproduct Δ is one, and adjoining the unit element is one, too. Now

$$\begin{aligned}
& m_t \circ (i_{t',t} \otimes i_{t',t}) \\
= \; & \psi_t^{-1} \circ m^{\otimes|t|} \circ (\psi_t \otimes \psi_t) \circ ((\psi_t^{-1} \circ j_{t',t} \circ \psi_{t'}) \otimes (\psi_t^{-1} \circ j_{t',t} \otimes \psi_{t'})) \\
= \; & \psi_t^{-1} \circ m^{\otimes|t|} \circ (j_{t',t} \otimes j_{t',t}) \circ (\psi_{t'} \otimes \psi_{t'}) \\
= \; & \psi_t^{-1} \circ j_{t',t} \circ m^{\otimes|t'|} \circ (\psi_{t'} \otimes \psi_{t'}) \\
= \; & i_{t',t} \circ m_{t'}.
\end{aligned}$$

∎

This algebra plays the rôle of the canonical representation of the process. The maps $i_{(t)} : \mathcal{B} \cong \mathcal{A}_{(t)} \to \tilde{\mathcal{A}}$, $t \in \mathbb{R}_+$, embed $\mathcal{B}$ into $\tilde{\mathcal{A}}$. This is the process we want to construct, we set $k_t = i_{(t)}$ for $t \in \mathbb{R}_+$. Then the increments are given by

$$k_{st} = (k_s \circ S) \star k_t = i_{(s,t)} \circ (S \otimes \mathrm{id}) \circ \Delta,$$

for $0 \leq s \leq t$. If $\mathcal{B}$ has a $*$-structure, then $\tilde{\mathcal{A}}$ inherits a $*$-structure also and the maps k_{st}, $0 \leq s \leq t$, are even $*$-algebra homomorphisms.

Now we have to define a functional on $\tilde{\mathcal{A}}$. Let $(\varphi_t)_{t \in \mathbb{R}_+}$ be a convolution semi-group. Then we define

$$\tilde{\Phi}_t(a) = \varphi_{t_1} \otimes \varphi_{t_2 - t_1} \otimes \cdots \otimes \varphi_{t_n - t_{n-1}}(\psi_t(a)) \tag{4.7}$$

for $a \in \mathcal{A}_t$, $t \in T$.

Proposition 4.5.3 *The functionals $\{\tilde{\Phi}_t; t \in T\}$ determine a unique functional $\tilde{\Phi}$ on $\tilde{\mathcal{A}}$ such that $\tilde{\Phi} \circ i_t = \tilde{\Phi}_t$.*

Proof: We have to show that $\tilde{\Phi}_{t'} = \tilde{\Phi}_t \circ i_{t',t}$ for all $t' \subseteq t$. Since the functionals φ_t form a convolution semi-group, i. e., $(\varphi_t \otimes \varphi_s) \circ \Delta = \varphi_{t+s}$, and, more generally, for $t = (t_1, \ldots, t_n)$ and $t' = (t_{\nu_1}, \cdots, t_{\nu_k}) \subseteq t$,

$$\varphi_{t_{\nu_1}} \otimes \cdots \otimes \varphi_{t_{\nu_k} - t_{\nu_{k-1}}} = \varphi_{t_1} \otimes \varphi_{t_2 - t_1} \otimes \cdots \otimes \varphi_{t_n - t_{n-1}} \circ j_{t',t}.$$

The proposition now follows from the definition of $\tilde{\Phi}_t = (\varphi_{t_1} \otimes \cdots \otimes \varphi_{t_n - t_{n-1}}) \circ \psi_t$ and of $i_{t',t}$. ∎

If there is a $*$-structure defined on $\mathcal{B}$ and if the functionals φ_t, $t \in \mathbb{R}_+$ are states, then one immediately sees that $\tilde{\Phi}$ is also a state and thus that $(k_t)_{t \in \mathbb{R}_+}$ is a quantum stochastic process, i. e., it is a $*$-algebra homomorphism from a $*$-algebra $\mathcal{B}$ to the $*$-algebra $\tilde{\mathcal{A}}$ with the state $\tilde{\Phi}$. The one-dimensional distributions of this process are exactly the functionals $\varphi_t = \tilde{\Phi} \circ k_t$.

By the construction of the state the expectation factorises and the increments k_{st}, $0 \leq s \leq t$ are independent and stationary (or pseudo-independent and stationary, if there is no $*$-structure and no positivity).

The following theorem summarises the main results of this section.

Theorem 4.5.4 $\{k_t : \mathcal{B} \to (\tilde{\mathcal{A}}, \tilde{\Phi}); t \in \mathbb{R}_+\}$ *is a quantum process with independent and stationary increments, i. e., it satisfies*

1. $k_t : \mathcal{B} \to \tilde{\mathcal{A}}$ *is a unital algebra homomorphism for all* $t \in \mathbb{R}_+$,

2. *the functionals* $\varphi_t = \tilde{\Phi} \circ k_t = \tilde{\Phi} \circ k_{s,s+t}$ *form a convolution semi-group, i. e.,*

$$(\varphi_t \star \varphi_s)(a) = (\varphi_t \otimes \varphi_s)(\Delta a) = \varphi_{t+s}(a)$$

for all $a \in \mathcal{A}$, $t, s \in \mathbb{R}_+$, *and* $\lim_{t \searrow 0} \varphi_t = \varepsilon$.

Example: Take the q-affine group (see Subsections 3.2.7 and 3.3.1). We set $a_t = k_t(a)$ and $b_t = k_t(b)$. The algebra $\tilde{\mathcal{A}}$ is spanned by

$$\{a_{t_1}^{n_1} b_{t_1}^{m_1} \cdots a_{t_p}^{n_p} b_{t_p}^{m_p}; p \in \mathbb{N}, t_1 < t_2 < \ldots < t_p \in \mathbb{R}_+, n, m \in \mathbb{N}^p\}$$

The commutation relations between elements corresponding to the same time t remain unchanged, those between elements corresponding to different times $s < t$ are determined by

$$\begin{aligned} a_t \cdot a_s &= a_s \cdot a_t, & b_t \cdot a_s &= a_s \cdot b_t + h b_s, \\ a_t \cdot b_s &= b_s \cdot a_t - h b_s, & b_t \cdot b_s &= q^{-1} b_s \cdot b_t + (1 - q^{-1}) b_s^2. \end{aligned}$$

4.6 MULTIPLICATIVE STOCHASTIC INTEGRALS

Motivated by McKean's [McK69] stochastic product integrals on Lie groups (see also Section 2.7 and [FS89a, FS89b]) we define a limiting procedure to obtain functionals, that, if their limit exists, converge to a one-parameter semi-group of functionals that will be considered as the analogous multiplicative process on the quantum group. By the previous section these semi-groups determine a unique pseudo-Lévy process.

Let $(W_t)_{t\in\mathbb{R}_+} = (W_t^1, \cdots, W_t^d)_{t\in\mathbb{R}_+}$ be a stochastic process with values in $\mathbb{R}^d$ with independent increments, and assume that all moments of W_t are finite. Then a functional $\hat{\Phi}_{t_1,t_2}$ on $\mathbb{R}[x_1,\ldots,x_d]$, corresponding to an increment $W_{t_2} - W_{t_1}$, is defined by $\hat{\Phi}_{t_1,t_2}(x_1^{n_1}\cdots x_d^{n_d}) = \mathbb{E}((W_{t_2}^1 - W_{t_1}^1)^{n_1}\cdots(W_{t_2}^d - W_{t_1}^d)^{n_d})$. We can identify $\mathcal{A}$ with $\mathbb{R}[x_1,\ldots,x_d]$ as a vector space, if we fix a Poincaré-Birkhoff-Witt (PBW) basis $\{A_n; n\in\mathbb{N}^d\}$ for $\mathcal{A}$ and set $\imath(A_n) = x^n$. We denote the functional on $\mathcal{A}$ obtained in this way also by $\hat{\Phi}_{t_1,t_2}$.

We suppose that $\imath$ is chosen such that the functionals $\hat{\Phi}_{t_1,t_2}$ are positive. For example for the q-affine group this is the case for $\imath$ defined by $\imath(\frac{a^nb^m}{n!q_m!}) = \frac{x_1^nx_2^m}{n!m!}$ with an appropriate definition of positivity[2]. We do not know how to define $\imath$ in general to guarantee the positivity of $\hat{\Phi}_{t_1,t_2}$. Nevertheless, the results below hold regardless of the positivity hypothesis.

The following construction will start with these functionals, and then recover their properties with respect to the coproduct of $\mathcal{A}$.

We suppose that W_t has independent increments. For the functional $\hat{\Phi}_t$ on $\mathbb{R}[x_1,\ldots,x_d]$ this means that

$$\hat{\Phi}_{t_1,t_2}\otimes\hat{\Phi}_{t_2,t_3}(\Delta z) = \hat{\Phi}_{t_1,t_3}(z), \qquad \text{for } z\in\mathbb{R}[x_1,\ldots,x_d]$$

where the coproduct of $\mathbb{R}[x_1,\ldots,x_d]$ is defined by $\Delta x_i = x_i\otimes 1 + 1\otimes x_i$. We want to construct a functional Φ_t on $\mathcal{A}$ that satisfies the same relation with respect to the coproduct of $\mathcal{A}$. To this end we define a sequence of functionals $\Phi_t^{(N)}$, and take its limit for Φ_t, if it exists. Let

$$\Phi_{s,t}^{(N)}(a) = \hat{\Phi}_{s,s+(t-s)/N}\otimes\cdots\otimes\hat{\Phi}_{s+(t-s)(N-1)/N,t}\left(\Delta^{N-1}(a)\right) \qquad \text{for } N>1, \quad a\in\mathcal{A}.$$

Loosely speaking, this corresponds to decomposing the desired process into its increments via the coproduct, and approximating its expected value by the expected value with respect to $\hat{\Phi}$ in each increment. We define

$$\Phi_{s,t}(a) = \lim_{N\to\infty}\Phi_{s,t}^{(N)}(a) \qquad \text{for } a\in\mathcal{A}, \qquad s<t\in\mathbb{R}_+, \tag{4.8}$$

if this limit exists.

Definition 4.6.1 *Let W_t be a stochastic process on $\mathbb{R}^n$ with independent increments*

[2]If we define the positive elements of $\mathcal{A}$ as the inverse image under $\imath$ of the positive elements of $\mathbb{R}[x_1,x_2]$.

$W_{s,t} = W_t - W_s$, and all moments finite. Let further $\mathcal{A}$ be a Hopf algebra and $\imath : \mathcal{A} \to \mathbb{R}[x_1, \dots, x_n]$ a vector space isomorphism. We call

$$\Phi_{s,t}(a) = \lim_{N\to\infty} \mathbb{E}(\imath^{\otimes N}(\Delta^{N-1}(a))(W_{s,s+(t-s)/N}, \dots, W_{s+(t-s)(N-1)/N,t}))$$

the $\imath$-analogue of W_t on $\mathcal{A}$, if this limit exists.

In our applications we will consider processes whose increments are stationary, i. e., the functionals $\hat{\Phi}_{s,t}$ and $\Phi_{s,t}$ depend only on the difference $t - s$, in this case we can also write $\hat{\Phi}_t$ and Φ_t.

In the following section we have a class of examples where this limit exists, and the expected value of the dual pairing, which in this context should be understood as a generating function, is found. For this it is necessary to extend the definition of the functional Φ_t to infinite series thus:

$$\Phi_t(\sum_{n=0}^{\infty} a_n) = \sum_{n=0}^{\infty} \Phi_t(a_n),$$

if the left-hand-side converges absolutely.

On Lie groups the multiplicative stochastic integrals define stochastic processes. Unfortunately in our case the positivity of the constructed process is not *a priori* clear, but depends on the choice of the identification of $\mathcal{A}$ and $\mathbb{R}[x_1, \dots, d]$. Nevertheless, the proof of the Feynman-Kac formula in the next section doesn't require positivity. Finding good criteria for which cases the obtained process will be positive is an open problem, for some results about positive functionals on quantum groups see [Koo91].

4.7 FEYNMAN-KAC FORMULA

On Lie groups one can obtain Feynman-Kac type formulae with Trotter's product formula

$$\lim_{N\to\infty} \left(e^{X_1/N} \cdots e^{X_d/N}\right)^N = e^{X_1 + \cdots + X_d}.$$

Let W_t be a stochastic process on $\mathbb{R}^d$ with independent and stationary increments and independent components, and

$$\mathbb{E}\left(e^{W^1_{s,t} X_1} \cdots e^{W^d_{s,t} X_d}\right) = e^{L_1(X_1)(t-s)} \cdots e^{L_d(X_d)(t-s)}$$

be the expectation value of an increment, i. e., L_i is the generator of $W_i(t)$. Then, if we approximate the process $g(\tilde{W}(t); X)$ on the Lie group defined by McKean's stochastic product integral [McK69] by $g^{(n)}(t)$,

$$\begin{aligned}
\mathbb{E}\left(g^{(n)}(t)\right) &= \mathbb{E}\left(\prod e^{w_1 X_1} \cdots e^{w_d X_d}\right) = \prod \mathbb{E}\left(e^{w_1 X_1} \cdots e^{w_d X_d}\right) \\
&= \left(e^{L_1(X_1)\Delta t} \cdots e^{L_d(X_d)\Delta t}\right)^n \underset{n\to\infty}{\longrightarrow} e^{t(L_1(X_1) + \cdots + L_d(X_d))}.
\end{aligned}$$

If we take the limit on both sides we get

$$\mathbb{E}\Big(g(\tilde{W}(t);X)\Big) = e^{t(L_1(X_1)+\cdots+L_d(X_d))},$$

see also Subsection 2.7.1. Note that this formula still contains Trotter's product formula for the special case of a deterministic process, i. e., if $L_i(X_i) = X_i$ for $i = 1, \ldots, d$.

Theorem 4.7.1 (Feynman-Kac formula) *Let $\mathcal{A}$ and $\mathcal{U}$ be a quantum group and a quantum algebra that are in duality, with generators $a_1, \ldots, a_d$ and $X_1, \cdots, X_d$, respectively, and dual pairing*

$$g(a;X) = e_{q_1}^{a_1X_1}\cdots e_{q_d}^{a_dX_d}, \qquad q_i \in \mathbb{C}.$$

Let ρ be a representation of $\mathcal{U}$ by operators that are bounded with respect to some norm $||\cdot||$. We set

$$p_i = \begin{cases} \max\{p; \rho(X_i)^p \neq 0\} & \text{if } \rho(X_i) \text{ is nilpotent, i.e. } \rho(X_i)^{r+1} = 0 \text{ for some integer } r, \\ \infty & \text{otherwise.} \end{cases}$$

Let $\imath : \mathcal{A} \to \mathbb{R}[x_1, \ldots, x_n]$ be defined by $\imath(a^n) = x^n$

Suppose further that we are given a d-dimensional stochastic process $W(t)$ with independent and stationary increments and independent components, and moment generating functions

$$\mathbb{E}(e^{uW^i(t)}) = e^{tL_i(u)}, \qquad L_i(u) = \sum_{k=1}^{\infty} \frac{l_k^{(i)}u^k}{k!} \text{ analytic,}$$

and moments $m_k(W^i(t)) = \left(\frac{\mathrm{d}}{\mathrm{d}u}\right)^k e^{tL_i(u)}\Big|_{u=0}$, such that the series

$$\tilde{L}_i(u) = \sum_{k=1}^{p_i} \frac{l_k^{(i)}u^k}{(q_i)_k!}, \qquad \text{and} \qquad \sum_{k=0}^{p_i} \frac{m_k(W^i(t))\rho(X_i)^k}{(q_i)_k!}$$

are well-defined and converge (in particular, $(q_i)_k \neq 0$, for $1 \le k \le p_i$).

Then we have for the $\imath$-analogue of $A(t)$ on the associated quantum group $\mathcal{A}$

$$(\Phi_t \otimes \mathrm{id})\left(\prod_{i=1}^{d} e_{q_i}^{a_i\rho(X_i)}\right) = \exp\left[t\sum_{i=1}^{d}\tilde{L}_i(\rho(X_i))\right].$$

Remark: If $\mathcal{U}$ has a sufficiently rich class of representations that satisfy the conditions of the theorem, then this relation allows us to read off all moments of the Brownian motion.

Proof: According to Definition 4.6.1 we have to consider

$$\Phi_t^{(N)}(g(a;\rho(X))) = \hat{\Phi}_{t/N}^{\otimes N}(g(\Delta^{N-1}a;\rho(X)))$$

where $\hat{\Phi}_{t/N}$ is the functional corresponding to an increment of W_t for $\Delta t = t/N$. By the identification procedure outlined in the previous section this gives $\hat{\Phi}_{t/N}(a^n) = \prod_i m_{n_i}(W_i(t/N))$. If we take into account that (cf. Equation (3.15))

$$g(\Delta a; X) = g(a'; X) g(a''; X),$$

where $a' = a \otimes 1$, $a'' = 1 \otimes a$, then we get

$$\Phi_t^{(N)}\left(g(a; \rho(X))\right) = \hat{\Phi}_{t/N}^{\otimes N}(g(a^{(1)}; X) \cdots g(a^{(N)}; X))$$

where $a^{(k)} = 1^{\otimes k-1} \otimes a \otimes 1^{N-k}$, and thus

$$\begin{aligned}
\Phi_t^{(N)}\left(g(a; \rho(X))\right) &= \hat{\Phi}^{\otimes N}\left(\prod_{\nu=1}^{N}\prod_{i=1}^{d} e_{q_i}^{a_i \rho(X_i)}\right) \\
&= \prod_{\nu=1}^{N}\prod_{i=1}^{d}\left(1 + \sum_{k=1}^{p_i} \frac{m_k(W^i(t/N))\rho(X_i)^k}{(q_i)_k!}\right)
\end{aligned}$$

Note that for $k \geq 1$ each $m_k(W^i(\tau))$ can be written as

$$m_k(W^i(\tau)) = \tau l_k^{(i)} + \tau^2 R_{ik}(\tau)$$

therefore

$$\sum_{k=1}^{p_i} \frac{m_k(W^i(t/N))\rho(X_i)^k}{(q_i)_k!} = \frac{t}{N}\tilde{L}_i(\rho(X_i)) + \frac{t^2}{N^2} R_i(t/N)$$

where $R_i(t/N)$ is some bounded operator. Taking the product we get

$$\prod_{i=1}^{d}\left(1 + \sum_{k=1}^{p_i} \frac{m_k(W^i(t/N))\rho(X_i)^k}{(q_i)_k!}\right) = 1 + \frac{t}{N}\sum_{i=1}^{d} \tilde{L}_i(\rho(X_i)) + \frac{t^2}{N^2} R(t/N),$$

where $R(t/N)$ is also bounded.

Using $\prod_{\nu=1}^{N} a_\nu - \prod_{\nu=1}^{N} b_\nu = \sum_{\nu=1}^{N} a_1 \cdots a_{\nu-1}(a_\nu - b_\nu) b_{\nu+1} \cdots b_N$, we have

$$\begin{aligned}
&\left\| \prod_{\nu=1}^{N}\prod_{i=1}^{d}\left(1 + \sum_{k=1}^{p_i} \frac{m_k(W^i(t/N))\rho(X_i)^k}{(q_i)_k!}\right) - \prod_{\nu=1}^{N}\left(1 + \frac{t}{N}\sum_{i=1}^{d} \tilde{L}_i(\rho(X_i))\right) \right\| \\
= &\left\| \sum_{\nu=1}^{N}\left(\prod_{i=1}^{d} 1 + \sum_{k=1}^{p_i} \frac{m_k(W^i(t/N))\rho(X_i)^k}{(q_i)_k!}\right)^{\nu-1} \right. \\
&\left. \cdot \frac{t^2 R(t/N)}{N^2}\left(1 + \frac{t}{N}\sum_{i=1}^{d} \tilde{L}_i(\rho(X_i))\right)^{N-\nu} \right\| \\
\leq &\ \frac{\text{constant}}{N} \to 0
\end{aligned}$$

for $N \to \infty$. This concludes the proof, since $\prod_{\nu=1}^{N}\left(1+\frac{t}{N}\sum_{i=1}^{d}\tilde{L}_i(\rho(X_i))\right)$ converges to $\exp\left(t\sum_{i=1}^{d}\tilde{L}_i(\rho(X_i))\right)$. ∎

We can now obtain a Trotter-type product formula for q-exponentials.

Proposition 4.7.2 (Trotter's product formula) *Let X_i, $i=1,\dots,d$ be bounded operators and assume that for each $i=1,\dots,d$ one of the following two conditions is satisfied:*

1. $q_i \in \{1\} \cup \{z \in \mathbb{C}; |z| \neq 1\}$,
2. *the operator X_i is nilpotent of some order p_i, i. e., $X_i^{p_i+1}=0$, and $q_i \in \{1\} \cup \{z \in \mathbb{C}; z^\nu \neq 1$ for $\nu = 1,\dots,p_i\}$.*

Then we have

$$\prod_{\nu=0}^{N}\prod_{i=1}^{d} e_{q_i}^{X_i/N} \underset{N\to\infty}{\longrightarrow} \exp\sum_{i=1}^{d} X_i. \qquad \textit{(norm convergence)}$$

Remark: For the second case, the q-exponential can be defined by the finite sum $e_{q_i}^{X_i}=\sum_{\nu=0}^{p_i}\frac{X_i^\nu}{q_\nu!}$ since $q_\nu! \neq 0$ for $\nu \leq p_i$ and $X_i^\nu = 0$ for $\nu > p_i$.

Proof: Take a deterministic process with $L_i(X_i)=X_i$, $i=1,\dots,d$, and apply Theorem 4.7.1. ∎

The following proposition can also be presented as a consequence of Theorem 4.7.1.

Proposition 4.7.3 *The functionals defined as the $\imath$-analogue of a stochastic process with independent and stationary increments and independent components form a convolution semi-group.*

4.8 TIME-REVERSAL, DUALITY, AND R-MATRICES

In this section we show some constructions for obtaining new Lévy processes from a given Lévy process.

4.8.1 *Time-reversal*

If $(X_t)_{t\in[0,T]}$ is a classical right Lévy process with values in a group, then one defines the time-reversed process by

$$\hat{X}_t = X_T^{-1}X_{T-t}, \qquad \text{for } t\in[0,T].$$

The increments of this process are given by $\hat{X}_{st}=\hat{X}_s^{-1}\hat{X}_t = X_{T-s}^{-1}X_{T-t} = (X_{T-t,T-s})^{-1}$ for $0\leq s\leq t\leq T$, and it is obvious that we have again a right Lévy process.

If $(X_t)_{t\in[0,T]}$ takes values in $\mathbb{R}^n$ and has characteristic exponent Φ, i.e. if $\mathbb{E}(e^{iuX_t}) = e^{t\Phi(u)}$, then the characteristic exponent $\hat{\Phi}$ of the time-reversed process $(\hat{X}_t)_{t\in[0,T]}$ is given by $\hat{\Phi}(u) = \Phi(-u)$. This follows from

$$\mathbb{E}(e^{iu\hat{X}_t}) = \mathbb{E}(e^{iu(X_{T-t}-X_T)}) = \mathbb{E}(e^{-iu(X_T-X_{-t})}) = \mathbb{E}(e^{-iuX_t}).$$

Our first construction of a time-reversed Lévy process is taken from [Sch93, Proposition 1.9.4].

Proposition 4.8.1 *Let $\mathcal{B} = (\mathcal{B}, m, e, \Delta, \varepsilon, \Psi, *)$ be a braided $*$-bialgebra and let $(j_{st})_{0\leq s\leq t\leq T}$ be a Lévy process on $\mathcal{B}$. Then*

$$\mathcal{B}^{\rm cop} = (\mathcal{B}, m, e, \Delta^{\rm cop} = \Psi^{-1}\circ\Delta, \varepsilon, \Psi^{-1}, *)$$

is also a braided $*$-bialgebra and the time-reversed process

$$j_{st}^{\rm cop} = j_{T-t,T-s}$$

is a Lévy process on $\mathcal{B}^{\rm cop}$. Moreover, (j_{st}) and $(j_{st}^{\rm cop})$ have the same generator and the same marginal distribution.

Proof: That $(\mathcal{B}, m, e, \Delta^{\rm cop} = \Psi^{-1}\circ\Delta, \varepsilon, \Psi^{-1})$ is a braided bialgebra was shown in [Maj93a, Lemma 4.6], so we only need to show that $\Delta^{\rm cop} = \Psi^{-1}\circ\Delta$ is a $*$-algebra homomorphism (where the $*$-structure in $\mathcal{B}^{\rm cop}\otimes\mathcal{B}^{\rm cop}$ is defined by $*_\otimes = \Psi^{-1}\circ(*\otimes *)\circ\tau$). We have

$$\begin{aligned}
\Delta^{\rm cop}\circ * &= \Psi^{-1}\circ\Delta\circ * = \Psi^{-1}\circ\Psi\circ(*\otimes *)\circ\tau\circ\Delta \\
&= (*\otimes *)\circ\tau\circ\Delta = \Psi^{-1}\circ(*\otimes *)\circ\tau\circ\Psi^{-1}\circ\Delta \\
&= *_\otimes\circ\Delta^{\rm cop},
\end{aligned}$$

where we used $(*\otimes *)\circ\tau = (\Psi\circ(*\otimes *)\circ\tau\circ\Psi)^{-1} = \Psi^{-1}\circ(*\otimes *)\circ\tau\circ\Psi^{-1}$.

The rest of the proof is identical to the proof of [Sch93, Proposition 1.9.4]. ∎

For classical Lévy processes this construction does not give an essentially new process, it only amounts to interpreting a right Lévy process as a left Lévy process with values in the (semi-)group with the opposite multiplication.

To define a time-reversed process in an analogous manner to the classical case we need the antipode.

Proposition 4.8.2 *Let $(\mathcal{B}, m, e, \Delta, \varepsilon, \Psi, S, *)$ be a braided $*$-Hopf algebra with invertible antipode S and let $(j_{st})_{0\leq s\leq t\leq T}$ be a Lévy process on $\mathcal{B}$. Then $\mathcal{B}^{\rm op} = (\mathcal{B}, m^{\rm op} = m\circ\Psi^{-1}, e, \Delta, \varepsilon, \Psi^{-1}, S^{-1}, S^{-1}\circ *\circ S)$ is also a braided $*$-Hopf algebra and the time-reversed process*

$$\hat{j}_{st} = j_{T-t,T-s}\circ S$$

is a Lévy process on $\mathcal{B}^{\rm op}$. Moreover, if L is the generator of (j_{st}), then $\hat{L} = L\circ S$ is the generator of $(\hat{j}_{st})$

Proof: It is not difficult to check that $\mathcal{B}^{\rm op} = (\mathcal{B}, m^{\rm op} = m \circ \Psi^{-1}, e, \Delta, \varepsilon, \Psi^{-1}, S^{-1})$ is a braided Hopf algebra and that it is actually isomorphic to $\mathcal{B}^{\rm cop}$, the isomorphism being the antipode $S : \mathcal{B}^{\rm op} \to \mathcal{B}^{\rm cop}$. This also shows that $\mathcal{B}^{\rm op}$ is a braided $*$-Hopf algebra with the involution $S^{-1} \circ * \circ S$. Writing $(\hat{\jmath}_{st})$ as $j_{st}^{\rm cop} \circ S$, the rest of the proposition becomes obvious. ∎

We can also combine both constructions. Then we get the Lévy processes

$$j_{st}^{\rm cop\,op} = \widehat{j_{st}^{\rm cop}} = j_{st} \circ S^{-1}, \qquad \text{for } 0 \le s \le t \le T,$$

on $\mathcal{B}^{\rm cop\,op} = (\mathcal{B}, m \circ \Psi, e, \Psi^{-1} \circ \Delta, \varepsilon, S, \Psi, S \circ * \circ S^{-1})$ and

$$j_{st}^{\rm op\,cop} = \hat{\jmath}_{st}^{\rm cop} = j_{st} \circ S, \qquad \text{for } 0 \le s \le t \le T,$$

on $\mathcal{B}^{\rm op\,cop} = (\mathcal{B}, m \circ \Psi^{-1}, e, \Psi \circ \Delta, \varepsilon, S, \Psi, S^{-1} \circ * \circ S)$. More generally, we get the two families

$$j_{st}^{k\times\rm cop\,op} = j_{st} \circ S^{-k}, \qquad \text{for } 0 \le s \le t \le T,$$

on $\mathcal{B}^{k\times\rm cop\,op} = (\mathcal{B}, m \circ \Psi^k, e, \Psi^{-k} \circ \Delta, \varepsilon, S, \Psi, S^k \circ * \circ S^{-k})$ and

$$\hat{\jmath}_{st}^{k\times\rm cop\,op} = j_{T-t,T-s} \circ S^{k+1}, \qquad \text{for } 0 \le s \le t \le T,$$

on $\mathcal{B}^{\rm op\,(k\times cop\,op)} = (\mathcal{B}, m \circ \Psi^{-k-1}, e, \Psi^k \circ \Delta, \varepsilon, S, \Psi^{-1}, S^{-k-1} \circ * \circ S^{k+1})$ for $k \in \mathbb{Z}$.

4.8.2 *Lévy coprocesses*

Lévy processes can be defined on bialgebras, because the coproduct allows to compose representations. Similarly, the multiplication can be used to compose corepresentations, this leads to the definition of infinitely divisible corepresentations or Lévy coprocesses. We will consider only symmetric bialgebras (i.e. $\Psi = \tau$) in this subsection and also neglect $*$-structures, i.e. stay at the level of pseudo (co-) processes.

Let $(\mathcal{A}, \Phi)$ be a coalgebra with some fixed element Φ satisfying $\varepsilon(\Phi) = 1$. A *coprocess* on a coalgebra $\mathcal{C}$ over $(\mathcal{A}, \Phi)$ is a family of coalgebra homomorphisms $(j_t)_{t\in I} : \mathcal{A} \to \mathcal{C}$ indexed by some set I.

First we have to dualize the definition of independence.

Definition 4.8.3 *A tuple of coalgebra homomorphisms $(j_1, \ldots, j_n)$ from $\mathcal{A}$ to $\mathcal{C}$ is called independent (with respect to $\Phi \in \mathcal{A}$), if the following two conditions are satisfied:*

(i)

$$(j_1 \otimes \cdots \otimes j_n) \circ \Delta_{\mathcal{A}}^{(n)}(\Phi) = j_1(\Phi) \otimes \cdots \otimes j_n(\Phi)$$

where $\Delta^{(1)} = \mathrm{id}$, $\Delta^{(2)} = \Delta$, *and* $\Delta^{(k+1)} = (\Delta \otimes \mathrm{id}^{\otimes(k-1)}) \circ \Delta^{(k)}$ *for* $k = 3, 4, \ldots$

(ii)

$$\tau \circ (j_k \otimes j_l) \circ \Delta_{\mathcal{A}} = (j_l \otimes j_k) \circ \Delta_{\mathcal{A}}$$

for $k, l = 1, \ldots, n$.

Let $\mathcal{B}$ be a bialgebra. The operation that we will use to compose increments is the same as for Lévy processes, i.e. for two linear maps $j_1, j_2 : \mathcal{A} \to \mathcal{B}$ we set

$$j_1 \star j_2 = m_{\mathcal{B}} \circ (j_1 \otimes j_2) \circ \Delta_{\mathcal{A}},$$

only the roles of $\mathcal{A}$ and $\mathcal{B}$ are switched.

Definition 4.8.4 *Let $\mathcal{B}$ be a bialgebra. A coprocess $\{j_{st} | 0 \leq s \leq t \leq T\}$ on $\mathcal{B}$ over $(\mathcal{A}, \Phi)$ is called a (pseudo)* Lévy coprocess *if the following conditions are satisfied.*

1. *(increment property)*

$$\begin{aligned} j_{rs} \star j_{st} &= j_{rt} \quad \text{for all } 0 \leq r \leq s \leq t \leq T, \\ j_{tt} &= e_{\mathcal{B}} \circ \varepsilon_A \quad \text{for all } 0 \leq t \leq T, \end{aligned}$$

2. *(independence of increments) the tuples $(j_{s_1 t_1}, \ldots, j_{s_n t_n})$ are independent for all $n \in \mathbb{N}$ and all $0 \leq s_1 \leq t_1 \leq s_2 \leq \cdots \leq t_n \leq T$,*

3. *(stationarity of increments) the element $\varphi_{st} = j_{st}(\Phi)$ depends only on the difference $t - s$,*

4. *("weak" continuity) $h \circ j_{st}(\Phi)$ converges to $h \circ j_{ss}(\Phi) = h(e_{\mathcal{B}})$ as $t \searrow s$ for all linear functionals $h : \mathcal{B} \to \mathbb{C}$.*

The following result shows that these Lévy coprocesses are really the dual notion of pseudo-Lévy processes.

Theorem 4.8.5 *Let $\mathcal{B}_1$, $\mathcal{B}_2$ be two dually paired bialgebras and let $(j_{st})_{0 \leq s \leq t \leq T}$ be a Lévy coprocess on $\mathcal{B}_1$. Then the process $j^*_{st} : \mathcal{B}_2 \to \mathcal{A}^*$ defined by*

$$j^*_{st}(b) = \langle j_{st}(\cdot), b \rangle, \quad \text{for } b \in \mathcal{B}_2,$$

is a pseudo-Lévy process on $\mathcal{B}_2$.

Remarks: We can easily recover a pseudo-Lévy process from Lévy coprocess, because $\mathcal{A}^*$ is an algebra (as the dual of a coalgebra). We can not go in the opposite direction, because, in general, the dual of an infinite-dimensional algebra does not have a natural coalgebra structure.

Proof: The properties 1. to 4. of the definition of a (pseudo-) Lévy process follow immediately from the corresponding conditions in the definition of a Lévy coprocess. We have, e.g., for the increment property

$$\begin{aligned} (j^*_{st} \star j^*_{tu})(b) &= m_{\mathcal{A}^*} \circ (j^*_{st} \otimes j^*_{tu}) \circ \Delta_{\mathcal{B}_2}(b) = b \circ m_{\mathcal{B}_1} \circ (j_{st} \otimes j_{tu}) \circ \Delta_{\mathcal{A}} \\ &= b \circ (j_{st} \star j_{tu}) = b \circ j_{su} = j^*_{su}(b) \end{aligned}$$

and

$$j_{tt}^*(b) = b \circ j_{tt} = b \circ e_{\mathcal{B}_1} \circ \varepsilon_{\mathcal{A}} = e_{\mathcal{A}^*} \circ \varepsilon_{\mathcal{B}_2}(b)$$

for all $b \in \mathcal{B}_2$.

We omit the proof of the other properties. ∎

4.8.3 *Constructions using the universal R-matrix*

In this subsection we will only deal with symmetric bialgebras. As in Subsection 4.8.1 we can construct the bialgebras $\mathcal{B}^{\text{cop}} = (\mathcal{B}, m, e, \Delta^{\text{cop}} = \tau \circ \Delta, \varepsilon)$, $\mathcal{B}^{\text{op}} = (\mathcal{B}, m^{\text{op}} = m \circ \tau, e, \Delta, \varepsilon)$, and $\mathcal{B}^{\text{cop op}} = (\mathcal{B}, m^{\text{op}} = m \circ \tau, e, \Delta^{\text{cop}} = \tau \circ \Delta, \varepsilon)$ from a given bialgebra $\mathcal{B} = (\mathcal{B}, m, e, \Delta, \varepsilon)$, but further iteration will not lead to anything else, since we have $\tau^2 = \text{id}$. If $\mathcal{B}$ is a Hopf algebra and if its antipode is invertible, then $\mathcal{B}^{\text{op}}$, $\mathcal{B}^{\text{cop}}$, and $\mathcal{B}^{\text{op cop}}$ are Hopf algebra, too. If, furthermore, $\mathcal{B}$ is a $*$-Hopf algebra, then $\mathcal{B}^{\text{cop}}$, $\mathcal{B}^{\text{op}}$, and $\mathcal{B}^{\text{cop op}}$ are $*$-Hopf algebras, too, with the involutions $*^{\text{cop}} = *$, $*^{\text{op}} = S^{-1} \circ * \circ S$ and $*^{\text{cop op}} = S \circ * \circ S^{-1}$, respectively.

Let $\mathcal{B}_1$ and $\mathcal{B}_2$ be two dually paired bialgebras and suppose that $\mathcal{B}_1$ is quasi-triangular, i.e. that there exists an invertible element $R \in \mathcal{B}_1 \otimes \mathcal{B}_1$ (called universal R-matrix) such that

$$\tau \circ \Delta(a) = R\Delta(a)R^{-1}, \qquad \text{for all } a \in \mathcal{A}$$

and that satisfies $(\Delta \otimes \text{id})(R) = R_{13}R_{23}$ and $(\text{id} \otimes \Delta)(R) = R_{13}R_{12}$, see [Dri87, Section 10] or Section 3.7.

With the universal R-matrix we can build two linear maps from $\mathcal{B}_2$ to $\mathcal{B}_1$, they are defined by

$$_R\lambda(b) = (b \otimes \text{id})(R), \quad \text{and} \quad \lambda_R(a) = (\text{id} \otimes b)(R), \qquad \text{for } b \in \mathcal{B}_2,$$

where we identified elements $b \in \mathcal{B}_2$ with the linear functionals on $\mathcal{B}_1$ defined by $b : \mathcal{B}_1 \ni a \mapsto \langle a, b \rangle \in \mathbb{C}$. As a map between $\mathcal{B}_2$ and $\mathcal{B}_1$ the map λ_R is a coalgebra homomorphism and an algebra anti-homomorphism, and $_R\lambda$ is a coalgebra anti-homomorphism and an algebra homomorphism, see [Kas95, Proposition VIII.2.5]. Passing to opposites or co-opposites we can turn them into homomorphisms.

Proposition 4.8.6 *The maps $\lambda_R : \mathcal{B}_2^{\text{op}} \to \mathcal{B}_1$ and $_R\lambda : \mathcal{B}_2 \to \mathcal{B}_1^{\text{cop}}$ are bialgebra homomorphisms.*

Proof: (See also [Kas95, Proposition VIII.2.5]).

We have $\lambda_R(1) = (\text{id} \otimes \varepsilon)(R) = 1$ and thus

$$\varepsilon \circ \lambda_R(b) = (\varepsilon \otimes b)(R) = b(1) = \varepsilon(b),$$

for all $b \in \mathcal{B}_2$, i.e. λ_R preserves the counit, and

$$\begin{aligned} \lambda_R(m^{\text{op}}(b_1 \otimes b_2)) &= \lambda_R(b_2 b_1) = (\text{id} \otimes b_2 b_1)(R) = (\text{id} \otimes b_2 \otimes b_1) \circ (\text{id} \otimes \Delta)(R) \\ &= (\text{id} \otimes b_2 \otimes b_1)(R_{13}R_{12}) = \lambda_R(b_1)\lambda_R(b_2) \end{aligned}$$

and

$$\begin{aligned}(\lambda_R \otimes \lambda_R) \circ \Delta(b) &= (\mathrm{id} \otimes b^{(1)} \otimes \mathrm{id} \otimes b^{(2)})(R \otimes R) = (\mathrm{id} \otimes \mathrm{id} \otimes b)(R_{13}R_{23}) \\ &= (\mathrm{id} \otimes \mathrm{id} \otimes b) \circ (\Delta \otimes \mathrm{id})(R) = \Delta \circ \lambda_R(b)\end{aligned}$$

for all $b, b_1, b_2 \in \mathcal{B}_2$.

The second part of the proposition follows similarly. ∎

Proposition 4.8.7 *If $\mathcal{B}_1$ and $\mathcal{B}_2$ are even dually paired Hopf algebras with invertible antipodes, then $\lambda_R : \mathcal{B}_2^{\mathrm{op}} \to \mathcal{B}_1$ and ${}_R\lambda : \mathcal{B}_2 \to \mathcal{B}_1^{\mathrm{cop}}$ are Hopf algebra homomorphisms.*

Proof: Recall (see, e.g., [Kas95, Theorem VIII.2.4. (b)]) that the R-matrix of a Hopf algebra with invertible antipode satisfies

$$(S \otimes \mathrm{id})(R) = R^{-1} = (\mathrm{id} \otimes S^{-1})(R), \quad \text{and} \quad (S \otimes S)(R) = R.$$

Thus we have

$$\begin{aligned}\lambda_R \circ S^{-1}(b) &= (\mathrm{id} \otimes S^{-1}(b))(R) = (\mathrm{id} \otimes b) \circ (\mathrm{id} \otimes S^{-1})(R) \\ &= (\mathrm{id} \otimes b) \circ (S \otimes \mathrm{id})(R) = S \circ \lambda_R(b)\end{aligned}$$

and

$$\begin{aligned}{}_R\lambda \circ S(b) &= (S(b) \otimes \mathrm{id})(R) = (b \otimes \mathrm{id}) \circ (S \otimes \mathrm{id})(R) \\ &= (b \otimes \mathrm{id}) \circ (\mathrm{id} \otimes S^{-1})(R) = S^{-1} \circ {}_R\lambda(b)\end{aligned}$$

for all $b \in n\mathcal{B}_2$. ∎

Let us now check how λ_R and ${}_R\lambda$ behave with respect to the involution.

Proposition 4.8.8 *Let $\mathcal{B}_1$ and $\mathcal{B}_2$ be two dually paired $*$-Hopf algebras and suppose furthermore that $\mathcal{B}_1$ is quasi-triangular and that its R-matrix satisfies*

$$R^{*\otimes *} = R^{-1}.$$

Then $\lambda_R : \mathcal{B}_2^{\mathrm{op}} \to \mathcal{B}_1$ and ${}_R\lambda : \mathcal{B}_2 \to \mathcal{B}_1^{\mathrm{cop}}$ are $*$-Hopf algebra homomorphisms.

Proof: Write $R = R^{(1)} \otimes R^{(2)}$, then we get

$$\begin{aligned}\lambda_R\left(b^{*^{\mathrm{op}}}\right)^* &= \left((\mathrm{id} \otimes S^{-1}(S(b)^*))(R)\right)^* = \left(R^{(1)}\overline{\langle S(b)^*, S^{-1}(R^{(2)})\rangle}\right)^* \\ &= R^{(1)^*}\langle S(b), R^{(2)^*}\rangle = (\mathrm{id} \otimes b) \circ (\mathrm{id} \otimes S)(R^{*\otimes *}) \\ &= (\mathrm{id} \otimes b) \circ (\mathrm{id} \otimes S)(R^{-1}) = (\mathrm{id} \otimes b)(R) = \lambda_R(b),\end{aligned}$$

and

$$\begin{aligned}{}_R\lambda(b^*)^* &= \left(\langle b^*, R^{(1)}\rangle R^{(2)}\right)^* = \left(\overline{\langle b, S(R^{(1)})^*\rangle} R^{(2)}\right)^* \\ &= (b \otimes \mathrm{id}) \circ (* \otimes *) \circ (S \otimes \mathrm{id})(R) = (b \otimes \mathrm{id}) \circ (* \otimes *)(R^{-1}) \\ &= (b \otimes \mathrm{id})(R) = {}_R\lambda(b)\end{aligned}$$

for all $b \in \mathcal{B}_2$. ∎

We can now use these maps to obtain Lévy processes on $\mathcal{B}_2$ from Lévy processes on $\mathcal{B}_1$

Theorem 4.8.9 *(a) Let $\mathcal{B}_1$ and $\mathcal{B}_2$ be two dually paired Hopf algebras and suppose that $\mathcal{B}_1$ is quasi-triangular with the R-matrix R. If $(j_{st})_{0\le s\le t\le T}$ is a pseudo-Lévy process on $\mathcal{B}_1$, then*

$$j_{T-t,T-s} \circ \lambda_R \circ S, \quad \textit{and} \quad j_{T-t,T-s} \circ {}_R\lambda, \quad \textit{for} \quad 0 \le s \le t \le T,$$

are pseudo-Lévy processes on $\mathcal{B}_2$.

*(b) Let $\mathcal{B}_1$ and $\mathcal{B}_2$ be two dually paired *-Hopf algebras and suppose that $\mathcal{B}_1$ is quasi-triangular with the R-matrix R and that R satisfies $R^{*\otimes *} = R^{-1}$. If $(j_{st})_{0\le s\le t\le T}$ is a Lévy process on $\mathcal{B}_1$, then*

$$j_{T-t,T-s} \circ \lambda_R \circ S, \quad \textit{and} \quad j_{T-t,T-s} \circ {}_R\lambda, \quad \textit{for} \quad 0 \le s \le t \le T,$$

are Lévy processes on $\mathcal{B}_2$.

Proof:

(a) It follows from Proposition 4.8.6 that $j_{st} \circ \lambda_R$ is a pseudo-Lévy process on $\mathcal{B}_2^{\rm op}$. Using the time-reversal defined in Proposition 4.8.2 we then get the pseudo-Lévy process $j_{T-t,T-s} \circ \lambda_R \circ S$ on $\mathcal{B}_2$, all that has to be done is to check that Proposition 4.8.2 also holds for pseudo-Lévy processes, i.e. if no involution is defined.

Similarly, using first the time-reversal defined in Proposition 4.8.1 to obtain a Lévy process on $\mathcal{B}_1^{\rm cop}$ and then Proposition 4.8.6 we see that $j_{T-t,T-s} \circ {}_R\lambda$ also defines a pseudo-Lévy process on $\mathcal{B}_2$. Note that $j_{T-t,T-s} \circ {}_R\lambda$ can also be defined, if $\mathcal{B}_1$ and $\mathcal{B}_\in$ are only bialgebras.

(b) Follows by combining Proposition 4.8.8 with the time-reversals defined in Propositions 4.8.1 and 4.8.2.

∎

A "dual" theory of time-reversal for Lévy coprocesses can be developed also. The constructions on dually paired bialgebra using R-matrices have analogues using universal R-forms (see, e.g., [Kas95, Section VIII.5] for the definition of a universal R-form).

Examples

Group algebras. Let G be a finite group and let $\mathbb{C}G$ be defined as in Subsection 3.2.1. Then $\mathbb{C}G$ is quasi-triangular and its R-matrix is $R = b_e \otimes b_e$, where e denotes the neutral element of G. The *-Hopf algebras $\mathbb{C}G$ and Fun G are dually paired, so that we get two Lévy processes $j_{T-t,T-s} \circ \lambda_R \circ S$ and $j_{T-t,T-s} \circ R$ on Fun G

for every Lévy process j_{st} on $\mathbb{C}G$ by Theorem 4.8.9. But these processes are not very interesting, we only get the trivial Lévy process $e \circ \varepsilon$, since

$$j_{T-t,T-s} \circ \lambda_R \circ S(x) = (j_{T-t,T-s} \otimes S(x))(b_e \otimes b_e) = j_{T-t,T-s}(b_e)x(b_e) = \varepsilon(x)\,\mathrm{id},$$

and

$$j_{T-t,T-s} \circ {}_R\lambda(x) = (x \otimes j_{T-t,T-s})(b_e \otimes b_e) = \varepsilon(x)\,\mathrm{id}$$

for all $x \in \mathbb{C}G$.

Quantum double of group algebras. Drinfeld has given a construction of quasi-triangular Hopf algebras, cf. [Dri87, Section 13]. Let G be a finite group with unit element e, then the quantum double $D = D(\mathbb{C}G)$ of the group algebra $\mathbb{C}G$ has an R-matrix. The set $\{\mathbb{1}_g \otimes b_h; g, h \in G\}$ is a basis, and the Hopf algebra structure of D is given by

$$\begin{aligned}
e_D &= \sum_{g\in G} \mathbb{1}_g \otimes b_e \\
\mathbb{1}_{g_1} \otimes b_{h_1} \cdot \mathbb{1}_{g_2} \otimes b_{h_2} &= \delta_{g_1, h_1 g_2 h_1^{-1}} \mathbb{1}_{g_1} \otimes b_{h_1 h_2} \\
\varepsilon_D(\mathbb{1}_g \otimes b_h) &= \delta_{e,g} \\
\Delta_D \mathbb{1}_g \otimes b_h &= \sum_{g_1 g_2 = g} (\mathbb{1}_{g_2} \otimes b_h) \otimes (\mathbb{1}_{g_1} \otimes b_h) \\
S_D(\mathbb{1}_g \otimes b_h) &= \mathbb{1}_{h^{-1}g^{-1}h} \otimes b_{h^{-1}}
\end{aligned}$$

for $g, g_1, g_2, h, h_1, h_2 \in G$, see also [Kas95, Section IX.4]. Its R-matrix is given by

$$R = \sum_{g\in G} (\mathbb{1} \otimes b_g) \otimes (\mathbb{1}_g \otimes b_e),$$

where $\mathbb{1} = \sum_{g\in G} \mathbb{1}_g$ denotes the unit element of $\mathrm{Fun}\, G$.

Let D^* denote the dual of D, as a vector space it can be given as $D^* = \mathrm{span}\,\{b_g \otimes \mathbb{1}_h; g, h \in G\}$. The dual pairing $\langle \cdot, \cdot \rangle : D^* \times D \to \mathbb{C}$ is defined by

$$\langle b_{g_1} \otimes \mathbb{1}_{h_1}, \mathbb{1}_{g_2} \otimes b_{h_2} \rangle = \langle b_{g_1}, \mathbb{1}_{g_2} \rangle \langle \mathbb{1}_{h_1}, b_{h_2} \rangle = \delta_{g_1,g_2} \delta_{h_1,h_2},$$

and Hopf algebra structure of D^* is given by

$$\begin{aligned}
e_{D^*} &= \sum_{g\in G} b_g \otimes \mathbb{1}_e \\
b_{g_1} \otimes \mathbb{1}_{h_1} \cdot b_{g_2} \otimes \mathbb{1}_{h_2} &= \delta_{h_1,h_2} b_{g_2 g_1} \otimes \mathbb{1}_{h_1} \\
\varepsilon_{D^*}(b_g \otimes \mathbb{1}_h) &= \delta_{h,e} \\
\Delta_{D^*}(b_g \otimes \mathbb{1}_h) &= \sum_{h_1 h_2 = h} (b_g \otimes \mathbb{1}_{h_1}) \otimes (b_{h_1^{-1} g h_1} \otimes \mathbb{1}_{h_2}) \\
S_{D^*}(b_g \otimes \mathbb{1}_h) &= b_{h^{-1}g^{-1}h} \otimes \mathbb{1}_{h^{-1}}.
\end{aligned}$$

We get

$$\begin{aligned} j_{T-t,T-s} \circ \lambda_R \circ S_{(D^*)^{\mathrm{op}}}(b_g \otimes \mathbb{1}_h) &= \Big(j_{T-t,T-s} \otimes (b_g \otimes \mathbb{1}_h)\Big)(R^{-1}) \\ &= \delta_{h,e} j_{T-t,T-s}(\mathbb{1} \otimes b_{g^{-1}}), \\ j_{T-t,T-s} \circ {}_R\lambda(b_g \otimes \mathbb{1}_h) &= j_{T-t,T-s}(\mathbb{1}_h \otimes b_e), \end{aligned}$$

These processes can actually be interpreted as processes on $\mathbb{C}G$ and $\mathrm{Fun}\,G$, using the surjective Hopf algebra homomorphisms $p_{\mathbb{C}G} : D^* \to \mathbb{C}G$, $b_g \otimes \mathbb{1}_h \mapsto \delta_{e,h} b_{g^{-1}}$ and $p_{\mathrm{Fun}\,G} : D^* \to \mathrm{Fun}\,G$, $b_g \otimes \mathbb{1}_h \mapsto \mathbb{1}_h$.

REFERENCES

Lévy processes in non-commutative probability theory appeared first in [Wal73], the formulation of the algebraic framework was done in [ASW88]. For an accessible first introduction to these processes see [Mey93, Chapter VII], for a more complete account [Sch93]. The sections on multiplicative stochastic integrals and on the Feynman-Kac formula are taken from [FFS97]. Time-reversal has been considered in [Sch93, Proposition 1.9.4] and in [Maj93d], but Subsections 4.8.2 and 4.8.3 are new.

Chapter 5

Markov structure of quantum Lévy processes

In this chapter we will consider the question which quantum stochastic processes admit classical versions, i.e. for what families of operators $(X_t)_{t\in I}$ there exists a classical stochastic process $(\tilde{X}_t)_{t\in I}$ on some probability space $(\Omega, \mathcal{F}, P)$ such that all time-ordered moments agree, i.e.

$$\Phi(X_{t_1}^{k_1} \cdots X_{t_n}^{k_n}) = \mathbb{E}(\tilde{X}_{t_1}^{k_1} \cdots \tilde{X}_{t_n}^{k_n}), \tag{5.1}$$

for all $n, k_1, \ldots, k_n \in \mathbb{N}$, $t_1 \leq \ldots \leq t_n \in I$, for the (vacuum) state Φ.

A famous example of a classical version of a quantum Lévy process is the Azéma martingale [Azé85, Eme89, Par90, Sch91a, Sch93]. In this case we can choose an element x of the bialgebra on which the process is defined such that the process $(X_t = j_t(x))$ is actually commutative, and thus it is clear that it is equivalent to a classical process. This process turns out to have several surprising properties, it was the first example of a process that has the chaotic representation property, but is not a (classical) Lévy process, see the references cited above.

The approach presented here is based on the theory of quantum Markov processes. We show that every quantum Lévy process, i.e. every Lévy process on a (possibly braided) $*$-bialgebra has a natural Markov structure. Then we use this Markov structure to give sufficient conditions for the existence of classical versions of Lévy processes on $*$-bialgebras.

It is well known that on a commutative algebra the quantum Markov property is sufficient for the existence of classical versions, see e.g. [BKS96]. Therefore it is sufficient to find commutative $*$-subalgebras such that the restriction of our process is still Markovian. We will show that we can obtain quantum Markov processes on commutative subalgebras from quantum Lévy processes that are not commutative, and that there exist also commutative processes that are not Markovian.

We also show that this approach leads to a powerful tool for the explicit calculation of the classical generators or measures.

In Subsection 5.1, we construct a family of conditional expectations on the inductive limit representation (see Subsection 4.5) of a quantum Lévy process and show that these processes are always Markovian. We also give necessary and sufficient conditions for restrictions of Lévy processes to be also Markovian.

In the following subsections (Subsections 5.2, 5.3, and 5.4) we give a detailed study of several examples, including the Azéma martingale and a symmetrised Poisson process, and show how Hopf algebra duality can be used for explicit calculations.

5.1 CLASSICAL VERSIONS OF QUANTUM LÉVY PROCESSES

5.1.1 *Schürmann's condition*

Let us first recall a sufficient condition for the existence of a classical version that is due to M. Schürmann.

Theorem 5.1.1 *[Sch93, Proposition 4.2.3] Let $(j_t)_{t\in[0,T]}$ be a Lévy process on a (braided) bialgebra $\mathcal{B}$, and $b_1,\cdots,b_n \in \mathcal{B}$ be commuting elements of $\mathcal{B}$, i.e. $b_ib_k = b_kb_i$ for all $i,k = 1,\ldots,n$. If $b_i\otimes 1$ and Δb_k commute in $\mathcal{B}\otimes\mathcal{B}$ for all $i,k=1,\ldots,n$, then the family $(j_t(b_1),\ldots,j(b_n))_{t\in[0,T]}$ is a family of commuting operators.*

Remark: In [Sch93] this result is only stated for the symmetric case and for $n = 1$, but the generalisation is straightforward.
Proof: Let $0 \le s \le t \le T$, and let $\Delta(b_k) = \sum b_k^{(1)} \otimes b_k^{(2)}$. Then we have

$$\begin{aligned}
j_t(b_k)j_s(b_i) &= \sum j_s(b_k^{(1)})j_{st}(b_k^{(2)})j_s(b_i) \\
&= \sum m_{\mathcal{A}} \circ (j_s \otimes j_{st} \otimes j_s)(b_k^{(1)} \otimes b_k^{(2)} \otimes b_i) \\
&= \sum m_{\mathcal{A}} \circ (j_s \otimes j_s \otimes j_{st})(b_k^{(1)} \otimes \Psi(b_k^{(2)} \otimes b_i)) \\
&= m_{\mathcal{A}} \circ (j_s \otimes j_{st})(m_{\mathcal{B}\otimes\mathcal{B}}(\Delta b_k \otimes (b_i \otimes e)) \\
&= m_{\mathcal{A}} \circ (j_s \otimes j_{st})(m_{\mathcal{B}\otimes\mathcal{B}}((b_i \otimes e) \otimes \Delta(b_k)) \\
&= \sum m_{\mathcal{A}} \circ (j_s \otimes j_{st})(b_ib_k^{(1)} \otimes b_k^{(2)}) \\
&= \sum j_s(b_ib_k^{(1)})j_{st}(b_k^{(2)}) = \sum j_s(b_i)j_s(b_k^{(1)})j_{st}(b_k^{(2)}) \\
&= j_s(b_i)j_t(b_k).
\end{aligned}$$

■

5.1.2 *From quantum Lévy processes to quantum Markov processes*

We will take for $(\mathcal{A},\Phi)$ the quantum probability space $(\tilde{\mathcal{B}},\tilde{\Phi})$ obtained as inductive limit from finite tensor products of $\mathcal{B}$ and the marginal distribution $\varphi_t = \Phi\circ j_t$ by the construction described in Section 4.5.

For an interval $I \subseteq [0,T]$ we denote by $\mathcal{A}_I$ the $*$-subalgebra of $\mathcal{A}$ generated by $\bigcup_{\tau\in I} j_\tau(\mathcal{B})$. For singletons $\{t\} = [t,t]$ the algebra $\mathcal{A}_t = \mathcal{A}_{[t,t]}$ is isomorphic to $\mathcal{B}$, and we have the shift isomorphisms $T_{s,t} : \mathcal{A}_s \to \mathcal{A}_t$, $T_{s,t} = j_t \circ j_s^{-1}$. Let $\mathcal{F}_I$ be the $*$-subalgebra of $\mathcal{A}$ generated by the increments of subintervals of I, i.e. the algebra generated by $\bigcup_{[\tau_1,\tau_2]\subseteq I} j_{\tau_1,\tau_2}(\mathcal{B})$. Since $j_{t'} = j_t \star j_{t,t'}$ for $0 \le t \le t' \le T$, we see that $\mathcal{A}_{[t_1,t_2]}$ is exactly the algebra generated by $\mathcal{A}_{t_1} \cup \mathcal{F}_{[t_1,t_2]}$. In particular, we obviously have $\mathcal{F}_{[0,t]} = \mathcal{A}_{[0,t]}$ for all $t \in [0,T]$.

It is not essential to use this realisation, we could just as well use the realisation of the Lévy process on Bose Fock space defined by the representation theorem [Sch93, Theorem 2.5.3] (see also Section 4.4), in that case the decomposition $\Gamma(L^2([0,T],H)) \cong \Gamma(L^2([0,t_1],H)) \otimes \Gamma(L^2([t_1,t_2],H)) \otimes \Gamma(L^2([t_2,T],H))$ for $[t_1,t_2] \subseteq [0,T]$ can be used to define conditional expectations $P_{[t_1,t_2]}$. But since we want to treat the symmetric and the braided case simultaneously, this would require extending Schürmann's results to the braided case.

We want to define conditional expectations $P_I : \mathcal{A} \to \mathcal{F}_I$ for $I \subseteq [0,T]$, and show that a Lévy process satisfies $P_{[0,s]}(\mathcal{A}_t) \subseteq \mathcal{A}_s$ for all $0 \le s \le t \le T$, i.e. that it is a quantum Markov process. Remember that the factors in an element $b_1 \otimes \cdots \otimes b_n$ of $\mathcal{B}_{(t_1,\dots,t_n)}$ get mapped to $j_{t_i}(b_i) \in \mathcal{A}_{t_i}$, and that we can use the coproduct to 'decompose' them into their increments. This was done by the map

$$\psi_{(t_1,\dots,t_n)} = (\psi_2 \otimes \mathrm{id}_\mathcal{B} \otimes \cdots \otimes \mathrm{id}_\mathcal{B}) \circ (\mathrm{id}_\mathcal{B} \otimes \psi_2 \otimes \mathrm{id}_\mathcal{B} \otimes \cdots \otimes \mathrm{id}_\mathcal{B}) \circ \cdots \circ (\mathrm{id}_\mathcal{B} \otimes \cdots \otimes \mathrm{id}_\mathcal{B} \otimes \psi_2),$$

$\psi_{(t_1,\dots,t_n)} : \mathcal{B}_{(t_1,\dots,t_n)} \to \mathcal{B}^{\otimes n}$, where $\psi_2 = (m \otimes \mathrm{id}_\mathcal{B}) \circ (\mathrm{id}_\mathcal{B} \otimes \Delta)$. This motivates the following definition.

Let $\mathcal{B}_{(s_1,\dots,s_n)}$ be the 'approximation' of $\mathcal{A} = \tilde{\mathcal{B}}$ corresponding to the tuple $(s_1,\dots,s_k,\dots,s_n)$, $s_1 < \cdots < s_n$. Suppose $s_k = s$ for some k. Then we define $P_{[0,s]}^{(s_1,\dots,s_k,\dots,s_n)} : \mathcal{B}_{(s_1,\dots,s_n)} \to \mathcal{B}_{(s_1,\dots,s_k=s)} \subseteq \mathcal{A}_{[0,s]}$ by

$$P_{[0,s]}^{(s_1,\dots,s_k,\dots,s_n)} = \psi^{-1}_{(s_1,\dots,s_k)} \circ \left(\mathrm{id}_{\mathcal{B}^{\otimes k}} \otimes e \circ \varphi_{s_{k+1}-s_k} \otimes \cdots \otimes e \circ \varphi_{s_n - s_{n-1}}\right) \circ \psi_{(s_1,\dots,s_n)}$$

If we introduce $\Pi^{(s_1,\dots,s_k,\dots,s_n)}_{(s_1,\dots,s_k)} = \mathrm{id}_{\mathcal{B}^{\otimes k}} \otimes e \circ \varphi_{s_{k+1}-s_k} \otimes \cdots \otimes e \circ \varphi_{s_n-s_{n-1}}$, then we can also write $P_{[0,s]}^{(s_1,\dots,s_k,\dots,s_n)} = \psi^{-1}_{(s_1,\dots,s_k)} \circ \Pi^{(s_1,\dots,s_k,\dots,s_n)}_{(s_1,\dots,s_k)} \circ \psi_{(s_1,\dots,s_n)}$.

Proposition 5.1.2 *The maps* $P_{[0,s]}^{(s_1,\dots,s_k,\dots,s_n)} : \mathcal{B}_{(s_1,\dots,s_n)} \to \mathcal{A}_{[0,s]}$ *extend to a unique linear map* $P_{[0,s]} : \mathcal{A} \to \mathcal{A}_{[0,s]}$.

Proof: Let $i_{S',S} : \mathcal{B}_{S'} \to \mathcal{B}_S$ with $S = (s_1,\dots,s_n)$ and $S' \subseteq S$ be the inclusion maps from [FFS97, Proposition 8.2] (i.e. the maps that add the unit element in the factors that are missing in S'). To prove the proposition it is sufficient to show $P^S_{[0,s]} \circ i_{S',S} = P^{S'}_{[0,s]}$, where we can assume that s occurs in S' (and therefore also in S). We show the proof for the case $S = (s_1,\dots,s_k = s,\dots,s_n)$ and $S' = (s_1,\dots,s_k = s,\dots,s_{l-1},s_{l+1},\dots,s_n)$, the general case is similar. In this case the inclusion map acts as $i_{S',S}(b_1 \otimes \cdots \otimes b_{l-1} \otimes b_{l+1} \otimes \cdots \otimes b_n) = b_1 \otimes \cdots \otimes b_{l-1} \otimes e \otimes b_{l+1} \otimes \cdots \otimes b_n$ for all $b_1 \otimes \cdots \otimes b_{l-1} \otimes b_{l+1} \otimes \cdots \otimes b_n \in \mathcal{B}_{S'}$.

Therefore $\psi_S \circ i_{S',S} = (\mathrm{id}_{\mathcal{B}^{\otimes(l-1)}} \otimes \Delta \otimes \mathrm{id}_{\mathcal{B}^{\otimes(n-l-1)}}) \circ \psi_{S'}$, and thus

$$\begin{aligned} P^S_{[0,s]} \circ i_{S',S} &= \\ &= \psi^{-1}_{S\cap[0,s]} \circ \Big(\mathrm{id}_{\mathcal{B}^{\otimes k}} \otimes e \circ \varphi_{s_{k+1}-s_k} \otimes \cdots \otimes e \circ (\varphi_{s_l - s_{l-1}} \star \varphi_{s_{l+1}-s_l}) \otimes \\ &\qquad \cdots \otimes e \circ \varphi_{s_n - s_{n-1}}\Big) \circ \psi_{S'} \\ &= P^{S'}_{[0,s]}. \end{aligned}$$

■

Conditional expectations are usually studied on von Neumann algebras with a fixed faithful normal state (or on C^*-algebras), where they are defined as completely positive maps that preserve the unit and the state. Here, in our purely algebraic setting, we will call an element x of a $*$-algebra $\mathcal{A}$ positive if and only if it can be written in the form $x = \sum_{i=1}^n a_i^* a_i$ for some $n \in \mathbb{N}$ and $a_1, \ldots, a_n \in \mathcal{A}$. The set $M_n(\mathcal{A})$ of all $n \times n$ matrices with entries in $\mathcal{A}$ is again an involutive algebra with the obvious matrix multiplication and the involution $(a^*)_{ij} = a^*_{ji}$. A linear map $\varphi : \mathcal{A}_1 \to \mathcal{A}_2$ between two involutive algebras is called *completely positive*, if, for all $n \in \mathbb{N}$, the linear maps $\varphi_n : M_n(\mathcal{A}_1) \to M_n(\mathcal{A}_2)$ defined by $(\varphi_n(a))_{ij} = \varphi(a_{ij})$ preserve positivity.

We can now show that the maps $P_{[0,s]}$ really have the desired properties.

Proposition 5.1.3 *Let $s \in [0,T]$. The linear map $P_{[0,s]} : \mathcal{A} \to \mathcal{A}_{[0,s]}$ is completely positive and satisfies $P^2_{[0,s]} = P_{[0,s]}$, $\Phi \circ P_{[0,s]} = \Phi$, $P_{[0,s]}(1) = 1$, and $P_{[0,s]}(xyz) = xP_{[0,s]}(y)z$ for all $x, z \in \mathcal{A}_{[0,s]}$, $y \in \mathcal{A}$, i.e. it is a* conditional expectation. *Furthermore, we have $P_{[0,s_1]} \circ P_{[0,s_2]} = P_{[0,s_2]} \circ P_{[0,s_1]} = P_{[0,\min(s_1,s_2)]}$ for all $s_1, s_2 \in [0,T]$.*

Proof: $P^2_{[0,s]} = P_{[0,s]}$ is clear, because $P^S_{[0,s]} = \psi_S^{-1} \circ \mathrm{id}_{\mathcal{B}^{\otimes|S|}} \circ \psi_S = \mathrm{id}$ for all $S \subseteq [0,s]$. $\Phi \circ P_{[0,s]} = \Phi$ and $P_{[0,s]}(1) = 1$ follow also immediately from the definition.

We will show $P_{[0,s]}(xyz) = xP_{[0,s]}(y)z$ for the case $x, z \in \mathcal{A}_s$, $y \in \mathcal{A}_t$, $0 < s < t < T$, the general case is similar. For this case it is sufficient to take $S = (s,t)$, and we can write $\psi_S(x) = x \otimes e$, $\psi_S(y) = \Delta(y) = \sum y_i^{(1)} \otimes y_i^{(2)}$, $\psi_S(z) = z \otimes e$. Therefore $\psi_S(xyz) = \sum xy_i^{(1)} \Psi(y_i^{(2)} \otimes z)$, and $P_{[0,s]}(xyz) = \psi^{-1}_{(s)}\Big((\mathrm{id} \otimes \varphi_{t-s})(\sum xy_i^{(1)}\Psi(y_i^{(2)} \otimes z))\Big) = \sum xy_i^{(1)} z \varphi_{t-s}(y_i^{(2)}) = xP_{[0,s]}(y)z$, where we used the Ψ-invariance of φ_{t-s}.

In order to prove the complete positivity, it is sufficient to prove that the map $\Pi^{(s_1,\ldots,s_k,\ldots,s_n)}_{(s_1,\ldots,s_k)}$ is completely positive, since the ψ_S are $*$-algebra homomorphisms. But this is obvious from the way it is defined via the identity map and the states $\varphi_{s_{i+1}-s_i}$. ■

Theorem 5.1.4 *Every quantum Lévy process is a* quantum Markov process, *i.e. we have $P_{[0,s]}(\mathcal{A}_t) \subseteq \mathcal{A}_s$ for all $0 \leq s \leq t \leq T$.*

Proof: Since the one-dimensional distributions form a convolution semi-group, all $P_{[0,s]}^{(s_1,\ldots,s_k,\ldots,s_n)}$ for $n \geq 2$, $0 < s_k = s < s_n = t$ act in the same way on elements of $\mathcal{A}_t$, namely $P_{[0,s]}^{(s_1,\ldots,s_k,\ldots,s_n)}\Big|_{\mathcal{A}_t} = j_s \circ (\mathrm{id}_{\mathcal{B}} \otimes \varphi_{t-s}) \circ \Delta \circ j_t^{-1}$, and therefore this is also true for $P_{[0,s]}\big|_{\mathcal{A}_t}$. In particular, it follows $P_{[0,s]}(\mathcal{A}_t) \subseteq \mathcal{A}_s$ ∎

In this proof we have seen how $P_{[0,s]}$ can be calculated on $\mathcal{A}_t$.

Corollary 5.1.5 *For the Markov semi-group $P_{s,t} = P_{[0,s]} \circ T_{s,t}$ we have*

$$P_{s,t}(j_s(b)) = j_s(p_{t-s}(b)) \qquad \text{for all} \quad b \in \mathcal{B},$$

where $p_{t-s} = (\mathrm{id} \otimes \varphi_{t-s}) \circ \Delta$.

This result allows us to do all calculations directly on the Hopf algebra, without using any particular realisation of the process (j_t). One recognises that the map $p_t : \mathcal{B} \to \mathcal{B}$ is the right dual representation (cf. Section 3.3)) $\rho(\varphi_t)$ of the functional φ_t.

5.1.3 *From quantum Markov processes to classical Markov processes*

We know (see e.g. [BKS96]) that for every quantum Markov process on a commutative $*$-algebra (e.g. the subalgebra generated by one self-adjoint element) there exists a classical version, i.e. a classical stochastic process that has the same (time-ordered) joint moments as the quantum Markov process. The quantum Markov property is sufficient (not necessary) for the joint density associated to the joint time-ordered moments to be positive, and then the classical process exists by Kolmogorov's construction.

Thus we need to look for self-adjoint elements of $\mathcal{B}$ who generate a subalgebra such that the restriction of the quantum Lévy process to this algebra remains Markov.

We give two criteria that guarantee that the restriction of (j_t) to a subalgebra $\mathcal{B}_0$ remains Markovian.

Theorem 5.1.6

(a) Let $\mathcal{B}_0$ be a $*$-subalgebra of $\mathcal{B}$. If $\Delta(\mathcal{B}_0) \subseteq \mathcal{B}_0 \otimes \mathcal{B}$ (i.e. $\mathcal{B}_0$ is a right coideal), then the restriction $(j_t|_{\mathcal{B}_0})$ to $\mathcal{B}_0$ of every Lévy process (j_t) on $\mathcal{B}$ is a quantum Markov process.

(b) Let L be the generator of (j_t), and $\mathcal{B}_0 \subseteq \mathcal{B}$ a $*$-subalgebra of $\mathcal{B}$. Then the restriction of (j_t) to $\mathcal{B}_0$ is a quantum Markov process if and only if $\rho(L) = (\mathrm{id} \otimes L) \circ \Delta$ leaves $\mathcal{B}_0$ invariant, i.e. if $\rho(L)(\mathcal{B}_0) \subseteq \mathcal{B}_0$.

Proof:

a) If $\mathcal{B}_0$ is a right coideal, then $\rho(L)(\mathcal{B}_0) \subseteq \mathcal{B}_0$ for any functional L, and consequently $P_{[0,s]}(j_t(\mathcal{B}_0)) = j_s\Big(\rho(e^{(t-s)L})(\mathcal{B}_0)\Big) \subseteq j_s(\mathcal{B}_0)$.

b) That the condition is sufficient is clear, since $P_{[0,s]} = j_s \circ e^{(t-s)\rho(L)} \circ j_t^{-1}$ on $\mathcal{A}_t$. To see that it is necessary it is enough to differentiate the relation $\rho(e^{(t-s)L})(\mathcal{B}_0) \subseteq \mathcal{B}_0$ with respect to t and then set $t = s$.

■

In the symmetric case an Abelian subalgebra is a right coideal only if all its elements satisfy the condition of Theorem 5.1.1, so that Part a) of the preceding theorem does not enable us to find any classical versions that could not have been obtained with Schürmann's result.

Proposition 5.1.7 *Let $\mathcal{B}$ be a symmetric bialgebra. If $\mathcal{B}_0$ is an Abelian subalgebra and a right coideal of $\mathcal{B}$, then $a \otimes 1$ and Δb commute for all $a, b \in \mathcal{B}_0$.*

Proof: Let $\Delta b = \sum b_i^{(1)} \otimes b_i^{(2)}$, then $(a \otimes 1)\Delta b = \sum a b_i^{(1)} \otimes b_i^{(2)} = \sum b_i^{(1)} a \otimes b_i^{(2)} = \Delta a(b \otimes 1)$ since the $b_i^{(1)}$ are in $\mathcal{B}_0$ and commute therefore with a. ■

In Subsection 5.3.1, we will see that there exist commutative operator processes of the form $(j_t(x))$ where the quantum stochastic process $(j_t|_{\mathbb{C}[x]})$ is not Markovian, and also restrictions of Lévy processes to commutative $*$-subalgebras that are Markovian but not commutative, i.e. we can get processes that have a classical version with Part b) of Theorem 5.1.6, but that do satisfy the conditions of Theorem 5.1.1.

5.1.4 *Martingales*

Definition 5.1.8 *Let $(\mathcal{A}, \Phi)$ be a quantum probability space equipped with a family of conditional expectations $(P_{[0,s]})_{s\in[0,T]}$. An operator process $(X_t)_{t\in[0,T]} \subseteq \mathcal{A}$ is called a $(P_{[0,s]})_{s\in[0,T]}$ -martingale, if*

$$P_{[0,s]}(X_t) = X_s$$

for all $0 \leq s \leq t \leq T$.

It is often useful to find a function $f(x,t)$ for a given initial value $g(x)$ and a given stochastic process (X_t) such that $f(x,0) = g(x)$ and $f(X_t,t)$ is a martingale. For example, if $(X_t)_{t\in\mathbb{R}_+}$ is a classical Lévy process (with characteristic exponent L, i.e. $\mathbb{E}(e^{iuX_t}) = e^{tL(u)}$) and $g(x) = e^{iux}$ for some $u \in \mathbb{R}$, then $\left(e^{iuX_t - tL(u)}\right)_{t\in\mathbb{R}_+}$ is a martingale (w.r.t. the filtration of $(X_t)_{t\in\mathbb{R}_+}$), the so-called exponential martingale of (X_t).

We will show how solutions of this problem for quantum Lévy processes can be given.

Theorem 5.1.9 *Let $(j_t)_{t\in[0,T]}$ be a Lévy process on a $*$-bialgebra $\mathcal{B}$ (realized on the inductive limit quantum probability space $\tilde{\mathcal{B}}$) with generator L and let $b \in \mathcal{B}$. Then $\left(j_t\left(e^{-t\rho(L)}b\right)\right)_{t\in[0,T]}$ is a martingale.*

Proof: With Corollary 5.1.5 we get

$$\begin{aligned} P_{[0,s]}\left(j_t\left(e^{-t\rho(L)}b\right)\right) &= j_s\left(e^{(t-s)\rho(L)}e^{-t\rho(L)}b\right) \\ &= j_s\left(e^{-s\rho(L)}b\right), \end{aligned}$$

for all $0 \leq s \leq t \leq T$. ∎

5.2 EXAMPLES OF CLASSICAL VERSIONS OF LÉVY PROCESSES ON $\mathbb{C}_q\langle a, a^*\rangle$

Let us first consider the $*$-bialgebra on which the Azéma martingale is defined (cf. [Sch91a, Sch93]). It can be viewed as the free algebra with two generators a, a^* (and the obvious $*$-structure), we shall denote it here by $\mathbb{C}_q\langle a, a^*\rangle$. We will restrict ourselves to the case where $q \in \mathbb{R}\backslash\{0\}$. The set $\mathcal{B} = \{1, a, a^*, aa, aa^*, a^*a, \ldots\}$ of all words in the two letters a, a^* forms a basis of $\mathbb{C}_q\langle a, a^*\rangle$. The coproduct and counit are defined as

$$\Delta(a) = a + a', \qquad \Delta(a^*) = a^* + a^{*\prime}, \qquad \varepsilon(a) = \varepsilon(a^*) = 0$$

on the generators, where we used the shorthand notation $a = a \otimes 1$, $a' = 1 \otimes a$, $a^* = a^* \otimes 1$, $a^{*\prime} = 1 \otimes a^*$. Then extend as algebra homomorphisms, where the algebra structure of $\mathbb{C}_q\langle a, a^*\rangle \otimes \mathbb{C}_q\langle a, a^*\rangle$ is determined by the braid relations

$$\begin{aligned} a'a &= qaa', & a^{*\prime}a &= qaa^{*\prime}, \\ a'a^* &= q^{-1}a^*a' & a^{*\prime}a^* &= q^{-1}a^*a^{*\prime}. \end{aligned}$$

We need to construct the dual $\mathcal{U}$ of $\mathbb{C}_q\langle a, a^*\rangle$. For $q = 1$ this is the shuffle algebra, see e.g. [SS93, Section 3.8]. In the general case the dual might be called a q-shuffle algebra, the formulas of the shuffle algebra only have to be modified by some q-dependent combinatorial coefficients.

A functional on $\mathbb{C}_q\langle a, a^*\rangle$ is determined by its action on the basis chosen above, i.e. on the words in a, a^*. Thus it can be written as

$$u = \sum_{\chi \in \mathcal{X}} c_\chi \chi,$$

where $\mathcal{X}$ is the dual basis of $\mathcal{B}$, i.e. the set of all words in two letters (x, x^*), and $c_\chi \in \mathbb{C}$. As dual of a coalgebra this is an algebra, and if we restrict to finite linear combinations, then it is even a bialgebra, and the dual pairing

$$\langle \chi, \beta \rangle = \begin{cases} 1 & \text{if } \chi, \beta \text{ are identical modulo the substitution } a \leftrightarrow x,\ a^* \leftrightarrow x^*, \\ 0 & \text{else,} \end{cases}$$

$\chi \in \mathcal{X}$, $\beta \in \mathcal{B}$, is still non-degenerate.

The coproduct of a word in x, x^* is just the sum over the different ways to split the word in two, i.e. for a word with n letters there are exactly $n+1$ terms. Thus we have e.g. $\Delta x = x \otimes 1 + 1 \otimes x$, $\Delta x^* = x^* \otimes 1 + 1 \otimes x^*$, or $\Delta x x^* x = x x^* x \otimes 1 + x x^* \otimes x + x \otimes x^* x + 1 \otimes x x^* x$.

To compute the product[1] of two basis elements χ_1, χ_2 in the q-shuffle algebra, we have to look at all the ways in which the first word χ_1 can be 'shuffled' into the second. The first term is simply the concatenation $\chi_1\chi_2$. Now we form all combinations of the letters of χ_1 and χ_2 where the order of the letters of χ_1 (and χ_2, resp.) remains unchanged, and add a factor q every time an x (from χ_1) is moved to the right across a letter of χ_2, and a factor q^{-1} every time an x^* (from χ_1) is moved to the right across a letter of χ_2. Thus, e.g.

$$\begin{aligned} x \cdot x &= xx + qxx = (1+q)xx, & x \cdot x^* &= xx^* + qx^*x, \\ x^* \cdot x &= x^*x + q^{-1}xx^*, & x^* \cdot x^* &= x^*x^* + q^{-1}x^*x^* = (1+q^{-1})x^*x^*, \end{aligned}$$

or

$$\begin{aligned} &\overbrace{xx^*} \cdot \underbrace{xx^*} \\ &= \overbrace{xx^*}\,\underbrace{xx^*} + q^{-1}\,\overbrace{x\,\underbrace{xx^*}}\,\underbrace{x^*} + \underbrace{x\,\overbrace{xx^*}\,x^*} + q^{-2}\,\overbrace{x\,\underbrace{xx^*}\,x^*} + q^{-1}\,\underbrace{x\,\overbrace{x}}\overbrace{x^*\,x^*} + \underbrace{xx^*}\,\overbrace{xx^*} \\ &= 2xx^*xx^* + (1 + 2q^{-1} + q^{-2})xxx^*x^*. \end{aligned}$$

WARNING: xx, xx^*, xx^*xx^*, etc. means concatenation, in this section the multiplication in $\mathcal{U}$ is *always* indicated by a dot.

The dual action or right regular representation eliminates the letters corresponding to those of χ, if they are in the same order, with possibly additional letters in between, and adds a factor q (q^{-1}, resp.) for every letter to the left of an a (a^*, resp.) that is suppressed. E.g., on an element $a_1 a_2 \cdots a_n \in \mathbb{C}_q\langle a, a^*\rangle$, with $a_i \in \{a, a^*\}$ for $i = 1, \ldots, n$,

$$\begin{aligned} \rho(x) : a_1 \cdots a_n &\mapsto \sum_{i:\, a_i = a} q^{i-1} a_1 \cdots \check{a}_i \cdots a_n \\ \rho(x^*) : a_1 \cdots a_n &\mapsto \sum_{i:\, a_i = a^*} q^{-i+1} a_1 \cdots \check{a}_i \cdots a_n \\ \rho(xx^*) : a_1 \cdots a_n &\mapsto \sum_{i<j:\, a_i = a,\, a_j = a^*} q^{i-j+1} a_1 \cdots \check{a}_i \cdots \check{a}_j \cdots a_n. \end{aligned}$$

5.2.1 *The Azéma martingale*

If we choose the generator $L = c_1 xx^* + c_2 x^*x$, with $c_1, c_2 \in \mathbb{R}_+$, then the Azéma martingale M_t (cf. [Sch91a, Sch93]) is a classical version of $\{j_t(a + a^*)\}$, i.e. the (finite) joint moments of $(j_t(a+a^*))_{t\in\mathbb{R}_+}$ agree with those of the (classical) Azéma martingale: $\mathbb{E}(M_{t_1}^{n_1} \cdots M_{t_k}^{n_k}) = \Phi\left((j_{t_1}(a+a^*))^{n_1} \cdots (j_{t_k}(a+a^*))^{n_k}\right)$ for

[1] Remember that we use the convention $\langle u \cdot v, \beta\rangle = \sum \langle u, \beta^{(1)}\rangle\langle v, \beta^{(2)}\rangle$, i.e. we define the dual pairing of the tensor product algebra with the flip automorphism τ: $\langle\cdot,\cdot\rangle_\otimes = (\langle\cdot,\cdot\rangle \otimes \langle\cdot,\cdot\rangle) \circ (\mathrm{id} \otimes \tau \otimes \mathrm{id})$.

all $t_1, \ldots, t_k \in \mathbb{R}_+$, $n_1, \ldots, n_k \in \mathbb{N}$, $k \in \mathbb{N}$. In order to find the generator of the Azéma martingale (as a classical Markov process) and to compute the moments, we need to know how L acts on the subalgebra generated by $z = a + a^*$.

Proposition 5.2.1 (Leibniz formulas). *Let $f(z)$ be a polynomials in z, i.e. $f(z) = \sum_{k=0}^{n} f_k z^k$ with $f_k \in \mathbb{C}$ and $z = a + a^*$. Then*

$$\begin{aligned}
\rho(x)(zf(z)) &= f(z) + qz\rho(x)f(z), \\
\rho(x^*)(zf(z)) &= f(z) + q^{-1}z\rho(x^*)f(z), \\
\rho(xx^*)(zf(z)) &= \rho(x^*)f(z) + z\rho(xx^*)f(z), \\
\rho(x^*x)(zf(z)) &= \rho(x)f(z) + z\rho(x^*x)f(z).
\end{aligned}$$

Proof: This follows from the Lemma 3.8.1. We have $\rho(u)(\alpha\beta) = \sum \alpha^{(1)} \cdot (\mathrm{id} \otimes u^{(1)})(\Psi(\alpha^{(2)} \otimes \rho(u^{(2)})\beta))$ (where we used Sweedler's notation for the coproduct). Applying the braid relations and recalling that z, x, x^* are primitive, while $\Delta(xx^*) = xx^* \otimes 1 + x \otimes x^* + 1 \otimes xx^*$, $\Delta(x^*x) = x^*x \otimes 1 + x^* \otimes x + 1 \otimes x^*x$, we get the desired formulas. ∎

Setting $f = z^n$, we obtain recurrence relations that allow us to determine the operators on polynomials in z,

$$\begin{aligned}
&\rho(x) : z^n \mapsto q_n z^n && \Rightarrow \quad (\rho(x)f)(z) = \delta_q f(z) = \tfrac{f(qz)-f(z)}{z(q-1)} \\
&\rho(x^*) : z^n \mapsto (q^{-1})_n z^n && \Rightarrow \quad (\rho(x)f)(z) = \delta_{1/q} f(z) = \tfrac{f(q^{-1}z)-f(z)}{z(q^{-1}-1)} \\
&\rho(xx^*) : z^n \mapsto \begin{cases} \frac{(q^{-1})_n - n}{q^{-1}-1} z^{n-2} & n \geq 2, \\ 0 & n = 0,1 \end{cases} && \Rightarrow \quad (\rho(xx^*)f)(z) = \tfrac{\delta_{1/q}f(z) - f'(z)}{z(q^{-1}-1)} \\
&\rho(x^*x) : z^n \mapsto \begin{cases} \frac{q_n - n}{q-1} z^{n-2} & n \geq 2, \\ 0 & n = 0,1 \end{cases} && \Rightarrow \quad (\rho(xx^*)f)(z) = \tfrac{\delta_q f(z) - f'(z)}{z(q-1)}
\end{aligned}$$

for $q \neq 1$, and $\rho(x)f = \rho(x^*)f = f'$, $\rho(xx^*)f = \rho(x^*x)f = \frac{1}{2}f''$ for $q = 1$. Here we used the q-numbers $q_n = \sum_{\nu=0}^{n-1} q^\nu = \frac{q^n - 1}{q-1}$ for $q \neq 1$ and $q_n = n$ for $q = 1$.

Using these operators to calculate the Appell polynomials defined by

$$h_n(z;t) = e^{t\rho(L)} z^n,$$

we can also calculate the moments of the Azéma martingale, $\mathbb{E}(M_t^n) = h_n(0,t)$. We summarise the results in the following theorem.

Theorem 5.2.2 *Let $(j_t)_{t \in \mathbb{R}_+}$ be the quantum Lévy process on $\mathbb{C}_q\langle a, a^* \rangle$ with generator $L = c_1 xx^* + c_2 x^*x$, $c_1, c_2 \in \mathbb{R}_+$, and $(M_t)_{t \in \mathbb{R}_+}$ a classical version of $j_t(a + a^*)$ (i.e. the Azéma martingale). Then the generator L_{a+a^*} of M_t (as a classical Markov process) acts as*

$$L_{a+a^*} f(z) = \begin{cases} \frac{c_2 \delta_q f(z) - qc_1 \delta_{1/q} f(z) - (c_2 - qc_1) f'(z)}{z(q-1)} & q \neq 1 \\ \frac{c_1 + c_2}{2} f''(z) & q = 1 \end{cases}$$

on polynomials $f(z) = \sum_{k=0}^{n} f_k z^n$, $f_0, f_1, \ldots, f_n \in \mathbb{C}$. *The moments of this process are*

$$\mathbb{E}(M_t^n) = \begin{cases} \frac{k_{2m}k_{2m-2}\cdots k_2}{m!} t^m & n = 2m \text{ even,} \\ 0 & n \text{ odd,} \end{cases}$$

where $k_n = c_1 \frac{(q^{-1})_n - n}{q^{-1} - 1} + c_2 \frac{q_n - n}{q-1}$ *for* $q \neq 1$, *and* $k_n = (c_1 + c_2)\frac{n(n-1)}{2}$ *for* $q = 1$.

5.2.2 *Other processes on* $\mathbb{C}_q\langle a, a^*\rangle$

If we want to obtain other classical processes we can either change the quantum process, i.e. choose a different generator, or use a different commutative subalgebra of $\mathbb{C}_q\langle a, a^*\rangle$ that satisfies the conditions discussed in Subsection 5.1.3. Let us briefly look at the second possibility. We want an element of $u \in \mathcal{B} = \mathbb{C}_q\langle a, a^*\rangle$ such that $\Delta\,\mathbb{C}[u] \subseteq \mathbb{C}[u] \otimes \mathbb{C}_q\langle a, a^*\rangle$. A possible choice is $u = a^*a - qaa^*$. In fact, $\Delta u^n = \sum_{\nu=0}^{n} \begin{pmatrix} n \\ \nu \end{pmatrix} u^\nu (u')^{n-\nu}$, so that u actually generates a Hopf subalgebra. But, since this Hopf subalgebra is isomorphic to the Hopf algebra of polynomials in $\mathbb{R}$, we find exactly the classical Lévy processes whose moments are finite.

Let us now look at other generators. Since no algebraic relations are imposed on a and a^*, we can take any operator X acting on some Hilbert space $\mathcal{H}$ to define a representation ρ_X of $\mathbb{C}_q\langle a, a^*\rangle$, simply represent a by X and a^* by its adjoint. To get a positive functional on $\mathbb{C}_q\langle a, a^*\rangle$ we now simply fix an element $h \in \mathcal{H}$ and set $\tilde{\psi}_{h,X}(u) = < h, \rho_X(u)h >_{\mathcal{H}}$. Then $\psi_{h,X} = \tilde{\psi}_{h,X} - \tilde{\psi}_{h,X}(1)\varepsilon$ is conditionally positive, and, if $\psi_{h,X}$ also satisfies the invariance condition, then there exists a Lévy process with generator $\psi_{h,X}$. The invariance condition in this case means simply that $\psi_{h,X}$ vanishes on 'words' that do not have the same number of a's and a^*'s. Let $\mathcal{H} = \mathbb{C}^2$,

$$X_\alpha = \begin{pmatrix} 0 & \alpha \\ 0 & 0 \end{pmatrix}, \qquad h_1 = \begin{pmatrix} 1 \\ 0 \end{pmatrix}, \qquad h_2 = \begin{pmatrix} 0 \\ 1 \end{pmatrix},$$

then $\psi_{i,\alpha} = \psi_{h_i,X_\alpha}$ are generators of Lévy processes on $\mathbb{C}_q\langle a, a^*\rangle$. One verifies that they can be written as

$$\begin{aligned} \psi_{1,\alpha} &= |\alpha|^2 xx^* + |\alpha|^4 xx^*xx^* + |\alpha|^6 xx^*xx^*xx^* + \cdots \\ \psi_{2,\alpha} &= |\alpha|^2 x^*x + |\alpha|^4 x^*xx^*x + |\alpha|^6 x^*xx^*xx^*x + \cdots, \end{aligned}$$

i.e. $\psi_{1,\alpha}$ ($\psi_{2,\alpha}$, resp.) is the sum over all concatenations of xx^* (x^*x, resp.) with itself, with coefficient $|\alpha|^l$, where l is the length of the 'word'. The functional $\psi_{1,\alpha}$, e.g., gives zero on all words in a, a^*, except for those that start with an a, never have two identical letters next to each other, and have an even number of letters. On those words it gives

$$\psi_{1,\alpha}(\underbrace{aa^*a\cdots a^*aa^*}_{l \text{ letters}}) = |\alpha|^l.$$

Introducing

$$\begin{aligned}\phi_{1,\alpha} &= |\alpha|^2 x^* + |\alpha|^4 x^* x x^* + |\alpha|^6 x^* x x^* x x^* + \cdots = |\alpha|^2 x^*(1+\tilde{\psi}_{1,\alpha})\\ \phi_{2,\alpha} &= |\alpha|^2 x + |\alpha|^4 x x^* x + |\alpha|^6 x x^* x x^* x + \cdots = |\alpha|^2 x(1+\tilde{\psi}_{2,\alpha}),\end{aligned}$$

(Concatenation!) we can state the Leibniz formulas that allow to determine the action of $\psi_{1,\alpha}$ and $\psi_{2,\alpha}$ on polynomials in $z = a + a^*$.

Proposition 5.2.3 (Leibniz formula) *Let $f(z) = \sum_{k=0}^n f_k z^k$. We have*

$$\begin{aligned}\rho(\psi_{1,\alpha})(zf(z)) &= \rho(\phi_{1,\alpha})f(z) + z\rho(\psi_{1,\alpha})f(z)\\ \rho(\phi_{1,\alpha})(zf(z)) &= |\alpha|^2\rho(1+\psi_{1,\alpha})f(z) + q^{-1}z\rho(\phi_{1,\alpha})f(z)\\ \rho(\psi_{2,\alpha})(zf(z)) &= \rho(\phi_{2,\alpha})f(z) + z\rho(\psi_{2,\alpha})f(z)\\ \rho(\phi_{2,\alpha})(zf(z)) &= |\alpha|^2\rho(1+\psi_{2,\alpha})f(z) + qz\rho(\phi_{2,\alpha})f(z)\end{aligned}$$

Proof: Similarly as in the proof of Proposition 5.2.1. We only need the first two terms of $\Delta(\psi_{i,\alpha})$, since the others vanish when applied to 1 or z. ■

We set $P_n = \rho(\psi_{1,\alpha})z^n$, and combine the recurrence relations above to

$$P_{n+1}(z) = (1+q^{-1})zP_n(z) + (|\alpha|^2 - q^{-1}z^2)P_{n-1} + |\alpha|^2 z^{n-1},$$

for $n \geq 1$, $P_1(z) = P_0(z) = 0$. $Q_n = \rho(\psi_{2,\alpha})z^n$ satisfies the same relation with q instead of q^{-1}.

We use the generating function

$$G(t,z) = \sum_{n=0}^{\infty} t^n Q_n(z) = \sum_{n=0}^{\infty} \rho(\psi_{2,\alpha})z^n,$$

this is (formally) the image of $\frac{1}{1-zt}$ under $\rho(\psi_{2,\alpha})$. Using the recurrence relation we get

$$G(t,z) = (1+q)tzG(t,z) + (|\alpha|^2 - qz^2)t^2G(t,z) + \frac{|\alpha|^2t^2}{1-zt},$$

and thus

$$\begin{aligned}G(t,z) &= \frac{|\alpha|^2t^2}{(1-(1+q)tz - (|\alpha|^2-qz^2)t^2)(1-zt)}\\ &= \frac{A_1}{1-z_1t} + \frac{A_2}{1-z_2t} - \frac{1}{1-zt}\end{aligned}$$

where

$$\begin{aligned}z_1 &= \tfrac{1}{2}\left((1+q)z + \sqrt{(1-q)^2z^2+4|\alpha|^2}\right), & A_1 &= \tfrac{1}{2}\frac{(1-q)z+\sqrt{(1-q)^2z^2+4|\alpha|^2}}{\sqrt{(1-q)^2z^2+4|\alpha|^2}},\\ z_2 &= \tfrac{1}{2}\left((1+q)z - \sqrt{(1-q)^2z^2+4|\alpha|^2}\right), & A_2 &= \tfrac{1}{2}\frac{-(1-q)z+\sqrt{(1-q)^2z^2+4|\alpha|^2}}{\sqrt{(1-q)^2z^2+4|\alpha|^2}},\end{aligned}$$

We thus have the following result.

Theorem 5.2.4 *The generator $\rho(\psi_{2,\alpha})$ acts on $\mathbb{C}[z]$ as*

$$Lf(z) = A_1 f(z+\Delta_1) + A_2 f(z+\Delta_2) - f(z)$$

where

$$\begin{aligned}\Delta_1 &= \tfrac{1}{2}\left((q-1)z + \sqrt{(1-q)^2z^2+4|\alpha|^2}\right), & A_1 &= \tfrac{1}{2}\frac{(1-q)z+\sqrt{(1-q)^2z^2+4|\alpha|^2}}{\sqrt{(1-q)^2z^2+4|\alpha|^2}}, \\ \Delta_2 &= \tfrac{1}{2}\left((q-1)z - \sqrt{(1-q)^2z^2+4|\alpha|^2}\right), & A_2 &= \tfrac{1}{2}\frac{-(1-q)z+\sqrt{(1-q)^2z^2+4|\alpha|^2}}{\sqrt{(1-q)^2z^2+4|\alpha|^2}}.\end{aligned}$$

The action of $\rho(\psi_{1,\alpha})$ is given by the same expressions, if we replace q by q^{-1}.

The two generators $\rho(\psi_{2,\alpha})$, $\rho(\psi_{1,\alpha})$ on $\mathbb{C}[z]$ are the generators of classical real-valued jump processes (or rather their restrictions to the algebra of polynomials). These classical processes therefore have the same moments as $(j_t(z))$ and are thus classical versions of $(j_t(z))$.

Remarks:

1. For $q = 1$ this generator is simply $Lf(z) = \frac{1}{2}\{f(z+|\alpha|) + f(z-|\alpha|) - 2f(z)\}$, and the process is a difference of two independent Poisson processes with jumps of size $|\alpha|$ and same intensity, $Z_t = \hat{N}_t = N_t^{(1)} - N_t^{(2)}$. From these we can build any symmetric compound Poisson process by taking the generator $L_\mu = \int_{\mathbb{R}} \psi_{1,\alpha} d\mu(\alpha)$, and thus we have 'q-analogues' of symmetric compound Poisson processes at hand.

2. Another special case where the description of Z_t becomes a lot simpler is $q = -1$. In this case $Z^2 = [Z,Z]_t + 2\int_0^t Z_{s-}dZ_s = [\hat{N},\hat{N}]_t$ is a Poisson process, i.e. Z_t takes values in $\{\pm\sqrt{n}|\alpha|; n \in \mathbb{N}\}$. Jumps are only possible from $\sqrt{n}|\alpha|$ to $\pm\sqrt{n+1}|\alpha|$, the transition probabilities are given by $A_{1,1} = \frac{1}{2}\left(1 \pm \frac{\sqrt{n}}{\sqrt{n+1}}\right)$.

3. The jumps Δ_1, Δ_2 are the solutions of the quadratic equation

$$\Delta^2 = (q-1)z\Delta + |\alpha|^2,$$

this implies that Z_t is a solution of

$$[Z,Z]_t = (q-1)\int_0^t Z_{s-}dZ_s + [\hat{N},\hat{N}]_t,$$

i.e. a *structure equation* where t, which is the quadratic variation of a Brownian motion, has been replaced by the quadratic variation of a symmetric Poisson process.

5.3 EXAMPLES OF CLASSICAL VERSIONS OF LÉVY PROCESSES ON $U_q(sl(2))$

The Hopf algebra $U_q(sl(2))$ is generated by three elements e, k, k^{-1}, f with the relations [Dri87]

$$kk^{-1} = k^{-1}k = 1, \quad ke = qek, \quad fk = qkf, \quad ef - fe = \frac{k^2 - k^{-2}}{q - q^{-1}},$$
$$\Delta k = k \otimes k, \quad \Delta e = e \otimes k + k^{-1} \otimes e, \quad \Delta f = f \otimes k + k^{-1} \otimes f,$$
$$S(k) = k^{-1}, \quad S(q) = -qe, \quad S(f) = -q^{-1}f, \quad \varepsilon(k) = 1, \quad \varepsilon(e) = \varepsilon(f) = 0,$$

where q is a complex parameter, $q \neq 0, \pm 1$. The Casimir element of $U_q(sl(2))$ is

$$C = \frac{qk^2 + q^{-1}k^{-2} - 2}{(q - q^{-1})^2} + fe,$$

it is a generator of the center of $U_q(sl(2))$.

The dual of $U_q(sl(2))$ is generated by four functionals x, y, u, v which we will write as a matrix $X = \begin{pmatrix} x & u \\ v & y \end{pmatrix} : U_q(sl(2)) \to \mathbb{C}^{2\times 2}$, see [MMN+90a]. On the generators they are defined by

$$X(k) = \begin{pmatrix} x(k) & u(k) \\ v(k) & y(k) \end{pmatrix} = \begin{pmatrix} q^{1/2} & 0 \\ 0 & q^{-1/2} \end{pmatrix},$$
$$X(e) = \begin{pmatrix} x(e) & u(e) \\ v(e) & y(e) \end{pmatrix} = \begin{pmatrix} 0 & 1 \\ 0 & 0 \end{pmatrix},$$
$$X(f) = \begin{pmatrix} x(f) & u(f) \\ v(f) & y(f) \end{pmatrix} = \begin{pmatrix} 0 & 0 \\ 1 & 0 \end{pmatrix},$$

and extended to general elements by

$$X(ab) = X(a)X(b) \quad \text{for } a, b \in U_q(sl(2)).$$

This implies that the *Leibniz formula* for the right dual representation reads

$$\rho(X)(ab) = \Big(\rho(X)a\Big)\Big(\rho(X)b\Big) \quad \text{for } a, b \in U_q(sl(2))$$

in the same matrix notation. Therefore, to calculate the right dual action of x, y, u, v on polynomials in some element $z \in U_q(sl(2))$ it is sufficient to calculate $\rho(X)z$, from this follows immediately

$$\rho(X)p(z) = p(\rho(X)z), \quad \text{for } p(z) \in \mathbb{C}[z] \subset U_q(sl(2)).$$

There exist three inequivalent $*$-structures on $U_q(sl(2))$.

Proposition 5.3.1 *[MMN+ 90a, Proposition 3] Real forms of $U_q(sl(2))$ are classified as follows:*

- $U_q(su(2))$ *with* $q \in (-1,0) \cup (0,1)$ *(the compact real form): the involution is defined by*
$$k^* = k, \quad e^* = f, \quad f^* = e.$$
- $U_q(su(1,1))$ *with* $q \in (-1,0) \cup (0,1)$ *(a non-compact real form): the involution is defined by*
$$k^* = k, \quad e^* = -f, \quad f^* = -e.$$
- $U_q(sl(2,\mathbb{R}))$ *with* $|q| = 1$ *(a non-compact real form): the involution is defined by*
$$k^* = k, \quad e^* = -e, \quad f^* = -f.$$

5.3.1 *The compact real form* $U_q(su(2))$

Let us first consider the compact real form $U_q(su(2))$. The element $z = ek + kf$ is self-adjoint (i.e. $z^* = z$), and satisfies Schürmann's condition (see Theorem 5.1.1), since $\Delta z = z \otimes k^2 + 1 \otimes z$.

For the dual representation one obtains

$$\rho(X)z = \begin{pmatrix} qz & q^{-1/2} \\ q^{-1/2} & q^{-1}z \end{pmatrix}$$

and therefore

$$\rho(X)p(z) = p\begin{pmatrix} qz & q^{-1/2} \\ q^{-1/2} & q^{-1}z \end{pmatrix}$$

for all polynomials $p(z) \in \mathbb{C}[z]$. Since the functionals x, y, u, v generate the dual of $U_q(su(2))$ this implies that $\mathbb{C}[z] \subseteq U_q(su(2))$ is a right coideal and that $(j_t|_{\mathbb{C}[z]})$ is a quantum Markov process for any Lévy process (j_t) on $U_q(su(2))$.

Using $\frac{1}{1-\lambda z}$ as a generating function, we can get a more useful expression for $\rho(X)|_{\mathbb{C}[z]}$.

$$\begin{aligned}
\rho(X)\frac{1}{1-\lambda z} &= \left\{1 - \lambda\begin{pmatrix} qz & q^{-1/2} \\ q^{-1/2} & q^{-1}z \end{pmatrix}\right\}^{-1} \\
&= \frac{1}{1-(q+q^{-1})\lambda z + \lambda^2(z^2 - q^{-1})}\begin{pmatrix} 1 - q^{-1}\lambda z & q^{-1/2}\lambda \\ q^{-1/2}\lambda & 1 - q\lambda z \end{pmatrix} \\
&= \frac{1}{1-\lambda z_1}\begin{pmatrix} \frac{(q^2-1)z+\sqrt{Q}}{2\sqrt{Q}} & \frac{\sqrt{q}}{\sqrt{Q}} \\ \frac{\sqrt{q}}{\sqrt{Q}} & \frac{(q^2-1)z+\sqrt{Q}}{2\sqrt{Q}} \end{pmatrix} \\
&\quad + \frac{1}{1-\lambda z_2}\begin{pmatrix} \frac{-(q^2-1)z+\sqrt{Q}}{2\sqrt{Q}} & -\frac{\sqrt{q}}{\sqrt{Q}} \\ -\frac{\sqrt{q}}{\sqrt{Q}} & \frac{-(q^2-1)z+\sqrt{Q}}{2\sqrt{Q}} \end{pmatrix},
\end{aligned}$$

where $Q = (q^2-1)^2z^2 + 4q$, and

$$z_1 = \frac{(q^2+1)z + \sqrt{Q}}{2q}, \quad z_2 = \frac{(q^2+1)z - \sqrt{Q}}{2q}.$$

For example for $q > 0$ and $L = y - \varepsilon$, we get again a jump process that tends to a symmetric Poisson process in the classical limit $q \to 1$, as in Theorem 5.2.4.

Another element that satisfies Schürmann's condition is the Casimir element since $C \otimes 1$ is in the center of $U_q(su(2)) \otimes U_q(su(2))$. Therefore $(j_t(C))_{t\in\mathbb{R}_+}$ is commutative and has a classical version for any Lévy process $(j_t)_{t\in\mathbb{R}_+}$ on $U_q(su(2))$. But $\mathbb{C}[C]$ is not a right coideal and $(j_t|_{\mathbb{C}[C]})$ will in general not be a quantum Markov process, as we can see from

$$\rho(X)C = \begin{pmatrix} qC + \frac{(q^{-2}-1)k^{-2}+2(q-1)}{(q-q^{-1})^2} & q^{-1/2}k^{-1}f \\ q^{-1/2}ek^{-1} & q^{-1}C + \frac{(q^2-1)k^{-2}+2(q^{-1}-1)}{(q-q^{-1})^2} \end{pmatrix},$$

since $\rho(X)C \not\subseteq M_2(\mathbb{C}[C])$.

But if we take, e.g., $L = y - \varepsilon$, then the restriction of the corresponding process to the commutative $*$-subalgebra $\mathbb{C}[C,k]$ is a quantum Markov process, and has thus a classical version. The action of L on $\mathbb{C}[C,k]$ is determined by

$$\rho(y)\frac{p(k)}{1-\lambda C} = \frac{p(q^{-1/2}k)\left\{1-\lambda\left(qC + \frac{(1-q^2)k^{-2}+2(q-1)}{(q-q^{-1})^2}\right)\right\}}{1-\lambda\left((q+q^{-1})C + 2\frac{(q^{1/2}-q^{-1/2})^2}{(q-q^{-1})^2}\right) + \lambda^2\left(C - \frac{(q^{1/2}-q^{-1/2})^2}{(q-q^{-1})^2}\right)^2} \tag{5.2}$$

for $p(k) \in \mathbb{C}[k]$. But the process $(j_t|_{\mathbb{C}[C,k]})_{t\in\mathbb{R}_+}$ is not commutative, since

$$[k\otimes 1, \Delta(C)] = (q^{-1}-1)f\otimes ke + (1-q^{-1})e\otimes fk \neq 0$$

for $q \neq 1$, and, furthermore,

$$\begin{aligned}
&\Phi\Big(j_s(f)[j_s(k), j_t(C)]j_{ts}(e)\Big)\\
= \;& \varphi_s \otimes \varphi_{t-s}\Big((q^{-1}-1)f^2\otimes ke^2 + (1-q^{-1})fe\otimes kfe\Big)\\
= \;& (1-q^{-1})\varepsilon^{\otimes 2}\left(e^{s\rho(y-\varepsilon)}(fe)\otimes e^{(t-s)\rho(y-\varepsilon)}(kfe)\right)\\
= \;& \frac{\left(e^{(q^{-2}-1)s} - e^{(q^{-1}-1)s}\right)\left(e^{(q^{-5/2}-1)(t-s)} - e^{(q^{-3/2}-1)(t-s)}\right)}{1-q^{-4}}\\
\neq \;& 0 \qquad (\text{for } q \in (0,1)).
\end{aligned}$$

So we have here an example of a classical version whose existence does not follow from Schürmann's criterium (Theorem 5.1.1). From Equation (5.2) we can explicitly construct it.

Theorem 5.3.2 *Let $(j_t)_{t\in[0,T]}$ be a Lévy process on $U_q(su(2))$ with generator $L = y - \varepsilon$ and let $(N_t)_{t\in[0,T]}$ be a standard Poisson process. Then the classical Markov process*

$$\left(\left(\frac{q^{(N_t+1)/2} - q^{-(N_t+1)/2}}{q-q^{-1}}\right)^2, q^{-N_t/2}\right)_{t\in[0,T]}$$

is a classical version of $(j_t(C), j_t(k))_{t\in[0,T]}$.

Proof: From Equation (5.2) we deduce that $\rho(L)$ acts on polynomials in C and k as

$$(\rho(L)f)(C,k) = A_1 f(c_1, q^{-1/2}k) + A_2 f(c_2, q^{-1/2}k) - f(C,k),$$

for $f(C,k) \in \mathbb{C}[C,k]$, where

$$\begin{aligned} c_1 &= \frac{C(q^2+1)(q+1)^2 + 2q^2 + (q+1)^2\sqrt{C((q^2-1)^2C+4q^2)}}{2q(q+1)^2}, \\ c_2 &= \frac{C(q^2+1)(q+1)^2 + 2q^2 - (q+1)^2\sqrt{C((q^2-1)^2C+4q^2)}}{2q(q+1)^2}, \\ A_1 &= \frac{(1-q^2)C + 2\frac{1-qk^{-2}}{q^{-2}-1} + \sqrt{C((q^2-1)^2C+4q^2)}}{\sqrt{C((q^2-1)^2C+4q^2)}}, \\ A_2 &= \frac{-(1-q^2)C - 2\frac{1-qk^{-2}}{q^{-2}-1} + \sqrt{C((q^2-1)^2C+4q^2)}}{\sqrt{C((q^2-1)^2C+4q^2)}}. \end{aligned}$$

A Markov process $(\hat{C}_t, \hat{k}_t)$ is a classical version of $(j_t(C), j_t(k))$, if $(\hat{C}_0, \hat{k}_0) = (\varepsilon(C), \varepsilon(k))$ and if its generator acts on polynomials in the same way as $\rho(L)$, so we see that we have to take a jump process starting from

$$\left(\left(\frac{q^{1/2}-q^{-1/2}}{q-q^{-1}}\right)^2, 1\right).$$

We calculate that for $C = \left(\frac{q^{(n+1)/2}-q^{-(n+1)/2}}{q-q^{-1}}\right)^2$ we get $c_1 = \left(\frac{q^{(n+2)/2}-q^{-(n+2)/2}}{q-q^{-1}}\right)^2$ and $c_2 = \left(\frac{q^{n/2}-q^{-n/2}}{q-q^{-1}}\right)^2$. But, furthermore, it turns out that $A_2 = 0$ and $A_1 = 1$ for

$$(c,k) = \left(\left(\frac{q^{(n+1)/2}-q^{-(n+1)/2}}{q-q^{-1}}\right)^2, q^{-n/2}\right),$$

so that the probability to jump to c_2 is zero. ∎

This actually allows to characterise the representation j_t (for fixed t).

Corollary 5.3.3 *The non-commutative random variable j_t (in the state Φ) is equivalent to the representation*

$$\pi = \sum_{\nu=0}^{\infty} \pi_{\nu/2} \quad on \quad H = \bigoplus_{\nu=0}^{\infty} D_{\nu/2}$$

in the vacuum vector

$$\Omega_t = \sum_{nu=0}^{\infty} \omega_\nu(t)|\nu/2, -\nu/2\rangle,$$

with $|\omega_\nu(t)|^2 = P(N_t = \nu) = \frac{t^\nu}{\nu!}e^{-t}$, where $(\pi_{\nu/2}, D_{\nu/2})$ denotes the unitary irreducible $\nu+1$ dimensional representation of $U_q(su(2))$ and $|\nu/2, -\nu/2\rangle$ the lowest weight vector of $D_{\nu/2}$.

5.3.2 *The real form $U_q(su(1,1))$*

The element z is not self-adjoint for the $*$-structure of $U_q(su(1,1))$, but we can take, e.g., $\bar{z} = ek - kf$ instead. We get

$$\rho(X)\bar{z} = \begin{pmatrix} q\bar{z} & q^{-1/2} \\ -q^{-1/2} & q^{-1}\bar{z} \end{pmatrix},$$

and therefore $\mathbb{C}[\bar{z}]$ is also a right coideal.

To get conditionally positive functionals on $U_q(su(1,1))$ and to express them in terms of x, y, u, v the formulas for the matrix elements of the irreducible unitary representation of $U_q(su(1,1))$ given in [MMN+90b] can be used. Take for example $w_{00}^{(l)}$ with $0 < l < \frac{1}{2}$, as a diagonal element of a unitary representation (see [MMN+90b, Proposition 4]), it is a positive functional. Furthermore we can see that it converges weakly to ε as l goes to 0. Therefore $L = \lim_{l\searrow 0} \frac{w_{00}^{(l)} - \varepsilon}{l}$ is a conditionally positive functional. From [MMN+90b, Proposition 3] we get the following expression for L,

$$L = \sum_{n=1}^{\infty} \frac{2h(uv)^n}{q^n - q^{-n}},$$

where $h = \ln(q)$. To calculate $\rho(uv)$ we introduce

$$X^{(2)} = \begin{pmatrix} x^2 & xu & ux & u^2 \\ xv & xy & uv & uy \\ vx & vu & yx & yu \\ v^2 & vy & yv & y^2 \end{pmatrix},$$

on $\bar{z}$ we get

$$\rho(X^{(2)})\bar{z} = \begin{pmatrix} q^2\bar{z} & q^{1/2} & q^{-1/2} & 0 \\ q^{1/2} & \bar{z} & 0 & q^{-1/2} \\ q^{-1/2} & 0 & \bar{z} & q^{-3/2} \\ 0 & q^{-1/2} & q^{-3/2} & q^{-2}\bar{z} \end{pmatrix},$$

this allows to calculate $\rho(X^{(2)})p(\bar{z}) = p\left(\rho(X^{(2)})\bar{z}\right)$ for arbitrary polynomials $p(\bar{z}) \in \mathbb{C}[\bar{z}]$. $\rho(uv)$ is characterised by

$$\begin{aligned} &\rho(uv)\frac{1}{1-\lambda\bar{z}} \\ &= \frac{(1+q^{-2})\lambda^2}{(1-\lambda\bar{z})(1-(q^2+q^{-2})\lambda\bar{z}+\lambda^2(\bar{z}^2-q^{-1}(q+q^{-1})^2))} \\ &= \frac{q(q^2+1)}{\bar{z}^2q(q^2-1)^2+(q^2+1)^2}\left\{\frac{-(q^2-1)^2\bar{z}+\sqrt{S}}{2\sqrt{S}(1-\lambda\bar{z}_1)} + \frac{(q^2-1)^2\bar{z}+\sqrt{S}}{2\sqrt{S}(1-\lambda\bar{z}_2)} - \frac{1}{1-\lambda\bar{z}}\right\} \end{aligned}$$

where $S = (q^4-1)^2\bar{z}^2 + 4q(q^2+1)^2$ and

$$\bar{z}_1 = \frac{(q^4+1)\bar{z}+\sqrt{S}}{2q^2}, \quad \bar{z}_2 = \frac{(q^4+1)\bar{z}-\sqrt{S}}{2q^2}.$$

We see that $\rho(uv)$ is a difference operator where the differences depend on $\bar{z}$ and q. The action of L on polynomials can be written in the form

$$Lf(\bar{z}) = \sum_{n=0}^{\infty}\sum_{k=1}^{2^n} A_{nk} f(\bar{z}_{nk})$$

where the $\bar{z}_{nk}$ can be obtained by iterating $\bar{z}_1$ and $\bar{z}_2$, i.e. set $\bar{z}_{01} = \bar{z}$, $\bar{z}_{11} = \bar{z}_1(\bar{z}) = \bar{z}_1 = \frac{(q^4+1)\bar{z}+\sqrt{S}}{2q^2}$, $\bar{z}_{12} = \bar{z}_2(\bar{z}) = \frac{(q^4+1)\bar{z}-\sqrt{S}}{2q^2}$, $\bar{z}_{21} = \bar{z}_1(\bar{z}_{11})$, $\bar{z}_{22} = \bar{z}_2(\bar{z}_{11})$, $\bar{z}_{23} = \bar{z}_1(\bar{z}_{12})$, $\bar{z}_{24} = \bar{z}_2(\bar{z}_{12})$, and so forth. A classical version is given by the process (starting from 0) that jumps after an exponential time given by $-A_{01} = \sum_{n=1}^{\infty}\sum_{k=1}^{2^n} A_{nk}$ to one of the $\bar{z}_{nk}$ (with probability $-A_{nk}/A_{01}$).

For $q \to 1$ we get $\rho(uv)p(\bar{z}) = \frac{1}{4}\{p(\bar{z}+2) + p(\bar{z}-2) - 2p(\bar{z})\}$, and, using $e^{\lambda\bar{z}}$ as a generating function, we get for $\rho(L)$,

$$\begin{aligned}\rho(L)e^{\lambda\bar{z}} &= \sum_{n=1}^{\infty}\frac{(e^{2\lambda}+e^{-2\lambda}-2)^n}{4^n n}e^{\lambda\bar{z}} = e^{\lambda\bar{z}}\sum_{n=1}^{\infty}\sum_{\nu=0}^{2n}\frac{(-1)^\nu}{4^n n}\begin{pmatrix}2n\\ \nu\end{pmatrix}e^{2\lambda(n-\nu)}\\ &= -\ln(1-\sinh^2\lambda)e^{\lambda\bar{z}} = -\ln\left(\frac{3-\cosh(2\lambda)}{2}\right)e^{\lambda\bar{z}},\end{aligned}$$

from which we get

$$\rho(L)p(x) = \sum_{k=1}^{\infty}\frac{p(x+2k)+p(x-2k)}{(3+2\sqrt{2})^k k} - \log\left(\frac{3+2\sqrt{2}}{4}\right)p(x),$$

i.e. the process

$$\sum_{k\in\mathbb{Z},k\neq 0} 2kN^{(k)}_{t/[|k|(3+2\sqrt{2})^{|k|}]}$$

is a classical version of $j_t(\bar{z})$, if the $N_t^{(k)}$, $k \in \mathbb{Z}\backslash\{0\}$, are mutually independent Poisson processes. The marginal law of $\left(j_t(\bar{z})\right)_{t\in\mathbb{R}_+}$ (for $q=1$) is uniquely determined by its generating function

$$\Phi\left(e^{\lambda j_t(\bar{z})}\right) = \varepsilon\circ e^{t\rho(L)}\left(e^{\lambda\bar{z}}\right) = \left(\frac{2}{3-\cosh(2\lambda)}\right)^t.$$

5.4 LEVY PROCESSES ON $U_q(aff(1))$

We will now consider an example where the parameter q in not real, but of modulus 1, and not a root of unity. Instead of $U_q(sl(2,\mathbb{R}))$ we take a simpler Hopf algebra with only two generators that can also be obtained as a sub-Hopf algebra of $U_q(sl(2,\mathbb{R}))$, namely $U_q(aff(1))$. Recall this Hopf $*$-algebra is generated by two elements X, Y with the relations

$$\begin{aligned} XY - YX = Y, &\qquad X^* = -X, \quad Y^* = -Y\\ \Delta(X) = X\otimes 1 + 1\otimes X, &\qquad \Delta(Y) = Y\otimes q^X + 1\otimes Y\\ \varepsilon(X) = \varepsilon(Y) = 0, &\qquad S(X) = -X, \quad S(Y) = Yq^{-X},\end{aligned}$$

with $q \in \mathbb{C}$, $|q| = 1$. Let $h \in \mathbb{R}$ be such that $q = e^{ih}$. Positive functionals can be obtained from the trivial one-dimensional representation $X \mapsto ix$, $Y \mapsto 0$, for $x \in \mathbb{R}$, i.e. on the elements $\varphi_x : Y^m X^n \mapsto \delta_{m,0}(ix)^n$. Differentiating φ_x with respect to x at $x = 0$ we get the two conditionally positive functionals $\varphi^{(1)}$: $Y^m X^n \mapsto i\delta_{m,0}\delta_{n,1}$ and $\varphi^{(1)} : Y^m X^n \mapsto -\delta_{m,0}\delta_{n,2}$. But these functionals do not give any interesting processes because we always get $j_t(Y) = 0$.

Let X_1 and X_2 be the operators on $l^2(\mathbb{N})$ defined by

$$\begin{aligned} (X_1 u)_n &= n u_n + \sqrt{n(n+1)}\, u_{n+1} \\ (X_2 u)_n &= n u_n + \sqrt{n(n-1)}\, u_{n-1}. \end{aligned}$$

They are closed on their maximal domains and satisfy $[X_1, X_2] = X_1 + X_2$ and $X_1^* = X_2$, see [PS91].

If we set $X = \frac{1}{2}(X_1 - X_2)$ and $Y = i(X_1 + X_2)$, then this defines a $*$-representation of $U_q(aff(1))$. Let us compute the 'vacuum expectation' of $Y^n X^m$ in the state $\Omega = u_1 = (1, 0, \ldots)$.

Lemma 5.4.1 *For α, β sufficiently small we have*

$$e^{\beta Y} e^{\alpha X} = e^{b X_2} e^{a X_1}$$

with

$$a = -\log\left(\frac{e^\alpha + 1}{2} - i\beta\right) + \alpha, \qquad b = -\log\left(\frac{e^\alpha + 1}{2} - i\beta\right)$$

Proof: $\{X_1, X_2\}$ and $\{X, Y\}$ both form a basis of the Lie algebra $aff(1)$, so that the preceding formula is just a coordinate change for a neighbourhood of the unit element in the Lie group $Aff(1)$. ■

With this lemma we obtain

$$\begin{aligned} \langle e^{\beta Y} e^{\alpha X} \rangle &= \langle e_1, e^{b X_2} e^{a X_1} e_1 \rangle = \langle e^{\bar{b} X_2} e_1, e^{a X_1} e_1 \rangle \\ &= e^{a+b}, \end{aligned}$$

from which we can deduce

$$\langle Y^m e^{\alpha X} \rangle = (m+1)! \frac{(i)^m}{\cosh^{m+2}(\alpha/2)}, \tag{5.3}$$

which makes sense for $\alpha \in \mathbb{C}$, $\alpha \neq (2k+1)2i\pi$ for $k \in \mathbb{Z}$.

We can now define a conditionally positive functional on $U_q(aff(1))$ by

$$L(u) = \langle u \rangle - \varepsilon(u),$$

if q is not a root of unity.

The restrictions of this process to the subalgebras generated by X or Y turn out to be commutative and Markovian, so that there exist classical versions of $j_t(iX)$ and $j_t(iY)$, respectively.

Proposition 5.4.2 *Let $(\hat{X})_{t\geq 0}$ and $(\hat{Y})_{t\geq 0}$ be classical versions of $(j_t(iX))_{t\geq 0}$ and $(j_t(iY))_{t\geq 0}$, respectively. Then the action of the generators L_X and L_Y of $(\hat{X})_{t\geq 0}$ and $(\hat{Y})_{t\geq 0}$ on polynomials is determined by*

$$\begin{aligned} L_X e^{iux} &= -\tanh^2(u/2)e^{iux}, \\ L_Y y^n &= \sum_{\nu=0}^{n} \frac{(-1)^{n-\nu}(n-\nu+1)! \left[\begin{array}{c} n \\ \nu \end{array} \right]_q}{\cos^{n-\nu+2}\left(\frac{\nu h}{2}\right)} y^\nu - y^n, \end{aligned}$$

($u \in \mathbb{R}$, $n \in \mathbb{N}$).

Proof: This follows directly from

$$\Delta Y^m X^n = \sum_{\mu=0}^{m}\sum_{\nu=0}^{n} \left[\begin{array}{c} m \\ \mu \end{array} \right]_q \binom{n}{\nu} Y^\mu X^\nu \otimes Y^{m-\mu}X^{n-\nu}q^{\nu X},$$

and Equation (5.3). ∎

Remark: In the limit $h \to 0$ we obtain for L_Y,

$$L_Y(e^{iuy}) = \left\{ \frac{1}{(1-iu)^2} - 1 \right\} e^{iuy}.$$

REFERENCES

This chapter is based on the article [Fra99]. For an introduction to the theory of quantum Markov processes see, e.g., [Küm88]. Biane has shown that additive and multiplicative free Lévy processes are also Markov processes, see [Bia98], but no such results exist for more general free and boolean Lévy processes (in the sense of [Sch95b]).

Chapter 6

Diffusions on braided spaces

The applications of diffusions in physics go far beyond the description of the physical phenomenon they are named after. Functional integrals can be used to solve partial differential equations, cf. the celebrated Feynman-Kac formula. Wiener integrals are very close to Feynman path integrals. Another interesting application of diffusions is the stochastic mechanics of Nelson [Nel66] (see also [Nel88]).

Diffusions on manifolds are characterised by two properties, the first being their Markov property, i.e. that at every instant t they start again, and their evolution does not depend on their history, but only on their distribution at time t. This gives rise to a semi-group and this property also plays a central role for the study of diffusions on braided spaces. The other property, i.e. that they have continuous sample paths, does not have a direct counterpart on braided spaces. It is replaced by the condition that diffusions can be obtained as the limit of a (simple) random walk.

The motivations for this approach come from two directions. First, there is Majid's random walk approach to Brownian motion on the braided line and on any-spaces [Maj93d, MRP94]. In [FS98a] his definition is extended to (pseudo-) diffusions on multi-dimensional braided spaces and, using coalgebraic limit theorems due to M. Schürmann [Sch93], their explicit form is given. This allows to consider semi-groups of functionals and Markovian transition operators, as well as the associated heat equations, and to introduce Appell systems as their polynomial solutions. Heat kernels, i.e. the densities of the functionals are also considered.

The second ingredient, M. Schürmann's theory of quantum Lévy processes, comes into play to assure the existence of the associated processes, e.g. as operators on a Hilbert space. For this is was necessary to use a definition of $*$-structures for braided Hopf algebras that differs from the one due to S. Majid.

To distinguish the two levels of the constructions, we call the functionals and semi-groups obtained via S. Majid's construction pseudo-diffusions, and reserve the name diffusion for the case where the processes can be realised as operators on a Hilbert space and are Lévy processes (i.e. independent increment processes) in the sense of M. Schürmann.

Section 6.1 shows how (pseudo-) diffusion processes can be constructed on generic braided spaces associated to a pair of matrices (R, R').

Section 6.2 investigates the relation between the diffusion processes and associated evolution equations. These equations are of the form $(\partial_t - L)u = 0$, where L is an operator consisting of linear and quadratic terms in the braided-partials ∂^i. Appell systems (introduced as shifted moment polynomials) are shown to be solutions of the evolution equation. We investigate their properties.

In Section 6.3 we give a brief discussion of how density functions can be introduced.

6.1 A CONSTRUCTION OF (PSEUDO-) DIFFUSIONS ON BRAIDED SPACES

Following Majid we first define a quantum random walk. Consider a classical random walk on the lattice. We assume that the process can only jump from a lattice site x to a finite number of other sites $x + d_i$, $i = 1, \ldots, k$. If at time t the particle is located at x, then it will be at $x + d_i$ at time $t + \Delta t$ with a probability p_i, $0 \leq p_i \leq 1$, and naturally the probabilities add to one,

$$\sum_{i=1}^{k} p_i = 1.$$

In classical probability theory there corresponds an evolution operator to this, that maps a function (i.e. an observable) at time t to the corresponding function one time step later

$$f_t(x) \mapsto f_{t+\Delta t}(x) = \sum_{i=1}^{k} p_i f_t(x + d_i)$$

Let $\phi^{(d_i)}$ be the functional corresponding to evaluation of a function at d_i, i.e. $\phi^{(d_i)} : f(x) \mapsto f(d_i)$, then the above equation can be written as

$$f_t(x) \mapsto f_{t+\Delta t}(x) = (\phi \otimes \mathrm{id}) \circ \Delta f_t(x)$$

where $\phi = \sum p_i \phi^{(d_i)}$ and Δ stands for the usual cocommutative coproduct on $\mathbb{R}[x_1, \ldots, x_n]$. This can be done in the same manner on braided spaces and quantum groups, if one chooses an appropriate functional ϕ. The problem here (in dimension > 1) is that one can not define the $\phi^{(d_i)}$ simply as replacing the variables x_i by the components of the vectors $d_j = (d_{j1}, \ldots, d_{jn})$ due to the noncommutativity. One possibility is to bring the variables into some fixed order (i.e. fix a basis of monomials) and replace them then by the d_{ji}. We shall see later that in the limit only the choice on the quadratic and linear terms matters.

A functional $\phi^{(d)}$ that corresponds to evaluation at $d = (d_1, \ldots, d_n)$ should satisfy

F1: $\phi^{(d)}(1) = 1$, or equivalently $\phi^{(d)}|_{V^{\cdot}(R,R')^{(0)}} = \epsilon$,

F2: evaluation on the generators gives the components of d, i.e.

$$\phi^{(d)}(x_i) = d_i,$$

for $i = 1, \ldots, n$,

F3: it has the correct behaviour with respect to a rescaling of d, more precisely,

$$\phi^{(\lambda d)}(a) = \phi^{(d)}(s(\lambda)a) = \lambda^{\deg(a)}\phi^{(d)}(a)$$

for homogeneous $a \in V\check{}(R, R')$.

But these conditions do not determine $\phi^{(d)}$ uniquely.

A possible choice for $\phi^{(d)}$ with $d = (d_1, \cdots, d_n)$ is to use the braided-exponential

$$\phi^{(d)} = \exp(d|\mathbf{v}) = \sum_{a \in \mathcal{S}\{1,\ldots,n\}} d_a([m; R]!^{-1})^a_b \mathbf{v}^b. \tag{6.1}$$

This definition makes sense for all cases where the braided-integers $[m; R]$ are invertible, and can also be used if this is not the case, if an appropriate braided-exponential is chosen. For the commutative case we have $\phi^{(d)} f(\mathbf{x}) = f(d)$ by Taylor's theorem, and similarly for the braided line $\mathbb{R}_q$ by a q-Taylor theorem. On any-space this agrees with definition of the Dirac δ-function, see [MRP94].

Another possibility is to use (positive) Dirac functionals, defined as functionals φ for which there exists a ($*$-)algebra $\mathcal{A}$, a state Φ on $\mathcal{A}$, and a ($*$-)algebra homomorphism $j : V\check{}(R, R') \to \mathcal{A}$ such that $\varphi = \Phi \circ j$ and the pair (j, j) is independent, see Definition 8.1.17. Typically these Dirac functionals replace one generator by a real or complex number, and all other generators by zero, and lead to the same class of diffusions as the functional presented above.

Assume now that we have defined the functionals $\phi^{(d_i)}$, and chosen values for the transition probabilities p_i, i.e. fixed ϕ. Then this defines an evolution for functions, i.e. observables on our space by

$$f_t \mapsto f_{t+\Delta t} = (\phi \otimes \mathrm{id}) \circ \underline{\Delta} f_t$$

as before. As explained in [Maj93d] there is a one-to-one correspondence between functionals ϕ and transition operators T, to pass from one to the other take

$$\begin{aligned} \phi &\mapsto T_\phi = (\phi \otimes \mathrm{id}) \circ \underline{\Delta}, \\ T &\mapsto \phi_T = \epsilon \circ T. \end{aligned}$$

The transition operators form a (discrete) semi-group with respect to composition, and the functionals with respect to convolution. The passage from one to the other is in fact a homomorphism, i.e. $T_\phi \circ T_\psi = T_{\phi\psi}$ and $\phi_T \phi_S = \phi_{TS}$.

We study the limit that arises if the lattice parameters d_i and the time step Δt go to zero. The discrete semi-groups (indexed by $\mathbb{N}$) will then converge to continuous ones (indexed by $\mathbb{R}_+$).

Theorem 6.1.1 *Let $d_1, \ldots, d_k$ and $p_1^0, \ldots, p_k^0$ be chosen such that $\sum_{i=1}^k p_i^0 = 1$ and $\phi = \sum_{i=1}^k p_i^0 \phi^{(d_i)}$ vanishes on $V^{\check{}}(R, R')^{(1)}$, i.e. $\sum_{i=1}^k p_i^0 d_{ij} = 0$ for all $j = 1, \ldots, n$.*

Choose furthermore $p_1^1, \ldots, p_k^1$ such that $p_1(\mu) = p_1^0 + \mu p_1^1, \ldots, p_k(\mu) = p_k^0 + \mu p_k^1$ satisfy $0 \leq p_i \leq 1$ and $\sum_{i=1}^k p_i = 1$ for sufficiently small μ (this implies in particular $\sum_{i=1}^k p_i^1 = 0$).

Then for $\phi_N = \sum_{i=1}^k p_i(1/\sqrt{N}) \phi^{(d_i/\sqrt{N})}$ we have

$$\lim_{N \to \infty} (\phi_N)^N = \exp\left(\sum_{k,l=1}^n a_{kl} v^k v^l + \sum_{j=1}^n b_i v^i\right)$$

where $a_{kl} = \sum_{i=1}^k p_i^0 \sum_{r,s=1}^n d_{ir} d_{is} ([2; R]^{-1})_{kl}^{rs}$ and $b_j = \sum_{i=1}^k p_i^1 d_{ij}$.

Proof: We apply [Sch93, Theorem 6.1.1]. Let $k_N = N$, $\varphi_{Nk} = \phi_N$ for $k = 1, \ldots, N$. We have to check the conditions of the theorem.

(i) $\varphi_{N1}, \ldots, \varphi_{NN}$ commute since they are identical.

(ii) Let $c = \sum_{a \in S\{1,\ldots,n\}} c^a \mathbf{x}_a$ with some coefficients $c^a \in \mathbb{C}$. Then

$$\max_{1 \leq k \leq N} |(\varphi_{Nk} - \epsilon)(c)| = \phi_N(c) - c^{\emptyset} = O(1/\sqrt{N})$$

goes to zero as $N \to \infty$.

(iii)

$$\begin{aligned} \sum_{k=1}^N (\varphi_{Nk} - \epsilon)(c) &= N(\phi_N(c) - c^{\emptyset}) \\ &= \sum_{i=1}^k p_i^1 \sum_{j=1}^n d_{ij} c^{\{j\}} + \sum_{i=1}^k p_i^0 \sum_{r,s=1}^n d_{ir} d_{is} c^{rs} + O(1/\sqrt{N}) \end{aligned}$$

is bounded and tends to $(\sum_{k,l=1}^n a_{kl} v^k v^l + \sum_{j=1}^n b_i v^i)(c)$.

Thus $\lim_{N \to \infty} \left(\prod_{k=1}^N \varphi_{Nk}\right)(c) = \exp(\sum_{k,l=1}^n a_{kl} v^k v^l + \sum_{j=1}^n b_i v^i)(c)$. ■

We will consider functionals that can be obtained in this way as diffusions on braided spaces and call $L = \sum_{k,l=1}^n a_{kl} v_k v_l + \sum_{j=1}^n b_i v^i$ their generator.

Definition 6.1.2 *We will call a semi-group of functionals* $\exp(tL)$ *on $V^{\check{}}(R, R')$ a (homogeneous)* pseudo-diffusion, *if there exist parameters $d_i = (d_{i1}, \ldots, d_{in})$, p_i^0, p_i^1 $(i = 1, \ldots, k)$ such that the conditions of Theorem 6.1.1 are satisfied, and such that $L = \sum_{k,l=1}^n a_{kl} v^k v^l + \sum_{j=1}^n b_i v^i$ with $a_{kl} = \sum_{i=1}^k p_i^0 \sum_{r,s=1}^n d_{ir} d_{is} ([2; R]^{-1})_{kl}^{rs}$ and $b_j = \sum_{i=1}^k p_i^1 d_{ij}$.*

If the process associated to $\exp(tL)$ *is also a Lévy process, then we will call it a* diffusion.

Remarks:

1. We have formulated the limit theorem in terms of the functionals, analogous theorems for the associated transition operators hold also.

2. In the statements of Theorem 6.1.1 we assumed that the braided-integers are invertible. The theorem remains valid if this is not the case, as long as a braided-exponential $\exp(\mathbf{x}|\mathbf{v}) = \sum \mathbf{x}_a F(m;R)^a_b \mathbf{v}^b$ exists, the inverted braided-exponentials just have to be replaced by the $F(m;R)$.

6.1.1 *Examples of (pseudo-) diffusions*

Dimension $n = 1$: The braided line $\mathbb{R}_q$

Here $R = (q)$ and $R' = (1)$. This leads to the braiding $\Psi(x^m \otimes x^p) = q^{mp} x^p \otimes x^m$ (i.e. $x'x = qxx'$, where $x = x \otimes 1$, $x' = 1 \otimes x$). The diffusions on the braided line have the form

$$\phi_t = \exp\left[t\left(\frac{a^2}{1+q}v^2 + bv\right)\right].$$

Using the expansion $e^{-tu^2/2+xu} = \sum_{p=0}^{\infty} \frac{u^p}{p!} H_p(x,t)$ with the Hermite polynomials $H_p(x,t) = \sum_{k=0}^{[\frac{p}{2}]} \begin{pmatrix} p \\ 2k \end{pmatrix} \frac{(2k)!}{2^k k!} x^{p-2k}(-t)^k$ we find for the moments

$$\phi_t\left(\frac{x^p}{[p;q]!}\right) = \frac{H_p\left(bt, -\frac{2a^2t}{1+q}\right)}{p!}.$$

Dimension $n = 2$: The quantum plane $\mathbb{C}_q^{2|0}$

Here

$$R = \begin{pmatrix} q^2 & 0 & 0 & 0 \\ 0 & q & q^2-1 & 0 \\ 0 & 0 & q & 0 \\ 0 & 0 & 0 & q^2 \end{pmatrix}, \qquad R' = q^{-2}R.$$

This is the standard two-dimensional quantum plane. It has two generators x, y with relations $yx = qxy$ and braid statistics $x'x = q^2xx'$, $x'y = qyx'$, $y'y = q^2yy'$, $y'x = qxy' + (q^2-1)yx'$. For the second braided-integer we get

$$[2;R] = \begin{pmatrix} 1+q^2 & 0 & 0 & 0 \\ 0 & 1 & q & 0 \\ 0 & q & q^2 & 0 \\ 0 & 0 & 0 & 1+q^2 \end{pmatrix},$$

i.e. it is not invertible. There exists nonetheless a braided-exponential [Maj96, Example 5.4]

$$\exp(\mathbf{x}|\mathbf{v}) = \sum_{a\in S\{1,\dots,n\}} \frac{\mathbf{x}_a \mathbf{v}^a}{[|a|;q^2]!} = \sum_{p=0}^{\infty} \frac{(\mathbf{x}\cdot\mathbf{v})^p}{[p;q^{-2}]!},$$

where $[m;q^2] = \frac{1-q^{2m}}{1-q^2}$ and $[m;q^2]! = \prod_{\mu=1}^{m}[\mu;q^2]$. For the generator we get

$$L = \frac{1}{1+q^2}\sum a_{ij}\partial_i\partial_j + \sum b_i\partial_i.$$

The free braided-space

Here $R' = P$, there are no relations in the algebras $V(R,R')$ and $V\check{}(R,R')$ since the ideal generated by $R'_{12}\mathbf{v}_2\mathbf{v}_1 = \mathbf{v}_1\mathbf{v}_2$ (respectively $\mathbf{x}_1\mathbf{x}_2R'_{12} = \mathbf{x}_2\mathbf{x}_1$) is equal to $\{0\}$, i.e. we have the free algebra with n generators. $\{\mathbf{v}^a; a \in \mathcal{S}\{1,\ldots,n\}\}$ (respectively $\{\mathbf{x}_a; a \in \mathcal{S}\{1,\ldots,n\}\}$) is thus a basis of $V(R,R')$ (respectively $V\check{}(R,R')$).

We will assume that R is also equal to P, i.e. that the braiding is given by $\Psi(v^i \otimes v^j) = (v^i \otimes v^j)$ (respectively $\Psi(x_i \otimes x_j) = (x_i \otimes x_j)$). In this case the braided-integers $[m;P]^a_b = m\delta_{ab}$, $m = |a| = |b|$, are invertible and thus

$$\exp(\mathbf{x}|\mathbf{v}) = \sum_{k=0}^{\infty}\frac{1}{k!}\sum_{a\in\mathcal{S}^k\{1,\ldots,n\}}\mathbf{x}_a\mathbf{v}^a,$$

The dual action is

$$\rho(v^i)\mathbf{x}_a = \begin{cases} |a|\,\mathbf{x}_{a_2\cdots a_{|a|}} & \text{if } a_1 = i, \\ 0 & \text{else} \end{cases}$$

on the basis elements. Generators of diffusions have again the form

$$L = \sum_{k,l=1}^{n} a_{kl}\partial_k\partial_l + \sum_{j=1}^{n} b_j\partial_j \tag{6.2}$$

with symmetric a_{kl}.

6.1.2 *Examples of true diffusions*

A free two-dimensional braided plane: $V\check{} = \mathbb{R}_q * \check{\mathbb{R}}_q$

The algebra underlying this example can be considered as the free product of the braided line with its dual. As an algebra it is the free unital algebra $\mathbb{C}\langle x,p\rangle$ with two generators. Let x be the generator of the braided-covector space $\mathbb{R}_q$, and p that of the braided-vector space dual to it (i.e. another copy of $\mathbb{R}_q$). Let $q \in \mathbb{R}\backslash\{0\}$, so that $R = (q)$ is of real type I in the sense of [Maj95a]. Following Majid's definitions we have the braid relations

$$\begin{aligned} x'x &= qxx \qquad & x'p &= q^{-1}px' \\ p'x &= q^{-1}xp' \qquad & p'p &= qpp', \end{aligned}$$

and

$$\begin{aligned} *_{\mathbb{R}_q\underline{\otimes}\check{\mathbb{R}}_q} &: \mathbb{R}_q\underline{\otimes}\check{\mathbb{R}}_q \to \mathbb{R}_q\underline{\otimes}\check{\mathbb{R}}_q, & x^* &= p \\ *_{\mathbb{R}_q\underline{\otimes}\check{\mathbb{R}}_q\underline{\otimes}\mathbb{R}_q\underline{\otimes}\check{\mathbb{R}}_q} &: \mathbb{R}_q\underline{\otimes}\check{\mathbb{R}}_q\underline{\otimes}\mathbb{R}_q\underline{\otimes}\check{\mathbb{R}}_q \to \mathbb{R}_q\underline{\otimes}\check{\mathbb{R}}_q\underline{\otimes}\mathbb{R}_q\underline{\otimes}\check{\mathbb{R}}_q, & x^* &= p',\ (x')^* = p, \end{aligned}$$

turn $\mathbb{R}_q\underline{\otimes}\check{\mathbb{R}}_q$ and $\mathbb{R}_q\underline{\otimes}\check{\mathbb{R}}_q\underline{\otimes}\mathbb{R}_q\underline{\otimes}\check{\mathbb{R}}_q$ in into $*$-algebras such that $\underline{\Delta} : \mathbb{R}_q\underline{\otimes}\check{\mathbb{R}}_q \to \mathbb{R}_q\underline{\otimes}\check{\mathbb{R}}_q\underline{\otimes}\mathbb{R}_q\underline{\otimes}\check{\mathbb{R}}_q$ satisfies $\underline{\Delta} \circ *_{\mathbb{R}_q\underline{\otimes}\check{\mathbb{R}}_q} = *_{\mathbb{R}_q\underline{\otimes}\check{\mathbb{R}}_q\underline{\otimes}\mathbb{R}_q\underline{\otimes}\check{\mathbb{R}}_q} \circ \tau \circ \underline{\Delta}$. Even though this $*$-structure does not suit our purposes, at least not in this form, it does tell how we can guess a 'better' one.

The coproduct on $\mathbb{R}_q$ and its dual extends to a map $\underline{\Delta}_0 : \mathbb{R}_q * \check{\mathbb{R}}_q \to (\mathbb{R}_q\underline{\otimes}\mathbb{R}_q) * (\check{\mathbb{R}}_q\underline{\otimes}\check{\mathbb{R}}_q) = \mathbb{C}\langle p,p',x,x'\rangle/\mathcal{I}_0$, where $\mathcal{I}_0$ is the ideal generated by $x'x - qxx'$ and $p'p - qpp'$, and satisfies $\underline{\Delta} \circ *_{\mathbb{R}_q * \check{\mathbb{R}}_q} = *_{(\mathbb{R}_q\underline{\otimes}\mathbb{R}_q)*(\check{\mathbb{R}}_q\underline{\otimes}\check{\mathbb{R}}_q)} \circ \underline{\Delta}$, if we define the $*$-structures in those algebras by

$$\begin{aligned}
*_{\mathbb{R}_q * \check{\mathbb{R}}_q} \quad &: \quad \mathbb{R}_q * \check{\mathbb{R}}_q \to \mathbb{R}_q * \check{\mathbb{R}}_q, \\
& \qquad x^* = p, \\
_{(\mathbb{R}_q\underline{\otimes}\mathbb{R}_q)(\check{\mathbb{R}}_q\underline{\otimes}\check{\mathbb{R}}_q)} \quad &: \quad (\mathbb{R}_q\underline{\otimes}\mathbb{R}_q) * (\check{\mathbb{R}}_q\underline{\otimes}\check{\mathbb{R}}_q) \to (\mathbb{R}_q\underline{\otimes}\mathbb{R}_q) * (\check{\mathbb{R}}_q\underline{\otimes}\check{\mathbb{R}}_q), \\
& \qquad x^* = p', \ (x')^* = p.
\end{aligned}$$

We now note that we get $(\mathbb{R}_q * \check{\mathbb{R}}_q)\underline{\otimes}(\mathbb{R}_q * \check{\mathbb{R}}_q)$ if we quotient by the relations $qxp = px$ and $qp'x' = x'p'$, i.e. $(\mathbb{R}_q * \check{\mathbb{R}}_q)\underline{\otimes}(\mathbb{R}_q * \check{\mathbb{R}}_q) = (\mathbb{R}_q\underline{\otimes}\mathbb{R}_q) * (\check{\mathbb{R}}_q\underline{\otimes}\check{\mathbb{R}}_q)/\{qxp = pxp, qp'x' = x'p'\} = \mathbb{C}\langle p',x,p,x'\rangle/\mathcal{I}$, where $\mathcal{I}$ is the ideal generated by $x'x - qxx'$, $qxp - px$, $qp'x' - x'p'$, and $p'p - qpp'$. The canonical projection $\pi : (\mathbb{R}_q\underline{\otimes}\mathbb{R}_q) * (\check{\mathbb{R}}_q\underline{\otimes}\check{\mathbb{R}}_q \to (\mathbb{R}_q * \check{\mathbb{R}}_q)\underline{\otimes}(\mathbb{R}_q * \check{\mathbb{R}}_q)$ is a $*$-algebra homomorphism, and $\underline{\Delta} = \pi \circ \underline{\Delta}_0$ turns $V^{\check{}} = \mathbb{R}_q * \check{\mathbb{R}}_q$ into a braided $*$-Hopf algebra in the sense we need. In particular, the coproduct is a $*$-algebra homomorphism. Note that we now have $p_1 = p'$ and $x_1 = x$ in the first factor of the tensor product, and $p_2 = p$ and $x_2 = x'$ in the second, i.e. the braid relations take the form

$$\begin{aligned}
x_2x_1 &= qx_1x_2, & p_2x_1 &= q^{-1}x_1p_2, \\
x_2p_1 &= qp_1x_2, & p_2p_1 &= q^{-1}p_1p_2
\end{aligned}$$

and the $*$-structure becomes $x_i^* = p_i$, $i = 1, 2$.

The braiding of the real line, and thus also that of $V^{\check{}}$, are induced by an action and coaction of the group algebra $\mathbb{C}\mathbb{Z}$. $V^{\check{}}$ is the (free) braided plane associated to the matrices

$$R' = \begin{pmatrix} 1 & 0 & 0 & 0 \\ 0 & 0 & 1 & 0 \\ 0 & 1 & 0 & 0 \\ 0 & 0 & 0 & 1 \end{pmatrix}, \qquad R = \begin{pmatrix} q & 0 & 0 & 0 \\ 0 & q^{-1} & 0 & 0 \\ 0 & 0 & q & 0 \\ 0 & 0 & 0 & q^{-1} \end{pmatrix}.$$

The possible generators of pseudo-diffusions are of the same form as in Equation (6.2), but now we are interested in generators of 'true' diffusions, i.e. Ψ-invariant, hermitian, conditionally positive generators. Choose the set of all words $\{1 = \emptyset, x, p, xx, xp, px, pp, \ldots\}$ as basis of $V^{\check{}}$, then ϕ_{xp} and ϕ_{px} with

$$\phi_{xp}(u) = \begin{cases} 1 & \text{if } u = xp, \\ 0 & \text{else,} \end{cases} \qquad \phi_{px}(u) = \begin{cases} 1 & \text{if } u = px, \\ 0 & \text{else,} \end{cases}$$

on basis elements u are such generators. The processes associated to these generators have been studied by M. Schürmann[Sch93], we find the processes that have the Azéma martingales as classical version.

Let us now consider an example whose braiding does not come from a group, but from a non-commutative and non-cocommutative bialgebra.

Diffusions on the four-dimensional braided-covector space $V^{\check{}} = \mathbb{C}_q^{2|0} * \check{\mathbb{C}}_q^{2|0}$

Let again q in $\mathbb{R}\backslash\{0\}$, and set $V^{\check{}} = \mathbb{C}_q^{2|0} * \check{\mathbb{C}}_q^{2|0} = \mathbb{C}\langle v,w,x,y\rangle/\{yx-qxy, vw-qwv\}$. A similar construction as in the preceding example leads to a four-dimensional braided-covector space, where the braid relations between x_1, y_1 and x_2, y_2 are described by the R-matrix of the braided plane,

$$R = \begin{pmatrix} q^2 & 0 & 0 & 0 \\ 0 & q & q^2-1 & 0 \\ 0 & 0 & q & 0 \\ 0 & 0 & 0 & q^2 \end{pmatrix},$$

those between x_1, y_1 and v_2, w_2 by its inverse,

$$R^{-1} = \begin{pmatrix} q^{-2} & 0 & 0 & 0 \\ 0 & q^{-1} & q^{-2}-1 & 0 \\ 0 & 0 & q^{-1} & 0 \\ 0 & 0 & 0 & q^{-2} \end{pmatrix},$$

those between v_1, w_2 and x_2, y_2 again by R, and, finally, those between v_2, w_2 again by R^{-1}. The $*$-structure is defined by $x^* = v$ and $y^* = w$, compare also [Maj95a].

$\{x^n y^m | n,m \in \mathbb{N}\}$ and $\{w^m v^n | n,m \in \mathbb{N}\}$ are bases of the two components of the free product defining $V^{\check{}}$, and therefore we can take the finite, alternating sequences of elements of these bases as a basis of $V^{\check{}}$:
$\{1=\emptyset, x^{n_1}y^{m_1}, w^{m_1}v^{n_1}, x^{n_1}y^{m_1}w^{m_2}v^{n_2}, \ldots | n_1, n_2, \ldots, m_1, \ldots \in \mathbb{N}\}$. The functionals defined on this basis by

$$\phi_{xv}(u) = \begin{cases} 1 & \text{if } u = xv, \\ 0 & \text{else,} \end{cases} \qquad \phi_{vx}(u) = \begin{cases} 1 & \text{if } u = vx, \\ 0 & \text{else,} \end{cases}$$

$$\phi_{yw}(u) = \begin{cases} 1 & \text{if } u = yw, \\ 0 & \text{else,} \end{cases} \qquad \phi_{wy}(u) = \begin{cases} 1 & \text{if } u = wy, \\ 0 & \text{else,} \end{cases}$$

as well as linear combinations thereof with positive coefficients, are generators of diffusions.

6.2 APPELL SYSTEMS

Recall that Appell systems $\{h_k; k \in \mathbb{N}\}$ can be defined by the conditions

A1: h_k is a polynomial of degree k,

A2: $Dh_k = kh_{k-1}$,

where $D = \mathrm{d}/\mathrm{d}x$, see Subsection 2.7.2. Examples are furnished by shifted moment sequences of the form

$$h_k(x) = \int_{\mathbb{R}} (x+y)^k p(\mathrm{d}x),$$

where p is any probability measure on $\mathbb{R}$ with all moments finite. For example in the Gaussian case this gives the Hermite polynomials $H_n(x) = \frac{1}{\sqrt{2\pi}}\int_{\mathbb{R}}(x+iy)^n e^{-y^2/2}dy$. Recently analogous classes of 'moment polynomials' on Lie groups [FS92] and quantum groups [FFS95] have been introduced. They cease to be polynomials in general.

The definition on quantum groups is easily adapted to braided groups.

Definition 6.2.1 *We define the* Appell polynomials *on the braided covector space $V^{\check{}}(R,R')$ with respect to the functional ϕ by*

$$h_a = (\phi \otimes \mathrm{id}) \circ \underline{\Delta}\mathbf{x}_a,$$

i.e. $h_a = T_\phi(\mathbf{x}_a)$, for $a \in \mathcal{S}\{1,\ldots,n\}$.

We are in particular interested in the Appell systems with respect to diffusions.

Proposition 6.2.2 *Let ϕ_t be a diffusion on $V^{\check{}}(R,R')$ with generator L and let $\{h_a(t); a \in \mathcal{S}\{1,\ldots,n\}\}$ be the associated Appell system. Then we have the following:*

1. *$h_a(t) = \mathbf{x}_a + p_a(t)$, where $p_a(t) \in \bigoplus_{\alpha=0}^{|a|-1} V^{\check{}}(R,R')^{(\alpha)}$,*
2. *$D^i h_a = \sum_{b\in\mathcal{S}^{|a|-1}\{1,\ldots,n\}}[|a|;R]_a^{i.b} h_b$, where $D^i = \exp(tL)\partial^i\exp(-tL)$ and $i.b$ the concatenation of i with b, i.e. $i.b = ib_1\cdots b_{|a|-1}$,*
3. *the $h_a(t)$ satisfy the evolution equation $\partial_t u = Lu$.*

Proof: Note that 1. and 2. are the analogues of A1 and A2.

1. The braided-binomial theorem [Maj96, Equation (91)] tells us that

$$\underline{\Delta}\mathbf{x}_a = \sum_{i=0}^{|a|} \sum_{b\in\mathcal{S}^i\{1,\ldots,n\}, c\in\mathcal{S}^{|a|-i}\{1,\ldots,n\}} \begin{bmatrix} |a| \\ i \end{bmatrix}_a^{b.c} \mathbf{x}_b \otimes \mathbf{x}_c.$$

where the braided-binomial coefficients are defined by

$$\begin{aligned}
\begin{bmatrix} m \\ r \end{bmatrix};R &= (PR)_{r,r+1}\cdots(PR)_{m-1,m}\begin{bmatrix} m-1 \\ r-1 \end{bmatrix};R + \begin{bmatrix} m-1 \\ r \end{bmatrix};R,\\
\begin{bmatrix} m \\ 0 \end{bmatrix};R &= 1,\\
\begin{bmatrix} m \\ r \end{bmatrix};R &= 0 \qquad \text{for } r > m.
\end{aligned}$$

The statement follows now since $\phi(t)|_{V^{\check{}}(R,R')^{(0)}} = \epsilon$, i.e. $h_a(t) = \mathbf{x}_a + p_a(t)$ with

$$p_a(t) = \sum_{i=0}^{|a|-1} \sum_{b\in\mathcal{S}^{|a|-i}\{1,\ldots,n\}, c\in\mathcal{S}^{|i}\{1,\ldots,n\}} \begin{bmatrix} |a| \\ i \end{bmatrix}_a^{b.c} \phi_t(\mathbf{x}_b)\mathbf{x}_c.$$

Note that this formula can also be used to calculate the Appell systems, if the moments of the process are known.

2.

$$\begin{aligned} D^i h_a &= \exp(tL)\partial^i \exp(-tL)\exp(tL)\mathbf{x}_a \\ &= \exp(tL)\partial^i \mathbf{x}_a \\ &= \exp(tL) \sum_{b\in S^{|a|-1}\{1,\ldots,n\}} [|a|;R]_a^{i.b}\mathbf{x}_b \\ &= \sum_{b\in S^{|a|-1}\{1,\ldots,n\}} [|a|;R]_a^{i.b} h_b \end{aligned}$$

with [Maj96, Equation (77)].

3.

$$\begin{aligned} \partial_t h_a(t) &= \partial_t \exp(tL)\mathbf{x}_a = L\exp(tL)\mathbf{x}_a \\ &= Lh_a(t). \end{aligned}$$

■

Property 2 shows that the D^i act as lowering operators. We also define raising operators.

Proposition 6.2.3 *Let $X_i = \exp(tL)x_i\exp(-tL)$, then*

$$X_i h_a(t) = h_{i.a}(t).$$

Examples

6.2.1 *The braided line* $\mathbb{R}_q$

The Appell polynomials associated to the functional $\phi_t = \exp(tL)$ where $L = \frac{a^2}{1+q}v^2 + bv$ are

$$\begin{aligned} h_k(x;t) &= (\phi_t \otimes \mathrm{id}) \circ \underline{\Delta} x^k \\ &= \sum_{\nu=0}^{k} \begin{bmatrix} k \\ \nu \end{bmatrix} ; q \Big] \phi_t(x^\nu) x^{k-\nu} \\ &= \sum_{\nu=0}^{K} \frac{[k;q]!H_\nu\left(bt, -\frac{2a^2t}{1+q}\right)}{[k-\nu;q]!\nu!} x^{k-\nu}. \end{aligned}$$

These polynomials are solutions of

$$\partial_t u = \frac{a^2}{1+q}\delta^2 u + b\delta u$$

with $u(x,t=0) = x^k$ (where δ is the q-difference operator $(\delta f)(x) = \frac{f(qx)-f(x)}{x(q-1)}$).

For $b = 0$, $a = \sqrt{(1+q)/2}$ the Appell polynomials simplify to

$$h_k(x;t) = \sum_{\nu=0}^{[\frac{k}{2}]} \frac{[k;q]!t^\nu x^{k-2\nu}}{[k-2\nu;q]!2^\nu \nu!},$$

these polynomials are a q-analogue of the Hermite polynomials (see also [FFS95]).

6.2.2 *The quantum plane* $\mathbb{C}_q^{2|0}$

We choose $\phi_t = \exp tL$ with $L = \frac{1}{1+q^2}\left(\partial_1^2 + \partial_2^2\right)$. We get

$$\phi_t = \sum_{r=0}^{\infty} \frac{t^r}{(1+q^2)^r r!} \sum_{\nu=0}^{r} \begin{bmatrix} r \\ \nu \end{bmatrix} ; q^{-4} \Bigg] \partial_2^{2\nu} \partial_1^{2(r-\nu)}$$

since $\partial_2\partial_1 = q^{-1}\partial_1\partial_2$. This leads to the following formula for the Appell polynomials:

$$\begin{aligned} & h_{rm}(x, y; t) \\ = \; & \exp(tL) x^r y^m \\ = \; & \sum_{\nu=0}^{[\frac{r}{2}]} \sum_{\mu=0}^{[\frac{m}{2}]} \frac{\left[\begin{matrix} \mu+\nu \\ \nu \end{matrix} ; q^{-4}\right] [m; q^2]![r; q^2]! q^{2\mu(r-2\nu)} \; t^{\mu+\nu}}{[m-2\mu-1; q^2]![r-2\nu-1; q^2]!(1+q^2)^{\mu+\nu}(\mu+\nu)!} x^{r-2\nu} y^{m-2\mu}, \end{aligned}$$

where $\left[\begin{matrix} r \\ \nu \end{matrix} ; q^{-4}\right]$ are the braided-binomial coefficients with respect to $R = (q^{-4}) \in M(1) \otimes M(1)$. These polynomials solve the evolution equation

$$\partial_t u = \frac{1}{1+q^2}\left(\partial_1^2 + \partial_2^2\right) u$$

where ∂_1, ∂_2 can be defined by

$$\partial_1 f(x,y) = \frac{f(q^2x, y) - f(x,y)}{x(q^2-1)}, \qquad \partial_2 f(x,y) = \frac{f(qx, q^2y) - f(qx,y)}{y(q^2-1)}.$$

6.2.3 *The free braided-space*

We consider again the free braided-space associated to $R = R' = P$. It is straightforward to calculate the action of the generator of a diffusion on a basis element:

$$L\mathbf{x}_k = r(r-1)a_{k_1k_2}\mathbf{x}_{k_3\cdots k_r} + rb_{k_1}\mathbf{x}_{k_2\cdots k_r},$$

for $L = \sum_{k,l=1}^{n} a_{kl}\partial_k\partial_l + \sum_{j=1}^{n} b_j\partial_j$ and $k = (k_1, \ldots, k_r) \in \mathcal{S}\{1, \ldots, n\}$. Thus we get for the Appell polynomials

$$\begin{aligned} h_k &= \exp(tL)\mathbf{x}_k \\ &= \mathbf{x}_k + t(r(r-1)a_{k_1k_2}\mathbf{x}_{k_3\cdots k_r} + rb_{k_1}\mathbf{x}_{k_2\cdots k_r}) \\ &\quad + \frac{t^2}{2}\Big(r(r-1)(r-2)(r-3)a_{k_1k_2}a_{k_3k_4}\mathbf{x}_{k_5\cdots k_r} \\ &\quad + r(r-1)(r-2)(b_{k_1}a_{k_2k_3} + a_{k_1k_2}b_{k_3})\mathbf{x}_{k_4\cdots k_r} \\ &\quad + r(r-1)b_{k_1}b_{k_2}\mathbf{x}_{k_3\cdots k_r}\Big) + \cdots, \end{aligned}$$

If the drift term vanishes, i.e. $b_j = 0$ for $j = 1, \ldots, n$, then

$$h_k = \sum_{\rho=0}^{[\frac{r}{2}]} \frac{t^\rho}{\rho!} r(r-1)\cdots(r-2\rho+1)a_{k_1k_2}\cdots a_{k_{2\rho-1}k_{2\rho}}\mathbf{x}_{k_{2\rho+1}\cdots k_r}.$$

6.3 DENSITIES

There are two ways in which densities can be associated to the diffusions constructed here.

This first uses the (right or left) invariant integrals (see Section 3.11). Fix such a functional $\int : V^{\check{}}(R, R') \to \mathbb{C}$. One can look for an element $\rho_t \in V^{\check{}}(R, R')$ such that

$$\phi_t(a) = \int (\rho_t a), \qquad \left(\text{or } \phi_t(a) = \int (a\rho_t)\right)$$

is satisfied for all $a \in V^{\check{}}(R, R')$ (where $\phi_t(a) =< \phi_t, a >$ denotes the evaluation map between $V^{\check{}}(R, R')$ and $V(R, R')$). This defines a (non-commutative) left (or right) density. For e.g. any-space this leads to the results exhibited in [MRP94]. To calculate it the Fourier transform (w.r.t. to the integral considered) can be used. For a braided-Hopf algebra B with dual B^*, exponential exp and invariant integral $\int : B \to \mathbb{C}$ the *Fourier transform* $\mathcal{F} : B \to B^*$ is defined by $\mathcal{F} = (\int \otimes \mathrm{id}) \circ (m \otimes \mathrm{id}) \circ (\mathrm{id} \otimes \exp)$, see [KM94]. Let now $\int$ be a left invariant integral on $V^{\check{}}(R, R')$ and $\int^*$ a right invariant integral on $V(R, R')$. Then, since $(\mathrm{id} \otimes < \cdot, \cdot >) \circ (\exp \otimes \mathrm{id}) = \mathrm{id}$,

$$\int (\rho a) =< \mathcal{F}(\rho), a >$$

for all $a \in V^{\check{}}(R, R')$ and $\rho \in V(R, R')$. Thus we need to invert the Fourier transform $\mathcal{F}$ to find the density corresponding to the functional ϕ_t. By [KM94, Proposition 6.4], $\mathcal{F}^* \circ \mathcal{F} = \mathrm{vol}\,\underline{S}$, where vol is a constant and $\mathcal{F}^*$ the Fourier transform w.r.t. $\int^*$, i.e. $\mathcal{F}^* = (\mathrm{id} \otimes \int^* \circ m) \circ (\Psi \otimes \mathrm{id}) \circ (\mathrm{id} \otimes \exp)$. Thus

$$\rho_t = \frac{1}{\mathrm{vol}} \underline{S}^{-1} \circ \mathcal{F}^*(\phi_t).$$

Take, e.g., the braided-line and $\phi_t = \exp(\frac{tv^2}{2})$. We shall use the Jackson integral $\int_{-\gamma.\infty}^{+\gamma.\infty} f(x) \mathrm{d}_q x$ where γ is a parameter. For $f(x) = \sum f_n x^n$ we set $f(\gamma) = \sum f_n \gamma^n$ and

$$\int_{-\gamma.\infty}^{+\gamma.\infty} f(x) \mathrm{d}_q x = (1-q) \sum_{k=-\infty}^{\infty} \left(f(q^k\gamma) + f(-q^k\gamma)\right) q^k \gamma$$

if the right-hand-side converges. We will express the density in terms of the discrete q-Hermite II polynomials

$$\tilde{h}_n(x; q) = (1-q)^n [n; q]! \sum_{k=0}^{[\frac{n}{2}]} \frac{(-1)^k q^{-2nk} q^{k(2k+1)} x^{n-2k}}{(1-q^2)^k [k; q^2]! (1-q)^{n-2k} [n-2k; q]!}.$$

With the orthogonality relations for the discrete q-Hermite II polynomials [Koo95, Equation (8.14)] and

$$\phi_t(\tilde{h}_n(x; q)) = \begin{cases} 0 & \text{if } n \text{ is odd,} \\ (1-q)^n [n; q]! \sum_{k=0}^{[\frac{n}{2}]} \frac{(-1)^k q^{-2nk} q^{k(2k+1)} t^{\frac{n}{2}-k}}{(1-q^2)^k [k; q^2]! (1-q)^{n-2k} 2^{\frac{n}{2}-k}} & \text{if } n \text{ is even,} \end{cases}$$

we get

$$\rho_t = \frac{1}{c_q(\gamma)} \sum_{m=0}^{\infty} q^{4m^2} \sum_{k=0}^{m} \frac{(-1)^k q^{-4mk} q^{k(2k+1)} t^{m-k}}{(1-q^2)^k [k;q^2]! 2^{m-k}} \tilde{h}_{2m}(x;q) e_{q^2}(-(1-q^2)x^2), \tag{6.3}$$

where $e_{q^2}(z) = \sum_{n=0}^{\infty} \frac{z^n}{[n;q^2]!}$, and $c_q(\gamma)$ is a constant depending only on q and γ (for its value see [Koo95, Equation (8.15)]).

The other way, commonly used in quantum probability (see e.g. [Mey93]) leads to a density on the usual real line. Consider for example a diffusion $\phi_t = \exp(tv^2/2)$ on the braided line, and the generator x, then $\phi_t(e^{itx})$ can be interpreted as the Fourier transform of the density of x in the state ϕ_t. Let $0 < q < 1$ and $Y = (1-q)\sum_{i=0}^{\infty} X_i$, where the X_i are independent, and exponentially distributed with respective means q^i (see [Fei87]). Then $E(e^{iuY}) = \prod_{i=0}^{\infty} \left(1 - iu(1-q)q^i\right)^{-1} = \sum_{n=0}^{\infty} (iu)^n/[n;q]! = e_q^{iu}$. Thus $\phi_t(E(e^{iuYx})) = \phi_t(e_q^i ux) = e^{-tu^2/2}$, and it follows that Yx has a Gauss distribution. This determines the distribution of x uniquely, see [Fei87, Theorem 3].

REFERENCES

This chapter is based on the articles [FS98a, FSS98].

Chapter 7

Evolution equations and Lévy processes on quantum groups

In this chapter we consider stochastic processes on quantum groups that are related to evolution equations of the form

$$\partial_t u = Lu,$$

with some difference-differential operator L. For the equations considered in Section 7.1, u is an element of a quantum or braided group $\mathcal{A}$. We recall that solutions of these equations can be given as Appell systems or shifted moments of the associated process, and show how these can be calculated explicitly on the q-affine group and on a braided analogue of the Heisenberg-Weyl group.

In Section 7.2, we define a Wigner map from functionals on a quantum group or braided group to a "Wigner" density on the undeformed space. We prove that the densities associated in this way to (pseudo-) Lévy processes satisfy a Fokker-Planck type equation. In the one-dimensional case these coincide with the evolution equations of Section 7.1, but in the general case we get new equations.

7.1 APPELL SYSTEMS

We want to consider equations of the form

$$\partial_t u = Lu \tag{7.1}$$

where $L : \mathcal{A} \to \mathcal{A}$ is a differential operator, independent of t, e.g.

$$\begin{aligned} \partial_t u &= (a\delta_q^2 + b\delta_q)u \qquad \text{on } \mathbb{R}_q \\ \partial_t u &= \left(\partial_1^2 + \partial_2^2\right) u \qquad \text{on } \mathbb{C}_q^{2|0} \end{aligned}$$

$$\begin{aligned} \partial_t u &= \left(\rho(X)^2 + \rho(Y)^2\right) u && \text{on } \mathrm{Aff}_q \\ \partial_t u &= \left(\rho(x)^2 + \rho(z)^2\right) u && \text{on } \mathrm{HW}_q \end{aligned}$$

In the first equation, L is a general second order q-difference operator, but for the explicit calculations we shall assume that a and b are constants.

In the second equation we have an analogue of the Laplacian, the operator in the third equation is related to the Gegenbauer or ultraspherical polynomials, see e.g. [FF]. In the fourth equation we have an analogue of the Kohn-Laplacian on the Heisenberg group.

An equation of the form (7.1) gives rise to a transition operator, formally written as e^{tL}. If the functional $\varepsilon \circ L$ is conditionally positive (w.r.t. an involution on $\mathcal{A}$), then $\varepsilon \circ e^{tL}$ defines a convolution semi-group of states and thus a Lévy process (if the braiding is different from the twist map, then we have to impose the additional condition that $\varepsilon \circ L$ is Ψ-invariant, i.e. that $(\varepsilon \circ L \otimes \phi) \circ \Psi = \phi \otimes \varepsilon \circ L$ for all $\phi \in \mathcal{A}^*$). We can still associate a process to L, even if $\varepsilon \circ L$ is not conditionally positive, but in this case the state fails to be positive, see the construction of Section 4.5 or [FFS97].

Appell systems on Lie groups have been studied in [FS92], quantum groups were considered in [FFS97]. Appell systems on quantum or braided groups are defined in the same way as on braided spaces, see Section 6.2, but in general they are no longer polynomials (see also [FFS97, FS98a]).

Definition 7.1.1 *We define the (left)* Appell system *on a braided group $\mathcal{A}$ with respect to a semi-group of functionals $\{\varphi_t\}$ by*

$$h_k = (\varphi_t \otimes \mathrm{id}) \circ \Delta a_k,$$

i.e. $h_k = U_t(\varphi)(a_k)$, for a fixed basis $\{a_k\}$ of $\mathcal{A}$.

If L is the generator of $\{\varphi_t\}$, then h_k solves

$$\partial_t h_k = L h_k.$$

For other interesting properties, e.g., raising operators, or relation to matrix elements, see Section 6.2 and [FFS97, FS98a].

Examples of Appell systems on the braided line $\mathbb{R}_q$ and the braided plane $\mathbb{C}_q^{2|0}$ were already considered in the Subsections 6.2.1 and 6.2.2.

7.1.1 *Example: The q-affine group*

We can use the generalised Gegenbauer polynomials defined in [FF]

$$C_n^h(x) = \sum_{\nu=0}^{[\frac{n}{2}]} \frac{(h)_{n-\nu}}{q_{n-2\nu}} \frac{(-1)^\nu}{\nu!} (2x)^{n-2\nu},$$

and the representation $\rho_h(X) = \alpha(x\partial_x + h)$, $\rho_h(Y) = i\alpha\delta_x$ to calculate the moments of $\Phi_t = \exp\left(\frac{t}{2}(X^2 + Y^2)\right)$. These polynomials are eigenfunctions of

$S_h = (x\partial_x + h)^2 - \delta_x^2 = \rho_h(X)^2 + \rho_h(Y)^2$, i.e. $S_h C_n^h(x) = (n+h)^2 C_n^h(x)$, and their inversion formula is

$$x^n = \frac{q_n!}{2^n} \sum_{k=0}^{[\frac{n}{2}]} \frac{h+n-2k}{(h)_{n-k+1} k!} C_{n-2k}^h(x).$$

Using the Feynman-Kac type formula (cf. [FFS97])

$$\Phi_t\left(e^{a\rho_h(X)} e_q^{b\rho_h(Y)} x^n\right) = e^{\frac{t}{2}(\rho_h(x)^2 + \rho_h(Y)^2)} x^n,$$

and comparing the coefficients of x^{n-2r}, we get

$$\begin{aligned} \Phi_t\left(e^{(n-2r+h)\alpha a} b^{2r}\right) &= \frac{q_{2r}}{\alpha^{2r}} \sum_{k=0}^{r} \frac{(h+n-2k)(h)_{n-r-k}(-1)^k}{4^r (h)_{n-k+1} k!(r-k)!} e^{(n-2k+h)^2\alpha^2 t/2} \\ & \qquad \text{for } n \geq 2r, \\ \Phi_t\left(e^{(n-2r+h)\alpha a} b^{2r+1}\right) &= 0 \qquad \text{for } n \geq 2r+1. \end{aligned}$$

Differentiating ν times w.r.t. h and setting $h = m - \mu + 2r - n$, we get all moments that are needed to calculate the Appell functions $h_{nm}(a, b, t) =$

$$\Phi_t\left(\Delta(a^n b^m)\right) = \sum_{\nu=0}^{n} \sum_{\mu=0}^{m} \binom{n}{\nu} \begin{bmatrix} m \\ \mu \end{bmatrix}_q \Phi_t\left(a^\nu e^{(m-\mu)\alpha a} b^\mu\right) a^{n-\nu} b^{m-\mu}.$$

7.1.2 *Example: The braided q-Heisenberg-Weyl group*

Consider $L = x^2 + z^2$. Then

$$e^{tL} = \sum_{\nu=0}^{n} \sum_{\kappa=0}^{2(n-\nu)\wedge 2\nu} \frac{C_{\nu,\kappa}^n t^n}{n!} z^{2(n-\nu)-\kappa} y^\kappa x^{2\nu-\kappa}$$

where the coefficients $C_{\nu,\kappa}^n$ are determined by the recurrence relations

$$C_{\nu,\kappa}^{n+1} = C_{\nu-1,\kappa}^n + q^{2\kappa} C_{\nu,\kappa}^n + q_{2\nu-\kappa+1} q_2 q^{\kappa-1} C_{\nu,\kappa-1}^n + q_{2\nu-\kappa+2} q_{2\nu-\kappa+1} C_{\nu,\kappa-2}^n.$$

For $\kappa = 0$ we have the binomial coefficients $C_{\nu,0}^n = \binom{n}{\nu}$. This allows to calculate

$$h_{nmr}(t) = e^{tL}(a^n b^m c^r),$$

using e.g. the dual pairing

$$\langle z^r y^m x^n, a^{n'} b^{m'} c^{r'} \rangle = \delta_{nn'} \delta_{mm'} \delta_{rr'} q_n! q_m! q_r!.$$

7.2 WIGNER-TYPE DENSITIES

For one single variable, or in the commutative case, one can use Bochner's theorem to associate a density to a quantum random variable, cf. [Mey93]. For the braided real line we have done this for one example in Section 6.3

We want to associate joint densities to several non-commutating variables, along the line of Wigner distributions [Wig32]. We will map functionals on an algebra with n generators to measures on $\mathbb{R}^n$. Equivalently, we can ask for a map from functions on $\mathbb{R}^n$ (e.g., polynomials) to elements of the algebra.

Consider the following diagram:

$$\begin{array}{ccccc} & & \text{Wigner} & & \\ & \text{QS} & \longrightarrow & \text{CS} & \\ \text{Duality} & \updownarrow & & \updownarrow & \text{Duality} \\ & \text{QO} & \longleftarrow & \text{CO} & \\ & & \text{Weyl} & & \end{array}$$

where

QS: (linear span of) the set of quantum states

CS: (linear span of) set of classical states

QO: set of quantum observables

CO: set of classical observables

We want the following similar diagram:

$$\begin{array}{ccccc} & & \text{“Wigner”} & & \\ & \mathcal{U} & \longrightarrow & \mathcal{M}(X) & \\ \text{q-Fourier} & \updownarrow & & \updownarrow & \text{Fourier} \\ & \mathcal{A} & \longleftarrow & C(X) & \\ & & \text{“Weyl”} & & \end{array}$$

where X is the undeformed space or group, and $\mathcal{M}(X)$ denotes the (convolution) algebra of (signed) measures on X, $C(X)$ the algebra of continuous functions on X, and $\mathcal{A}$ and $\mathcal{U}$ the quantum group and algebra. The *q-Fourier transformation* (with respect to an invariant integral $\int_{\mathcal{A}}$ on $\mathcal{A}$) is defined by

$$\mathcal{F}_{\mathcal{A}}(u) = \int_{\mathcal{A}} (u \exp)$$

where exp is the exponential or coevalution map of $\mathcal{A}$ and $\mathcal{U}$, see [KM94, Koo95]. We have $\int_{\mathcal{A}}(ab) = \langle \mathcal{F}_{\mathcal{A}}(a), b\rangle$, and thus in this setting a density (w.r.t. $\int_{\mathcal{A}}$) of a functional $\Phi \in \mathcal{A}^*$ can, at least in principle, be calculated with the inverse Fourier transform, $\rho_\Phi = \mathcal{F}_{\mathcal{A}}^{-1}(\Phi)$. A more detailed discussion, including an explicite example on $\mathbb{R}_q$, can be found in Section 6.3 or in [FS98a].

We use the right-hand-side of the diagram to introduce densities that 'live' on the classical, i.e. undeformed, group or space. Following Anderson[And69], we fix a set of generators $x_1, \ldots, x_n$ of $\mathcal{A}$ and define the *Weyl map* [Wey31] on polynomials $u_1^{k_1} \cdots u_n^{k_n}$ by

$$W(u_1^{k_1} \cdots u_n^{k_n}) = \frac{k_1! \cdots k_n!}{k!} \sum_{\pi \in S_k} x_{\pi(1)} \cdots x_{\pi(k)}$$

where $k = k_1 + \cdots + k_n$. Other definitions are possible, e.g. $W_W : u_1^{k_1} \cdots u_n^{k_n} \mapsto x_1^{k_1} \cdots x_n^{k_n}$ ("Wick"), or $W_{AW} : u_1^{k_1} \cdots u_n^{k_n} \mapsto x_n^{k_n} \cdots x_1^{k_1}$ ("Anti-Wick"), or also $W_{q-\exp}$ defined by $e^{u \cdot v} \mapsto \exp(x|v)$. But $W_{q-\exp}$ will not leave the marginal distributions unchanged. In fact, W is uniquely determined by the conditions $W(u_i) = x_i$ and $W\left((a_1 u_1 + \cdots + a_n u_n)^k\right) = (a_1 W(u_1) + \cdots + a_n W(u_n))^k$, and thus only W^* will give the correct marginal distributions for all linear combinations of the generators. Ordered monomials like the "Wick" or "Anti-Wick" map still lead to the right marginal distributions for the generators.

The *Wigner map* W^* is defined by the condition

$$\langle \Phi, W(u) \rangle = \int u \mathrm{d}W^*(\Phi),$$

i.e. as the dual of the Weyl map. The measure $W^*(\Phi)$ is called the *Wigner density* or *Wigner distribution* on X associated to the state Φ. The Fourier transform of the measure $W^*(\Phi)$ is

$$\begin{aligned} g_\Phi(v) &= \mathcal{F}(W^*(\Phi))(v) = \int e^{iu \cdot v} \mathrm{d}W^*(\Phi) \\ &= \langle \Phi, W(e^{iu \cdot v}) \rangle, \end{aligned}$$

where we assumed that we can interchange the limits involved, and that this series defines an analytic function. Note that here $\int$ and $\mathcal{F}$ denote integration and Fourier transform on X.

If the functionals Φ_t form a convolution semi-group, then the associated Wigner densities satisfy an evolution equation or *Fokker-Planck equation.*

Proposition 7.2.1 *Let $\{\Phi_t; t \in \mathbb{R}_+\}$ be a convolution semi-group with generator L, i.e. $\frac{\mathrm{d}}{\mathrm{d}t}\Phi_t = L\Phi_t = \Phi_t L$. Suppose further that W is invertible. Then the Wigner distribution of Φ_t satisfies*

$$\frac{\mathrm{d}}{\mathrm{d}t} W^*(\Phi_t) = \tilde{\rho}(L)^* W^*(\Phi_t)$$

with $\tilde{\rho}(X) = W^{-1} \circ \rho(X) \circ W$ and $\tilde{\rho}(X)^*$ defined by duality.

Proof: Differentiate

$$\langle \Phi_t, W(u) \rangle = \int u \mathrm{d}W^*(\Phi_t)$$

with respect to t, on the right hand side we get $\int u \mathrm{d}\frac{\mathrm{d}W^*(\Phi_t)}{\mathrm{d}t}$, while the left hand side gives

$$\begin{aligned} \frac{\mathrm{d}}{\mathrm{d}t}\langle \Phi_t, W(u) \rangle &= \langle L\Phi_t, W(u) \rangle = \langle \Phi_t, \rho(L) W(u) \rangle = \langle \Phi_t, W(\tilde{\rho}(L)u) \rangle \\ &= \int \tilde{\rho}(L) u \mathrm{d}W^*(\Phi_t) = \int u \mathrm{d}\left(\tilde{\rho}(L)^* W^*(\Phi_t)\right). \end{aligned}$$

7.2.1 *Example: The braided line*

Here we have only one variable, and the algebra is commutative, so the Weyl map is just $W : u^n \mapsto x^n$. The Fourier transform of the functional $\Phi = \sum_{n=0}^{\infty} a_n p^n$ is thus

$$g_\Phi(u) = \langle \Phi, e^{iux} \rangle = \sum_{n=0}^{\infty} \frac{a_n q_n!(iu)^n}{n!}$$

where we assumed that the regularity conditions necessary for interchanging the limits are satisfied. E.g. for $\Phi_a = e^{ap}$, i.e. the functional determined by $\Phi_a(e_q^{x|p}) = e^{ap}$ (where $e_q^{x|p} = \sum_{n=0}^{\infty} \frac{x^n p^n}{q_n!}$) we get

$$g_{\Phi_a}(u) = \sum_{n=0}^{\infty} \frac{q_n!(iau)^n}{(n!)^2}.$$

We need $\tilde{\rho}(p)^*$ to be able to give the form of the Fokker-Planck equation. Because of the simple form of W we have $\tilde{\rho}(p) = \delta$. The adjoint of the q-difference operator is a multiple of the q-difference with q replaced by q^{-1},

$$\delta^* f(x) = -\frac{1}{q} \frac{f(q^{-1}x) - f(x)}{x(q^{-1} - 1)} = -\frac{1}{q} \delta_{1/q} f(x)$$

so that the Wigner density $\eta_t(\mathrm{d}x) = \mathrm{d}W^*(\Phi_t)$ of the semi-group with generator $L = \sum c_n p^n$ satisfies

$$\partial_t \eta_t = \sum \frac{(-1)^n c_n}{q^n} \delta_{1/q}^n \eta_t.$$

7.2.2 *Example: The braided plane*

In order to determine W we calculate $(a_1 x + a_2 y)^N$, the coefficient of $a_1^n a_2^m$ is the image of $u_1^n u_2^m$. We get $W : u_1^n u_2^m \mapsto \dfrac{\begin{bmatrix} n+m \\ n \end{bmatrix}_q}{\begin{pmatrix} n+m \\ n \end{pmatrix}} x^n y^m$, and thus

$$g_\Phi(v_1, v_2) = \sum_{n,m=0}^{\infty} \frac{\begin{bmatrix} n+m \\ n \end{bmatrix}_q}{(n+m)!} i^{n+m} \langle \Phi, x^n y^m \rangle v_1^n v_2^m.$$

For the Gauss functionals in the sense of Bernstein ($\Phi(x^n y^m) = z^n \delta_{m,0}$ or $z^m \delta_{n,0}$, see [FNS97a]) we get $g_\Phi(v) = e^{izv_1}$ or $g_\Phi(v) = e^{izv_2}$, i.e. $W^*(\Phi)$ is a Dirac mass in $(z, 0)$ or $(0, z)$.

To write down the Fokker-Planck equations for Wigner distributions we need the representation $\tilde{\rho}$. For the two generators we get

$$\begin{aligned} \tilde{\partial}_1 u_1^n u_2^m &= n \frac{q_{n+m}}{n+m} u_1^{n-1} u_2^m \\ \tilde{\partial}_2 u_1^n u_2^m &= m q^n \frac{q_{n+m}}{n+m} u_1^n u_2^{m-1}. \end{aligned}$$

7.2.3 *Example: The q-affine group*

We get $W : u_1^n u_2^m \mapsto \sum_{\nu=0}^n \frac{n!m!}{(n+m)!} A_{n,\nu}^{n+m} \beta^{n-\nu} a^\nu b^m$, where the $A_{n\nu}^N$ are determined by $A_{n\nu}^N = 0$ if $n > N$ or $\nu > n$ or $n < 0$, $A_{00}^N = 1$, $A_{N\nu}^N = \delta_{N\nu}$, and

$$A_{n\nu}^{N+1} = A_{n-1,\nu-1}^N + (N+1-n)A_{n-1,\nu}^N + A_{n\nu}^N.$$

For the special case where the two lower indices are identical, we get $A_{nn}^N = \begin{pmatrix} N \\ n \end{pmatrix}$.

The Weyl map and its inverse are characterised by

$$\begin{array}{rcccc} W & : & e^{a_1 u_1 + a_2 u_2} & \mapsto & e^{a_1 a + a_2 b} = e^{a_1 a} e^{\frac{e^{a_1 \beta} - 1}{a_1 \beta} a_2 b} \\ W^{-1} & : & e^{b_1 a} e^{b_2 b} = e^{b_1 a + \frac{\beta b_1 b_2}{e^{b_1 \beta} - 1} b} & \mapsto & e^{b_1 u_1 + \frac{\beta b_1 b_2}{e^{b_1 \beta} - 1} u_2} \end{array}$$

This allows to write down $\tilde{\rho}$ and thus the Fokker-Planck equation for any Lévy process. For X we simply get $\tilde{\rho}(X) = \frac{\partial}{\partial u_1}$, the expression for $\tilde{\rho}(Y)$ is more complicated.

7.2.4 *Example: The q-Heisenberg-Weyl algebra*

The q-Heisenberg-Weyl algebra HW_q can be treated in the same way. Choose a, b, c as generators of HW_q, then the Weyl map is given by $W : u_1^n u_2^m u_3^r \mapsto \sum_{k=0}^{n \wedge r} \frac{n!m!r! D_{n,m,k}^{n+m+r}}{(n+m+r)!} a^{n-k} b^{m+k} c^{r-k}$, where the coefficients $D_{n,r,k}^{n+m+r}$ are defined by

$$\begin{array}{rcll} D_{n,r,k}^N & = & 0 & \text{if } n < 0 \text{ or } r < 0 \text{ or } n + r > N \text{ or } k < 0 \text{ or } k > n \wedge r, \\ D_{n,0,0}^N & = & \begin{pmatrix} N \\ n \end{pmatrix}, & \text{if } 0 \le n \le N \\ D_{0,r,0}^N & = & \begin{pmatrix} N \\ r \end{pmatrix}, & \text{if } 0 \le r \le N \end{array}$$

and the recurrence relation

$$D_{n,r,k}^{N+1} = D_{n,r,k}^N + D_{n,r-1,k}^N + q^{k-r} D_{n-1,r,k}^N + (q^{-1})_{r-k+1} \left(1 - \frac{1}{q}\right) D_{n-1,r,k-1}^N.$$

REFERENCES

The material in this chapter is mainly taken from [FS98b].

Chapter 8

Gauss laws in the sense of Bernstein on quantum groups

In this chapter we turn to the characterisation of certain probability laws and convolution semi-groups on nilpotent quantum groups and nilpotent braided groups.

In Section 8.1 we determine the functionals which satisfy an analogue of the Bernstein property, i.e. that the sum and difference of independent random variables are also independent, on several braided groups. This extends results obtained on Lie groups by D. Neuenschwander et al., cf. [Neu93], [NRS97]. As for Lie groups this class turns out to be too small to constitute a satisfactory definition of Gaussianity. Therefore we turn to convolution semi-groups. We show the uniqueness of embedding into continuous convolution semi-groups on nilpotent quantum groups and nilpotent braided groups in Section 8.2. In Section 8.3 we define Gaussian convolution semi-groups in the sense of Bernstein and calculate their generators on several nilpotent braided groups.

8.1 GAUSSIAN FUNCTIONALS IN THE SENSE OF BERNSTEIN

The Gauss law plays a central role in probability theory on Euclidean spaces. Bernstein's theorem gives a characterisation in terms of the group law of the underlying space.

Theorem 8.1.1 *Let P be a probability measure on $\mathbb{R}^n$. P is a Gauss law if and only if for all pairs of independent random variables X, Y with distribution P the random variables $X+Y$ and $X-Y$ are also independent.*

Note that it is not essential that the two random variables X and Y have the same law, we also have the following version of the above theorem.

Theorem 8.1.2 *Let P_X and P_Y be two probability measures on $\mathbb{R}^n$. If there exists a pair of independent random variables X, Y with distribution P_X, P_Y, respectively, such that the random variables $X+Y$ and $X-Y$ are also independent, then P_X and P_Y are Gauss laws (with identical covariance matrices).*

This characterisation can be used as a definition of Gaussian functionals on groups in general, if some care is taken concerning the order of the factors in non-commutative groups.

Definition 8.1.3 *Let G be a (locally compact topological) group. A probability measure μ on G is called Gaussian in the sense of Bernstein if there exists a probability space $(\Omega, \mathcal{B}, P)$ and a pair of G-valued r.v. X, Y, such that $\mu = X(P) = Y(P)$, and $(X \cdot Y, Y \cdot X)$ and $(X \cdot Y^{-1}, Y^{-1} \cdot X)$ are independent.*

D. Neuenschwander showed [Neu93] that the Gauss measures in the sense of Bernstein on the Heisenberg-Weyl group are exactly the Gauss measures concentrated on Abelian subspaces. In fact, this result holds on connected simply-connected nilpotent Lie groups in general [NRS97].

We will now give a definition of Gaussian functionals on (braided) Hopf algebras.

8.1.1 *Definition of Gaussian functionals in the sense of Bernstein on (braided) Hopf algebras*

We have to translate $X+Y$ and $X-Y$ into Hopf algebra language. Replacing the group operation by the coproduct and the inverse by the antipode we write $j_1 \star j_2 = m_{\mathcal{A}} \circ (j_1 \otimes j_2) \circ \Delta_{\mathcal{B}}$ and $j_1 \star (j_2 \circ S) = m_{\mathcal{A}} \circ (j_1 \otimes (j_2 \circ S)) \circ \Delta_{\mathcal{B}}$ for the sum and difference of two quantum random variables $j_1, j_2 : \mathcal{B} \to \mathcal{A}$, see also [ASW88, Sch93].

Definition 8.1.4 *A functional φ on the (braided) Hopf algebra $\mathcal{B}$ is called* right (left) Gaussian in the sense of Bernstein *if there exists a quantum probability space $(\mathcal{A}, \Phi)$ and two independent quantum random variables $j_1, j_2 : \mathcal{B} \to \mathcal{A}$ such that $\varphi = \Phi \circ j_1 = \Phi \circ j_2$ and the applications*

$$J = m_{\mathcal{A}} \circ (j_1 \otimes j_2) \circ \Delta_{\mathcal{B}} \quad \textit{and} \quad K_r = m_{\mathcal{A}} \circ (j_1 \otimes j_2) \circ (\mathrm{id} \otimes S) \circ \Delta_{\mathcal{B}}$$
$$\Big(J = m_{\mathcal{A}} \circ (j_1 \otimes j_2) \circ \Delta_{\mathcal{B}} \quad \textit{and} \quad K_l = m_{\mathcal{A}} \circ (j_1 \otimes j_2) \circ (S \otimes \mathrm{id}) \circ \Delta_{\mathcal{B}} \Big)$$

form also a pair of independent quantum random variables.

A Gaussian functional in the sense of Bernstein *is a functional that is right and left Gaussian in the sense of Bernstein.*

Here j_1 and j_2 are required to have the same law. We shall see that in the examples we consider this condition is too strong. We propose therefore also a weaker definition that is motivated by Theorem 8.1.2.

Definition 8.1.5 *A functional φ on a (braided) Hopf algebra $\mathcal{B}$ is called* weakly right (left) Gaussian in the sense of Bernstein *if there exists a quantum probability space $(\mathcal{A}, \Phi)$ and a pair of independent quantum random variables $j_1, j_2 : \mathcal{B} \to \mathcal{A}$ such that $\varphi = \Phi \circ j_1$ ($\varphi = \Phi \circ j_2$) and the applications*

$$J = m_{\mathcal{A}} \circ (j_1 \otimes j_2) \circ \Delta_{\mathcal{B}} \quad \textit{and} \quad K_r = m_{\mathcal{A}} \circ (j_1 \otimes j_2) \circ (\mathrm{id} \otimes S) \circ \Delta_{\mathcal{B}}$$
$$\Big(J = m_{\mathcal{A}} \circ (j_1 \otimes j_2) \circ \Delta_{\mathcal{B}} \quad \textit{and} \quad K_l = m_{\mathcal{A}} \circ (j_1 \otimes j_2) \circ (S \otimes \mathrm{id}) \circ \Delta_{\mathcal{B}}\Big)$$

form also a pair of independent quantum random variables.

8.1.2 *Preliminary results*

Before presenting explicit results on particular (braided) Hopf algebras, we give a lemma showing a link between left and right (weakly) Gaussian functionals in the sense of Bernstein.

Taking a Hopf algebra H whose antipode S is invertible, we can define three Hopf algebras H^{op}, H^{cop}, H^{opcop}, if we replace the multiplication or the comultiplication by their opposites $m^{\mathrm{op}} = m \circ \tau$ or $\Delta^{\mathrm{cop}} = \tau \circ \Delta$, and invert the antipode each time we take an opposite. Similarly, we can take opposites or co-opposites of braided Hopf algebras using the braiding or its inverse and inverting the braiding, i.e. the opposites lie in the braided category that has the opposite braiding Ψ^{-1}.

In particular, let $(\mathcal{B}, m, 1, \Delta, \varepsilon, S, \Psi)$ be a braided Hopf algebra with invertible antipode, and define $\Delta^{\mathrm{cop}} = \Psi^{-1} \circ \Delta$. Then $\mathcal{B}^{\mathrm{cop}} = (\mathcal{B}, m, 1, \Delta^{\mathrm{cop}}, \varepsilon, S^{-1}, \Psi^{-1})$ is also a braided Hopf algebra [Maj93a, Lemma 4.6]. Notice that we have to use the inverse of the braiding, $\tilde{\Delta}^{\mathrm{cop}} = \Psi \circ \Delta$ does not lead to a braided Hopf algebra. We can also define a braided Hopf algebra with opposite multiplication. Set $m^{\mathrm{op}} = m \circ \Psi^{-1}$, then $\mathcal{B}^{\mathrm{op}} = (\mathcal{B}, m^{\mathrm{op}}, 1, \Delta, \varepsilon, S^{-1}, \Psi^{-1})$ is a braided Hopf algebra. Combining the above operations we get the braided Hopf algebras $(\mathcal{B}^{\mathrm{cop}})^{\mathrm{op}} = (\mathcal{B}, m \circ \Psi, 1, \Psi^{-1} \circ \Delta, \varepsilon, S, \Psi)$ and $(\mathcal{B}^{\mathrm{op}})^{\mathrm{cop}} = (\mathcal{B}, m \circ \Psi^{-1}, 1, \Psi \circ \Delta, \varepsilon, S, \Psi)$, and, more generally, $\mathcal{B}^{k \times \mathrm{cop\,op}} = (\mathcal{B}, m \circ \Psi^k, 1, \Psi^{-k} \circ \Delta, \varepsilon, S, \Psi)$ for $k \in \mathbb{Z}$.. All these braided Hopf algebras are isomorphic, since the antipode is an isomorphism from $\mathcal{B}^{k \times \mathrm{cop\,op}}$ to $\mathcal{B}^{(k+1) \times \mathrm{cop\,op}}$. In the same way we can construct a chain of isomorphic braided Hopf algebras starting from $\mathcal{B}^{\mathrm{op}}$ or $\mathcal{B}^{\mathrm{cop}}$, in particular, $S : \mathcal{B}^{\mathrm{op}} \to \mathcal{B}^{\mathrm{cop}}$ is an isomorphism.

We have the following lemma.

Lemma 8.1.6 *Let $j_1, j_2 : \mathcal{B} \to \mathcal{A}$ be a pair of Ψ-independent quantum random variables on $\mathcal{B}$ over some quantum probability space $(\mathcal{A}, \Phi)$. Then the following statements are equivalent.*

1. *$J(j_1, j_2) = m_{\mathcal{A}} \circ (j_1 \otimes j_2) \circ \Delta$ and $K_r(j_1, j_2) = m_{\mathcal{A}} \circ (j_1 \otimes (j_2 \circ S)) \circ \Delta$ are independent (and thus $\varphi = \Phi \circ j_1$ is a (weakly) right Gaussian functional in the sense of Bernstein on $\mathcal{B}$),*

2. *$J^{\mathrm{cop}}(j_2, j_1) = m_{\mathcal{A}} \circ (j_2 \otimes j_1) \circ \Delta^{\mathrm{cop}}$ and $K_l^{\mathrm{cop}}(j_2, j_1) = m_{\mathcal{A}} \circ ((j_2 \circ S) \otimes j_1) \circ \Delta^{\mathrm{cop}}$ are independent,*

3. $J^{\mathrm{op}}(j_2 \circ S, j_1 \circ S) = m_{\mathcal{A}} \circ ((j_2 \circ S) \otimes (j_1 \circ S)) \circ \Delta$ *and* $K_r^{\mathrm{op}}(j_2 \circ S, j_1 \circ S) = m_{\mathcal{A}} \circ ((j_2 \circ S \circ S) \otimes (j_1 \circ S)) \circ \Delta$ *are independent.*

Proof: The equivalence of 2. and 3. is clear since $S : \mathcal{B}^{\mathrm{op}} \to \mathcal{B}^{\mathrm{cop}}$ is an isomorphism.

Let us show now that 1. and 2. are equivalent. The pair (j_1, j_2) is Ψ-independent if and only if (j_2, j_1) is Ψ^{-1}-independent.

We find

$$\begin{aligned} J(j_1, j_2) &= m_{\mathcal{A}} \circ (j_1 \otimes j_2) \circ \Delta \\ &= m_{\mathcal{A}} \circ (j_1 \otimes j_2) \circ \Psi \circ \Psi^{-1} \Delta \\ &= m_{\mathcal{A}} \circ (j_2 \otimes j_1) \circ \Delta^{\mathrm{cop}} \\ &= J^{\mathrm{cop}}(j_2, j_1), \end{aligned}$$

and

$$\begin{aligned} K_r(j_1, j_2) &= m_{\mathcal{A}} \circ (j_1 \otimes (j_2 \circ S)) \circ \Delta \\ &= m_{\mathcal{A}} \circ ((j_2 \circ S) \otimes j_1) \circ \Delta^{\mathrm{cop}} \\ &= K_l^{\mathrm{cop}}(j_2, j_1), \end{aligned}$$

so that $J(j_1, j_2)$ and $K_r(j_1, j_2)$ are independent exactly when $J^{\mathrm{cop}}(j_2, j_1)$ and $K_l^{\mathrm{cop}}(j_2, j_1)$ are independent. ∎

To use this lemma to relate (weakly) left Gaussian functionals in the sense of Bernstein to (weakly) right Gaussian functionals in the sense of Bernstein on the opposite or co-opposite quantum/braided group we have to deal with two technical difficulties. First, we have to assume that j_1, j_2 are Ψ-independent. In the examples considered we see that the condition that $j_1 \star j_2$ is also an algebra homomorphism is sufficient to show that j_1, j_2 satisfy the commutation relations for Ψ-independence.

Second, in the definition of $K_l^{\mathrm{op}}, K_l^{\mathrm{cop}}$ we have the original antipode. We should use the antipode of the opposite or co-opposite quantum/braided group though, i.e. S^{-1}. But in the applications we will see that the condition that $j_1 \star (j_2 \circ S)$ is an algebra homomorphism implies that there is no difference between $j_2 \circ S$ and $j_2 \circ S^{-1}$.

8.1.3 *The braided line*

If we look for left or right Gaussian functionals in the sense of Bernstein on the braided line $\mathbb{R}_q$ we find only the trivial solution $\varphi = \varepsilon$ (the counit). But, as the following theorem shows, we can find non-trivial weakly Gaussian functionals in the sense of Bernstein.

Theorem 8.1.7 *Let φ be a weakly right Gaussian functional in the sense of Bernstein on the braided line $\mathbb{R}_q$, $q \neq 0, 1$, and q not a root of unity. Then there exists a constant $a \in \mathbb{C}$ such that φ is given by*

$$\varphi(x^n) = a^n.$$

Conversely, if φ is of this form, then it is a weakly right Gaussian functional in the sense of Bernstein.

Proof: We shall determine the pairs of functionals (φ, ψ) on $\mathbb{R}_q$ such that there exists a quantum probability space $(\mathcal{A}, \Phi)$ and independent quantum random variables $j_1, j_2 : \mathbb{R}_q \to \mathcal{A}$ such that $\varphi = \Phi \circ j_1$ and $\psi = \Phi \circ j_2$ and the applications

$$J = m_{\mathcal{A}} \circ (j_1 \otimes j_2) \circ \Delta_{\mathbb{R}_q} \text{ and } K_r = m_{\mathcal{A}} \circ (j_1 \otimes j_2) \circ (\mathrm{id} \otimes S) \circ \Delta_{\mathbb{R}_q} \tag{8.1}$$

form a pair of independent quantum random variables. This way we will also see that there are no right Gaussian functionals in the sense of Bernstein.

Step 1. We can assume that $\mathcal{A}$ is of the form $\mathbb{R}_q \tilde{\otimes} \mathbb{R}_q / \{(x')^2 = 0\} = \mathbb{C} << x, x' >> / \{x'x = qxx', (x')^2 = 0\}$, where $\mathbb{C} << x, x' >>$ denotes the free complex algebra with two generators.

Lemma 8.1.8 *Suppose we are given an algebra C and two algebra homomorphisms $k_1, k_2 : \mathbb{R}_q \to C$ such that $k_1 \star k_2$ and $k_1 \star (k_2 \circ S)$ are also algebra homomorphisms. Then there exists an algebra homomorphism $\pi : \mathbb{R}_q \tilde{\otimes} \mathbb{R}_q / \{(x')^2 = 0\} \to C$ and $j_1, j_2 : \mathbb{R}_q \to \mathbb{R}_q \tilde{\otimes} \mathbb{R}_q / \{(x')^2 = 0\}$ with $j_1(x) = x$, $j_2(x) = x'$, and $k_i = \pi \circ j_i$, $i = 1, 2$.*

Proof of Lemma. Set $\pi(x) = k_1(x)$ and $\pi(x') = k_2(x)$. We can directly check the relations to verify that π is an algebra homomorphism. $\pi(x')\pi(x) = q\pi(x)\pi(x')$ follows since $k_1 \star k_2$ is an algebra homomorphism:

$$\begin{aligned} 0 &= k_1 \star k_2(x^2) - (k_1 \star k_2(x))^2 \\ &= \pi(x)^2 + (1+q)\pi(x)\pi(x') + \pi(x')^2 - (\pi(x) + \pi(x'))^2 \\ &= q\pi(x)\pi(x') - \pi(x')\pi(x), \end{aligned}$$

Similarly, checking on $x^2 = x \cdot x$ that $k_1 \star (k_2 \circ S)$ is an algebra homomorphism yields $\pi(x')^2 = 0$. ■

Step 2. If Φ is a functional on $\mathbb{R}_q \tilde{\otimes} \mathbb{R}_q / \{(x')^2 = 0\}$ such that j_1 and j_2 are independent, then Φ is of the form $\varphi \otimes \psi$, $\varphi : \mathbb{R}_q \to \mathbb{C}$, $\psi : \mathbb{R}_q / \{(x')^2 = 0\} \to \mathbb{C}$.

This follows from the factorisation property

$$\Phi(j_{\sigma(1)}(y) j_{\sigma(2)}(z)) = \Phi \circ j_{\sigma(1)}(y) \Phi \circ j_{\sigma(2)}(z)$$

(where $y, z \in \mathbb{R}_q$, $\sigma \in S_2$), i.e. taking $\sigma = \mathrm{id}$ we can set $\varphi = \Phi \circ j_1$, $\psi = \Phi \circ j_2$. Furthermore, setting $y = x$ and $z = x^n$ we see

$$\begin{aligned} \psi(x)\varphi(x^n) &= \Phi(j_2(x) j_1(x^n)) = \Phi(x' x^n) = q^n \Phi(x^n x') \\ &= q^n \psi(x)\varphi(x^n). \end{aligned}$$

Since we suppose that q is not a root of unity this means $\varphi(x^n) = 0$ for all $n \geq 1$ if $\psi(x) \neq 1$. Vice versa, if $\varphi(x^n) \neq 0$ for some $n \geq 1$, then $\psi(x) = 0$.

Step 3. The only pair (φ, ψ) such that $J = j_1 \star j_2$ and $K_r = j_1 \star (j_2 \circ S)$ are independent is φ as given in Theorem 8.1.7, and $\psi = \varepsilon$.

a. Let us first consider the case $\psi(x) = c \neq 0$. Then $\varphi = \varepsilon$. We check the factorisation property for J and K_r on $y = z = x$:

$$\Phi(J(x)K_r(x)) = \Phi((x+x')(x-x')) = \Phi(x^2 + (q-1)xx' + (x')^2) = 0,$$

but

$$\Phi(J(x))\Phi(K_r(x)) = \Phi(x+x')\Phi(x-x') = -c^2$$

so that the factorisation property is only satisfied for $c = 0$.

b. So let now $\psi(x) = 0$, $\varphi(x^n) = c_n$. We check the factorisation property on $y = x^n$, $z = x^m$. We have $J(x^n) = j_1 \otimes j_2 \left(\sum_{\nu=0}^{n} \begin{bmatrix} n \\ \nu \end{bmatrix}_q x^\nu (x')^{n-\nu} \right) = x^n + q_n x^{n-1}x'$ and $K_r(x^m) = x^m - q_m x^{m-1}x'$, since $(x')^2 = 0$. For the product we get $J(x^n)K_r(x^m) = x^{n+m} + (q_n q^m - q_m)x^{n+m-1}x'$. This implies

$$c_{n+m} = \Phi(J(x^n)K_r(x^m)) = \Phi(J(x^n))\Phi(K_r(x^m)) = c_n c_m,$$

and thus $c_n = a^n$ with $a = c_1$.

Step 4. One checks without difficulty that J and K_r verify the factorisation property for all $\sigma \in S_2$. But on $\mathbb{R}_q \tilde{\otimes} \mathbb{R}_q / \{(x')^2 = 0\}$ they don't satisfy the right commutation relations. To remedy this we quotient by $x' = 0$, this is possible without changing the functionals since $\psi(x) = 0$. We get thus $\tilde{A} = \mathbb{R}_q$, $\tilde{j}_1 = \mathrm{id} = J = K_r$, $\tilde{j}_2 = e \circ \varepsilon$ and $\varphi(x^n) = a^n$ for all n, and (J, K_r) satisfy (trivially) the conditions of Bose or symmetric independence. ∎

Corollary 8.1.9 *The weakly left Gaussian functionals on the braided line $\mathbb{R}_q$ are exactly the functionals that can be written as*

$$\varphi(x^n) = a^n$$

for some constant $a \in \mathbb{C}$.

Proof: Note that the first step of the proof of Theorem 8.1.7 shows that the quantum random variables j_1 and j_2 have to be Ψ-independent and that $j_2 \circ S$ and $j_2 \circ S^{-1}$ are identical.

We show that $\mathbb{R}_{q^{-1}}$ and $\mathbb{R}_q^{\mathrm{cop}}$ are isomorphic, and then apply Lemma 8.1.6.

The opposite coproduct gives $\Delta x = x + x'$, if we extend (using the inverse braiding) as an algebra homomorphism we get $\Delta x^n = \sum_{\nu=0}^{n} \begin{bmatrix} n \\ \nu \end{bmatrix}_{q^{-1}} x^\nu (x')^{n-\nu}$.

We see that the identity map is a braided Hopf algebra isomorphism between $\mathbb{R}_{q^{-1}}$ and $\mathbb{R}_q^{\mathrm{cop}}$. ∎

8.1.4 *The braided plane*

Theorem 8.1.10 *The weakly right Gaussian functionals in the sense of Bernstein on the braided plane $\mathbb{C}_q^{2|0}$ are exactly the functionals of the form*

$$\varphi(x^n y^m) = \delta_{m,0} z^n \quad \text{or} \quad \varphi(x^n y^m) = \delta_{n,0} z^m$$

where $z \in \mathbb{C}$ is some constant.

Proof: The proof is based on the same ideas as in the case of the braided line, we indicate only the results of each step.

Step 1. We can suppose $\mathcal{A} = \mathbb{C}_q^{2|0} \tilde{\otimes} \mathbb{C}_q^{2|0} / \{(x')^2 = 0, x'y' = 0, (y')^2 = 0\}$.

Step 2. We can suppose $\Phi = \varphi \otimes \psi$ and

$$\begin{aligned} \psi(x) \neq 0 \quad &\Rightarrow \quad \varphi(x^n y^m) = 0 \text{ if } n+m > 0 \\ \psi(y) \neq 0 \quad &\Rightarrow \quad \varphi(x^n y^m) = 0 \text{ if } n+m > 0 \end{aligned}$$

Step 3. We can show $\psi = \varepsilon$, and φ is of the form given in the theorem.

Step 4. We get $\tilde{\mathcal{A}} = \mathbb{C}_q^{2|0}$, $\tilde{\jmath}_1 = \mathrm{id}$, $\tilde{\jmath}_2 = 1 \circ \varepsilon$. ∎

Corollary 8.1.11 *A functional on the braided plane $\mathbb{C}_q^{2|0}$ is a weakly left Gaussian functional in the sense of Bernstein if and only if it is a weakly right Gaussian functional in the sense of Bernstein on $\mathbb{C}_q^{2|0}$.*

Proof: We use Lemma 8.1.6, as for the braided line. The isomorphism between $\left(\mathbb{C}_q^{2|0}\right)^{\mathrm{cop}}$ and $\mathbb{C}_{q^{-1}}^{2|0}$ is defined by $x^n y^m \mapsto y^n x^m$. ∎

8.1.5 *The braided q-Heisenberg algebra*

Theorem 8.1.12 *Let $q \in \mathbb{C}\backslash\{0\}$, $q \neq 1$, and q not a root of unity.*

The weakly right Gaussian functionals in the sense of Bernstein on the braided q-Heisenberg-Weyl algebra are exactly the functionals which can be defined by

$$\varphi(a^n d^r c^s) = m_a^n m_c^s \delta_{r0}$$

with constants $m_a, m_c \in \mathbb{C}$.

Proof: We proceed as for the braided line.

Step 1. We can assume that $\mathcal{A}$ is of the form $\mathrm{HW}_q \tilde{\otimes} \mathrm{HW}_q / \{c'a' - 2b' = 0, (a')^2 = 0, (c')^2 = 0\}$

Lemma 8.1.13 *Suppose we are given an algebra C and two algebra homomorphisms $k_1, k_2 : \mathrm{HW}_q \to C$ such that $k_1 \star k_2$ and $k_1 \star (k_2 \circ S)$ are also algebra homomorphisms. Then (in addition to the HW_q-relations) $k_i(a), k_i(b), k_i(c)$, $i = 1, 2$, satisfy the following relations:*

$$\begin{array}{lll} k_2(a)k_1(a) = qk_1(a)k_2(a) & k_2(a)k_1(b) = qk_1(b)k_2(a) & k_2(a)k_1(c) = k_1(c)k_2(a) \\ k_2(b)k_1(a) = k_1(a)k_2(b) & k_2(b)k_1(b) = qk_1(b)k_2(b) & k_2(b)k_1(c) = qk_1(c)k_2(b) \\ k_2(c)k_1(a) = \frac{1}{q}k_1(a)k_2(c) & k_2(c)k_1(b) = k_1(b)k_2(c) & k_2(c)k_1(c) = qk_1(c)k_2(c) \\ k_2(a)^2 = 0 & k_2(c)k_2(a) = 2k_2(b) & k_2(c)^2 = 0. \end{array}$$

Proof of Lemma: The first nine relations are exactly the braid relations, they follow from the condition that $k_1 \star k_2$ is an algebra homomorphism.

The last three relations follow if we also require $k_1 \star (k_2 \circ S)$ to be an algebra homomorphism. ∎

We shall from now on simply write a, b, c for $j_1(a), j_1(b), j_1(c)$ and a', b', c' for $j_2(a), j_2(b), j_2(c)$.

Step 2. If j_1 and j_2 are independent, then Φ can be written as the tensor product of two functionals, i.e. $\Phi = \varphi \otimes \psi$. Furthermore, we have the following restrictions

$$\begin{aligned} \psi(a) \neq 0 &\Rightarrow \varphi(a^n b^m c^r) = 0 \text{ if } n \neq 0 \text{ or } m \neq 0 \\ \psi(b) \neq 0 &\Rightarrow \varphi(a^n b^m c^r) = 0 \text{ if } m \neq 0 \text{ or } r \neq 0 \\ \psi(c) \neq 0 &\Rightarrow \varphi(a^n b^m c^r) = 0 \text{ if } n \neq r. \end{aligned}$$

To show this we apply the factorisation property twice, e.g.

$$\psi(c)\varphi(a^n b^m c^r) = \Phi(c' a^n b^m c^r) = q^{r-n}\Phi(a^n b^m c^r c') = q^{r-n}\varphi(a^n b^m c^r)\psi(c).$$

Step 3. $J = j_1 \star j_2$ and $K_r = j_1 \star (j_2 \circ S)$ only satisfy the factorisation property if $\psi = \varepsilon$ and φ has the form given in Theorem 8.1.12.

First we show $\psi = \varepsilon$. Suppose $\psi(a) \neq 0$. Then φ is of the form $\varphi(a^n b^m c^r) = \delta_{n0}\delta_{m0}m_r$. Checking the factorisation property for J and K_r on a leads to a contradiction:

$$\begin{aligned} 0 &= \Phi((a + a')(a - a')) = \Phi(J(a)K_r(a)) = \Phi(J(a))\Phi(K_r(a)) \\ &= \Phi(a + a')\Phi(a - a') = -\psi(a)^2, \end{aligned}$$

and therefore $\psi(a) = 0$. In the same way we obtain $\psi(b) = \psi(c) = 0$.

So we have seen that all non-trivial moments of ψ vanish. Let $\varphi(a^n d^m c^r) = m_{nmr}$, then

$$\begin{aligned} \varphi_J(a^n d^m c^r) &= \Phi \circ J(a^n d^m c^r) = m_{nmr} \\ \varphi_{K_r}(a^n d^m c^r) &= \Phi \circ K_r(a^n d^m c^r) = m_{nmr}, \end{aligned}$$

and the factorisation property implies

$$\begin{aligned} \varphi_{K_r}(a^{n_1} d^{m_1} c^{r_1})\varphi_J(a^{n_2} d^{m_2} c^{r_2}) &= \Phi(K_r(a^{n_1} d^{m_1} c^{r_1})J(a^{n_2} d^{m_2} c^{r_2})) \\ &= \Phi(a^{n_1} d^{m_1} c^{r_1} a^{n_2} d^{m_2} c^{r_2}). \end{aligned}$$

which leads to

$$\begin{aligned} m_{n_1 m_1 r_1} m_{n_2 m_2 r_2} &= \sum_{k=0}^{r_1 \wedge n_2} \frac{(q^{-1})_{r_1}!(q^{-1})_{n_2}!(-1)^k}{(q^{-1})_{r_1 - k}!(q^{-1})_{n_2 - k}!(q^{-1})_k!} \\ &\quad \times q^{-m_1(n_2 - k)} q^{m_2(r_1 - k)} m_{n_1 + n_2 - k, m_1 + m_2 + k, r_1 + r_2 - k} \end{aligned}$$

Taking only the n's different from zero (or the m's, or the r's), we immediately see that all moments are completely determined by the first order moments, $m_{n00} = (m_{100})^n$ ($m_{0m0} = (m_{010})^m$, $m_{00r} = (m_{001})^r$), and $m_{nmr} = m_{100}^n m_{010}^m m_{001}^r$.

Furthermore, choosing (100) and (001) we get

$$m_{101} = m_{100}m_{001} = m_{001}m_{100} = m_{101} - m_{010},$$

i.e. $m_{010} = 0$.

The general solution is thus given by

$$\varphi(a^n d^m c^r) = \delta_{m0} m_a^n m_c^r,$$

(one checks that this really defines a solution).

Step 4. We can now take $\mathcal{A} = \mathrm{HW}_q/\{d = 0\} = \mathbb{C}[a, c]$, i.e. the algebra of polynomials in two variables, and $j_1 = \pi = J = K_r$ and $j_2 = \pi \circ 1 \circ \varepsilon$ without changing the functionals. π is here the canonical projection $\pi : \mathrm{HW}_q \to \mathrm{HW}_q/\{d = 0\}$. ∎

Corollary 8.1.14 *The sets of weakly left and weakly right Gaussian functionals in the sense of Bernstein on the braided q-Heisenberg-Weyl group* HW_q *are identical.*

Proof: Notice that $a^n d^m c^r \mapsto (-1)^m c^n d^m a^r$ defines a braided Hopf algebra isomorphism between $\mathrm{HW}_q^{\mathrm{cop}}$ and $\mathrm{HW}_{q^{-1}}$, and apply Lemma 8.1.6. ∎

8.1.6 *Positivity*

The braided line. Let $q \in \mathbb{R}$. Then $x^* = x$ defines a $*$-structure on $\mathbb{R}_q$. One easily verifies the following result.

Proposition 8.1.15 *Let* $q \in \mathbb{R}$, $q \notin \{-1, 0, 1\}$.

The positive weakly right Gaussian functionals in the sense of Bernstein on the braided line $\mathbb{R}_q$ *are exactly the linear functionals* φ *such that*

$$\varphi(x^n) = c^n \qquad \text{for all } n \in \mathbb{N},$$

for some real constant $c \in \mathbb{R}$.

The braided plane. If q is real then we can define a $*$-structure on the braided plane $\mathbb{C}_q^{2|0}$ by $x^* = y$, $y^* = x$. The only weakly right Gaussian functional in the sense of Bernstein that is positive with respect to this $*$-structure is the counit ε. To see this evaluate a weakly right Gaussian functional in the sense of Bernstein φ_z (where $\varphi_z(x^n y^m) = z^n \delta_{m,0}$ or $z^m \delta_{n,0}$, $z \in \mathbb{C}$) on $a^* a$ for $a = x + y$ and $a = x - y$. We get

$$\begin{aligned} \varphi_z\Big((x+y)^*(x+y)\Big) &= \varphi_z(x^2 + (1+q)xy + y^2) = z^2, \\ \varphi_z\Big((x-y)^*(x-y)\Big) &= \varphi_z(-x^2 + (1+q)xy - y^2) = -z^2, \end{aligned}$$

so that φ_z can only be positive, if $z = 0$.

The braided q-Heisenberg-Weyl group. Let q be real, and consider the $*$-structure defined by $a^* = c$, $c^* = a$, $d^* = d$.

Proposition 8.1.16 *Let $q \in \mathbb{R}$, $q \notin \{-1, 0, 1\}$.*

The weakly right Gaussian functionals in the sense of Bernstein on the braided q-Heisenberg-Weyl group HW_q *are positive if and only if $m_a = \overline{m_c}$.*

Proof: In the proof of Theorem 8.1.12 we saw that we can assume that j_1, j_2, J, K_r take values in the algebra of polynomials in two variables, denoted e.g. by $\alpha = j_1(a), \gamma = j_1(c)$. For j_1, j_2, J, K_r to be $*$-algebra homomorphisms we need $\alpha^* = \gamma$ and $\gamma^* = \alpha$. Introduce $x = \alpha + \gamma$, $y = i(\alpha - \gamma)$, then $x^* = x$, $y^* = y$. In these coordinates φ is given by $\varphi(x^n y^m) = (m_a + m_c)^n i^m (m_a - m_c)^m$. We see that φ is positive if and only if $m_a + m_c$ and $i(m_a - m_c)$ are real, i.e. if $m_a = \overline{m_c}$. ∎

8.1.7 *Remarks*

The result of Section 8.1 has to be considered as a negative result. The only Gaussian functionals in the sense of Bernstein that we found, can be considered as deterministic or Dirac laws, in the sense of the following definition.

Definition 8.1.17 *A state ϕ on $\mathcal{B}$ is called* deterministic, *or a* Dirac law, *if there exists a quantum probability space $(\mathcal{A}, \Phi)$ and a quantum random variable $j : \mathcal{B} \to \mathcal{A}$ such that the pair (j, j) is independent and $\Phi \circ j = \phi$.*

The Bernstein functionals obtained here do not form convolution semi-groups. It is interesting to remark that they can be considered as quantum convolution semi-groups, though. Note that a continuous convolution semi-group (ccs) $\{\mu_t; t \in \mathbb{R}_+\}$ (see Section 8.2 for their definition) defines a map from $\mathcal{A}$ to the algebra of continuous functions on $\mathbb{R}_+$ via $f_a(t) = \mu_t(a)$ such that

$$f_a(t + t') = \sum f_{a^{(1)}}(t) f_{a^{(2)}}(t') \qquad \text{and} \qquad f_a(0) = \varepsilon(a),$$

where $\Delta a = \sum a^{(1)} \otimes a^{(2)}$ (Sweedler's notation). But $f(t) \mapsto f(t + t')$ and $f(t) \mapsto f(0)$ are nothing else but the coproduct and counit of the algebra of functions on the semi-group $(\mathbb{R}_+, +)$, i.e. the map $a \mapsto f_a(t)$ is a coalgebra homomorphism.

This suggests the following definition, if one wants to "deform" the t-dependence of convolution semi-groups.

Definition 8.1.18 *Let C be a coalgebra. A C-*quantum convolution semi-group *(C-qcs) on $\mathcal{A}$ is a coalgebra homomorphism from $\mathcal{A}$ to C.*

The family of weakly right Gaussian functionals in the sense of Bernstein on the braided plane $\mathbb{C}_q^{2|0}$ e.g. gives two families of $\mathbb{R}_q$-qcs $x^n y^m \mapsto \alpha^n z^n \delta_{m0}$ and $x^n y^m \mapsto \alpha^m z^m \delta_{n0}$, where z denotes now the generator of $\mathbb{R}_q$, and $\alpha \in \mathbb{C}$ is some constant.

The weakly right Gaussian functionals in the sense of Bernstein on the braided q-Heisenberg-Weyl group also give rise to $\mathbb{R}_q$-qcs if one of the constants is replaced by the generator of $\mathbb{R}_q$. The original functional can be recuperated, if the $\mathbb{R}_q$-qcs are composed with weakly right Gaussian functionals in the sense of Bernstein on $\mathbb{R}_q$.

The family of one-dimensional distributions of the quantum Brownian motion defined by Bozejko and Speicher [BS91] can also be interpreted as a $\mathbb{R}_q$-qcs (for $0 < q < 1$). These functionals can be represented on the braided q-Heisenberg-Weyl group as

$$\varphi_t(a^n b^m c^r) = t^m \delta_{n0}\delta_{r0}, \qquad \forall n, m, r \in \mathbb{N}.$$

To see this consider the unitary representation of HW_q on the Hilbert space $H = \mathrm{span}\{\psi_n; n \in \mathbb{N}\}$ with inner product $< \psi_n|\psi_m >= \delta_{nm} q_n!$, defined by

$$\begin{aligned}
\rho_t(a)\psi_n &= t\psi_{n+1} \\
\rho(c)\psi_n &= \begin{cases} 0 & \text{if } n = 0 \\ q_n\psi_{n-1} & \text{if } n > 0 \end{cases} \\
\rho(b)\psi_n &= t\psi_n.
\end{aligned}$$

Then in the state $\Phi =< \psi_0| \cdot \psi_0 >$ we get $\Phi \circ \rho_t = \varphi_t$.

Interpreting t as the generator of $\mathbb{R}_q$ we have a $\mathbb{R}_q$-qcs: $a^n b^m c^r \mapsto t^m \delta_{n0}\delta_{r0}$.

8.2 UNIQUENESS OF EMBEDDING

A *continuous convolution semi-group* (ccs) of functionals on a (braided) Hopf algebra $\mathcal{A}$ is a family of functionals $\{\varphi_t : \mathcal{A} \to \mathbb{C}; t \in \mathbb{R}_+\}$ indexed by $\mathbb{R}_+$ such that

1. $\varphi_0 = \varepsilon$ (the counit).
2. $\varphi_s \star \varphi_t = \varphi_{s+t}$ for all $s, t \in \mathbb{R}_+$. Here the convolution is defined by $\phi \star \psi = (\phi \otimes \psi) \circ \Delta$,
3. $\varphi_t(a)$ is continuous as a function of t for all $a \in \mathcal{A}$.

M. Schürmann [Sch93] has shown that every ccs on a (braided) Hopf algebra has a generator ψ and can be expressed as a (convolution) exponential $\varphi_t = \exp_\star(t\psi) = \varepsilon + t\psi + \frac{t^2}{2}\psi\star\psi + \cdots + \frac{t^n}{n!}\psi^{\star n} + \cdots$. Here we are interested in the question whether such a ccs is already uniquely determined by its value for one fixed value $t \neq 0$. D. Neuenschwander and others [Neu96, Bal85, Pap94] have shown on connected simply-connected nilpotent Lie groups that if a probability measure is embeddable, then this embedding is unique (under some regularity conditions that guarantee the existence of moments). In this section we will introduce a notion of nilpotency and show that the above uniqueness result can be extended to nilpotent (braided) Hopf algebras.

For Hopf algebras in general this result is not true, we can e.g. translate the counter-example for ccs on $\mathbb{Z}_3$ from Heyer's book [Hey77] to a counter-example on the Hopf algebra of functions on $\mathbb{Z}_3$. This counter-example also shows that our notion of nilpotency can not include all algebras of functions on nilpotent groups.

8.2.1 *Definition of nilpotency*

Nilpotent connected simply-connected Lie groups are diffeomorphic to some $\mathbb{R}^n$, and the group law can be brought into the following form (e.g. with the Campbell-Hausdorff formula)

$$(x_1, \ldots, x_n) \cdot (y_1, \ldots, y_n) = (z_1, \ldots, z_n)$$

where the z_i are of the form

$$z_i = x_i + y_i + p_i(x_1, \ldots, x_{i-1}, y_1, \ldots, y_{i-1})$$

where the p_i are polynomials. We will use the analogue of this property as a definition of nilpotency for (braided) Hopf algebras.

Definition 8.2.1 *Let $\mathcal{A}$ be a (braided) Hopf algebra with unit element* 1. *We say that $\mathcal{A}$ is* nilpotent *if there exists a well-ordered set $\mathcal{I}$ and a basis of the form $\{1\} \cup \mathcal{V}$ with $\mathcal{V} = \{a_\imath; \imath \in \mathcal{I}\}$ indexed by $\mathcal{I}$ such that for all $a_\imath \in \mathcal{V}$*

$$\Delta(a_\imath) = a_\imath \otimes 1 + 1 \otimes a_\imath + \sum_{\imath',\imath'' < \imath} C^{\imath}_{\imath',\imath''} a_{\imath'} \otimes a_{\imath''} \tag{8.2}$$

where (for fixed $\imath$) only finitely many $C^{\imath}_{\imath',\imath''} \neq 0$.

The above discussion shows that the Hopf algebras of polynomials on connected simply-connected nilpotent Lie groups are nilpotent in the sense of our definition.

For the "unbraided" or quantum group case (i.e. $\Psi = \tau$) there exists a nilpotent Lie algebra that is naturally associated to a nilpotent quantum group. It is obtained in the same way that can be used to reconstruct the Lie algebra of a Lie group from the Hopf algebra of smooth functions on the groups, i.e. as the space of derivations at the neutral element.

Proposition 8.2.2 *Let $\mathcal{A}$ be a nilpotent quantum group and suppose that*

$$L(\mathcal{A}) = \{f : \mathcal{A} \to \mathbb{C} \text{ linear} | f(ab) = f(a)\varepsilon(b) + \varepsilon(a)f(b)\}$$

is finite-dimensional. Then $L(\mathcal{A})$ is a nilpotent Lie algebra with the commutator $[f, g] = m(f \otimes g) - m(g \otimes f)$, $m = \Delta^*_{\mathcal{A}}$.

Proof: $L(\mathcal{A})$ is the set of primitive elements of the dual of $\mathcal{A}$, and thus a Lie algebra, see Page 49. Set $L^\imath = \{f \in L(\mathcal{A}) | f(a_{\imath'}) = 0 \text{ for } \imath' < \imath\}$. We have $[L(\mathcal{A}), L^\imath] \subseteq L^{\imath+1}$, since

$$\begin{aligned} [f, g](a_{\imath'}) &= (f \otimes g - f \otimes g) \left(\sum_{\imath'',\imath''' < \imath'} C^{\imath}_{\imath'',\imath'''} a_{\imath''} \otimes a_{\imath'''} \right) \\ &= 0 \end{aligned}$$

for $f \in L$, $g \in L^\imath$, $\imath' \leq \imath$. But since $L(\mathcal{A})$ is finite-dimensional, this implies that we actually have a finite sequence $L_0 = L(\mathcal{A}) \supset L^1 \supset \cdots \supset L^{r+1} = \{0\}$ of ideals, and thus that $L_0 = L(\mathcal{A})$ is nilpotent. ∎

There does not seem to be a natural generalisation of this result for braided groups, because the primitive elements of a braided Hopf algebra do not have an obvious Lie algebra structure.

8.2.2 *Poincaré-Birkhoff-Witt property and nilpotency*

We say that a (braided) Hopf algebra $\mathcal{A}$ has the Poincaré-Birkhoff-Witt (PBW) property if there exists a set of elements (called generators) $\{a_1, \ldots, a_d\}$ such that $\{a_1^{k_1} \cdots a_d^{k_d}; k = (k_1, \ldots, k_d) \in \mathbb{N}^d\}$ is a basis of $\mathcal{A}$. We shall also use the multi-index notation $a^k = a_1^{k_1} \cdots a_d^{k_d}$.

Proposition 8.2.3 *Let $\mathcal{A}$ have PBW property with generators $a_1, \ldots, a_d$. If for all $i, j = 1, \ldots, d$*

$$[a_i, a_j] = \sum_{k \in \mathbb{N}^{\max(i,j)-1}} P_{i,j,k} a^k, \tag{8.3}$$

$$\Delta(a_i) = a_i \otimes 1 + 1 \otimes a_i + \sum_{k,k' \in \mathbb{N}^{i-1}} D^i_{k,k'} a^k \otimes a^{k'} \tag{8.4}$$

then $\mathcal{A}$ is nilpotent.

Proof: We are going to use the index set $\mathcal{I} = \mathbb{N}^d$ with the lexicographic order and the PBW basis $\mathcal{V} = \{a_1^{k_1} \cdots a_d^{k_d}; k = (k_1, \ldots, k_d) \in \mathbb{N}^d\}$. Let $0 \leq \delta \leq d$ and define $\mathcal{A}^{(\delta)}$ to be the linear span of $\{a_1^{k_1} \cdots a_\delta^{k_\delta}; k = (k_1, \ldots, k_\delta) \in \mathbb{N}^\delta\}$. By the condition on the product, $\mathcal{A}^{(\delta)}$ is a subalgebra. Furthermore, we even have $[a_\delta, \mathcal{A}^{(\delta-1)}] \subseteq \mathcal{A}^{(\delta-1)}$. Thus

$$\begin{aligned}
\Delta(a_1^{k_1} \cdots a_\delta^{k_\delta}) &= \Delta(a_1)^{k_1} \cdots \Delta(a_\delta)^{k_\delta} \\
&= (a_1 \otimes 1 + 1 \otimes a_1)^{k_1} \left(a_2 \otimes 1 + 1 \otimes a_2 + \sum_{r,s} D^2_{r,s} a_1^r \otimes a_1^s \right)^{k_2} \cdots \\
&\qquad \left(a_\delta \otimes 1 + 1 \otimes a_\delta + \sum_{\imath',\imath'' \in \mathbb{N}^{\delta-1}} D^\delta_{\imath',\imath''} a^{\imath'} \otimes a^{\imath''} \right)^{k_\delta} \\
&= \sum_{k'+k''=k} \prod_{i=1}^{\delta} \binom{k_i}{k'_i} a^{k'} \otimes a^{k''} + K,
\end{aligned}$$

where $K \in \mathcal{A}^{(\delta-1)} \otimes \mathcal{A}^{(\delta-1)}$, so that $\Delta(a^k)$ is of the form required in Definition 8.2.1. ∎

Remark: A similar result for braided Hopf algebras with PBW property can be formulated if we require in addition that the braiding $a'_i a_j = \psi(a_i \otimes a_j) = \sum_{kl} R^{kl}_{ij} a_l \otimes a_k$ doesn't involve generators that come "after" a_i and a_j, i.e.

$$R^{kl}_{ij} = 0 \qquad \text{if} \qquad k, l \geq \max(i,j). \tag{8.5}$$

The following result is immediate from the formula for the coproduct of the HW_q algebra, see Equation (3.28).

Proposition 8.2.4 *The braided q-Heisenberg-Weyl algebra* HW_q *is nilpotent.*

8.2.3 *Uniqueness of embedding*

Lemma 8.2.5 *Let $\mathcal{A}$ be a nilpotent (braided) Hopf algebra and $\Phi : \mathcal{A} \to \mathbb{C}$ a normed functional on $\mathcal{A}$. Then there exists a unique normed functional $\Xi : \mathcal{A} \to \mathbb{C}$ such that*

$$\Phi = \Xi \star \Xi.$$

Proof: We can directly calculate the action of Ξ on the basis $\mathcal{V}$. On 1 we have

$$1 = \Phi(1) = (\Xi \otimes \Xi)(\Delta 1) = \Xi(1)^2.$$

so that Ξ can be chosen to be normed, i.e. $\Xi(1) = 1$. Now Ξ is uniquely determined by

$$\Phi(a_\imath) = (\Xi \otimes \Xi)(\Delta(a_\imath)).$$

In fact, suppose we have calculated $\Xi(a_\imath)$ already for all $\imath \in M \subseteq I$. Then

$$\Xi(a_\kappa) = \frac{1}{2}\left\{\Phi(a_\kappa) - \sum_{\imath',\imath''<\kappa} C^{\kappa}_{\imath',\imath''}\Xi(a_{\imath'})\Xi(a_{\imath''})\right\}$$

allows to calculate $\Xi(a_\kappa)$ for $\kappa = \min \mathcal{I} \backslash M$. ∎

Theorem 8.2.6 *If ϕ is normed, infinitely divisible, and embeddable, then the embedding is unique.*

Proof: Lemma 8.2.5 implies that the embedding is unique on the dyadic numbers. The rest follows from continuity. ∎

8.2.4 *Remark*

Note that e.g. on $\mathbb{R}$ we can also find a "root" for functionals that are not divisible (as probability measures). Take e.g. a symmetric Bernoulli law (concentrated on ± 1). The moments are

$$\Phi(x^n) = \begin{cases} 1 & \text{if } n \text{ is even,} \\ 0 & \text{if } n \text{ is odd.} \end{cases}$$

Then (by the lemma in Subsection 8.2.3) there exists a normed functional Ξ such that $\Phi = \Xi \star \Xi$. But Ξ does not correspond to a probability measure.

We can calculate $\Xi_n = \Xi(x^n)$ recursively by

$$\Xi_n = \frac{1}{2}\left\{1 - \sum_{\nu=1}^{n-1} \binom{n}{\nu} \Xi_\nu \Xi_{n-\nu}\right\} \qquad \text{if } n \text{ is even,}$$

$$\Xi_n = -\frac{1}{2}\sum_{\nu=1}^{n-1} \binom{n}{\nu} \Xi_\nu \Xi_{n-\nu} \qquad \text{if } n \text{ is odd.}$$

All odd moments vanish, and $\Xi_0 = 1$, $\Xi_2 = \frac{1}{2}$, $\Xi_4 = -1$, etc. We see that Ξ is not positive.

8.3 GAUSSIAN SEMI-GROUPS IN THE SENSE OF BERNSTEIN

Since the results of Section 8.1 must be interpreted as a rather "negative" result, the question arises if there are "better" extensions of the Bernstein property to quantum groups and braided groups.

Let us consider the situation on Lie groups. If from the beginning one suggests that the measure is embeddable into a continuous convolution semi-group (ccs) of probability measures on the Lie group G, then a natural version of the Bernstein theorem on G is given in Hazod ([Haz77], Satz III.3.3) and Heyer ([Hey77], Theorem 8.4) by considering the generating distribution of the ccs. A ccs $\{\mu_t\}_{t\geq 0} \subset M^1(G)$ is a topological monoid homomorphism

$$[0,\infty[\ni t \mapsto \mu_t \in (M^1(G), \delta_e, \stackrel{w}{\to}).$$

The generating distribution of $\{\mu_t\}_{t\geq 0}$ is given by

$$\mathcal{A}(f) := \lim_{t\to 0+} \frac{1}{t} \int_G [f(x) - f(0)]\mu_t(dx)$$

for bounded complex-valued C^∞-functions f on G (cf. Siebert [Sie81, p. 119]). We write $\mu_t = \exp t\mathcal{A}$. Then $\mathcal{A}$ has always the form

$$\mathcal{A}(f) = \sum_{i=1}^d a_i X_i f(0) + \sum_{i,j=1}^d b_{i,j} X_i X_j f(0) + \int_{G\setminus\{0\}} [f(x) - f(0) - \Phi(f,x)]\eta(dx)$$

where $a = (a_1, a_2, \ldots, a_d) \in I\!R^d$, $M = (b_{i,j})_{1\leq i,j\leq d}$ is a symmetric positive semidefinite $d \times d$-matrix, η is a Lévy measure (i.e. a non-negative measure on $G\setminus\{0\}$ such that $\int_{G\setminus\{0\}} x^2/(1+x^2)\eta(dx) < \infty$),

$$\Phi(f,x) := \begin{cases} \langle x, \nabla\rangle f(0) & : \ ||x|| \leq 1, \\ \langle \frac{x}{||x||}, \nabla\rangle f(0) & : \ ||x|| > 1 \end{cases}$$

and $X_1, X_2, \ldots, X_d$ is a fixed basis of the Lie algebra $\mathcal{G}$ of G. (Cf. Siebert [Sie73, Satz 1]) For short, we will write $\mathcal{A} = (a, M, \eta)$. The ccs $\{\exp t\mathcal{A}\}_{t\geq 0}$ is called

Gaussian if $\mathcal{A} = (a, M, 0)$. Now the "semi-group version" of the Bernstein theorem reads as follows: Let $H := G \times G$ and define the map $\varphi : H \to H$ by $\varphi(x,y) := (x \cdot y, x \cdot y^{-1})$.

Theorem 8.3.1 *Let $\mathcal{A}_0$ be a generating distribution on G and consider the generating distribution $\mathcal{A} := \mathcal{A}_0 \otimes \delta_0 + \delta_0 \otimes \mathcal{A}_0$ on H. Furthermore let $\mathcal{B}(f) := \mathcal{A}(f \circ \varphi^{-1})$ and define, for any generating distributions $\mathcal{B}_1, \mathcal{B}_2$ on G, the generating distribution $\mathcal{B}_3$ on H by $\mathcal{B}_3 := \mathcal{B}_1 \otimes \delta_0 + \delta_0 \otimes \mathcal{B}_2$. There exist $\mathcal{B}_1, \mathcal{B}_2$ such that $\mathcal{B} = \mathcal{B}_3$ iff $\mathcal{A}_0, \mathcal{A}, \mathcal{B}, \mathcal{B}_i$ $(1 \leq i \leq 3)$ are Gaussian.*

8.3.1 *Definition of Gaussian convolution semi-groups in the sense of Bernstein*

Schürmann [Sch93] tells us that every convolution semi-group on a quantum or braided group has a unique generator. We will use it to define Gaussian convolution semi-groups. The map $\varphi : (x, y) \mapsto (x + y, x - y)$ becomes $\Phi = (m \otimes m) \circ (\mathrm{id} \otimes \Psi \otimes S) \circ (\Delta \otimes \Delta)$ on (braided) Hopf algebras.

Definition 8.3.2 *The semi-group μ_t on $\mathcal{A}$ with generator A is called* Gaussian in the sense of Bernstein *if $A \otimes \varepsilon + \varepsilon \otimes A$ can be written in the form $(B_1 \otimes \varepsilon + \varepsilon \otimes B_2) \circ \Phi$ with some functional $B_1, B_2 : \mathcal{A} \to \mathbb{C}$. We will call μ_t* weakly Gaussian in the sense of Bernstein *if there exists another semi-group μ_t' with generator A' such that $A \otimes \varepsilon + \varepsilon \otimes A'$ can be written in the form $(B_1 \otimes \varepsilon + \varepsilon \otimes B_2) \circ \Phi$.*

If B_1 and B_2 are elements of a bialgebra that is dually paired with $\mathcal{A}$ and if we define $\Phi^\vee = (m_\mathcal{B} \otimes m_\mathcal{B}) \circ (\mathrm{id} \otimes \Psi_{\mathcal{B} \otimes \mathcal{B}} \otimes S_\mathcal{B}) \circ (\Delta_\mathcal{B} \otimes \Delta_\mathcal{B})$, then we can also write $\Phi^\vee(B_1 \otimes \varepsilon + \varepsilon \otimes B_2)$ instead of $(B_1 \otimes \varepsilon + \varepsilon \otimes B_2) \circ \Phi$.

8.3.2 *General results*

Since Φ is linear, it follows directly from the definition that the generators of Gaussian semi-groups in the sense of Bernstein and those of weakly Gaussian semi-groups in the sense of Bernstein form vector spaces. If $\mathcal{A}$ has an involution $*$ then we want to consider only semi-groups of positive functionals. If we restrict ourselves to α-invariant functionals, then $\mu_t = \exp_\star tA$ is positive for all $t \geq 0$, if and only if A is hermitian and conditionally positive, i.e. positive on the kernel of the counit ε, cf. [Sch93, Theorem 3.2.7].

The following proposition is the analogue of one direction of Theorem 8.3.1, if A is the sum of first and second order terms of primitive elements, then A is the generator of a weakly Gaussian convolution semi-group in the sense of Bernstein.

Proposition 8.3.3 *Let $\mathcal{A}$ be a (braided) Hopf algebra, and let $X_1, X_2 : \mathcal{A} \to \mathbb{C}$ be primitive linear functionals on $\mathcal{A}$, i.e. $X_i(ab) = X_i(a)\varepsilon(b) + \varepsilon(a)X_i(b)$, $i = 1, 2$, for all $a, b \in \mathcal{A}$. Then X_1, and X_2 are generators of Gaussian convolution semi-groups in the sense of Bernstein, and $X_1 X_2$ is a generator of a weakly Gaussian convolution semigroup in the sense of Bernstein.*

Proof: To show that $A \in \{X_1, X_2, X_1X_2\}$ is (weakly) Gaussian in the sense of Bernstein we have to show that there exist B_1, B_2 such that $(B_1 \otimes \varepsilon + \varepsilon \otimes B_2) \circ \Phi = A \otimes \varepsilon + \varepsilon \otimes A'$.

It suffices to use that $S(X_i) = -X_i$ since the X_i are primitive and $(\mathrm{id} \otimes S) \circ \Psi(X_2 \otimes X_1) = \Psi(S(X_2) \otimes X_1) = -\Psi(X_2 \otimes X_1)$ since we can pull the antipode across a braiding, and to take

$$(B_1, B_2) = \begin{cases} (X_1, 0) & \text{for } A = X_1, \\ (X_2, 0) & \text{for } A = X_2, \\ (\frac{1}{2}X_1X_2, \frac{1}{2}X_1X_2) & \text{for } A = X_1X_2. \end{cases}$$

E.g. for the last case we get

$$\begin{aligned} (B_1 \otimes \varepsilon + \varepsilon \otimes B_2) \circ \Phi &= \frac{1}{2}\Delta(X_1X_2) + \frac{1}{2}(\mathrm{id} \otimes S) \circ \Delta(X_1X_2) \\ &= X_1X_2 \otimes \varepsilon + \varepsilon \otimes \frac{1}{2}(X_1X_2 + S(X_1X_2)). \end{aligned}$$

■

Remark: We see that $S(X_1X_2) = m \circ \Psi(X_1 \otimes X_2)$ is also generator of a weakly Gaussian convolution semi-group.

M. Schürmann has proposed an alternative definition, see [Sch93, Proposition 5.1.1].

Definition 8.3.4 *A functional A on a bialgebra $\mathcal{A}$ is called* quadratic, *or a* Gaussian generator in the sense of Schürmann, *if for all $a, b, c \in \mathcal{A}$ with $\varepsilon(a) = \varepsilon(b) = \varepsilon(c) = 0$ follows*

$$A(abc) = 0.$$

This definition is more general in that it requires less structure on $\mathcal{A}$, only the counit and the multiplication are needed. But, if $\mathcal{A}$ is actually a Hopf algebra (or a braided Hopf algebra), then we can show that this property implies that A is a generator of a weakly Gaussian convolution semi-group in the sense of Bernstein.

Proposition 8.3.5 *Let $\mathcal{A}$ be a (braided) Hopf algebra, and $A : \mathcal{A} \to \mathbb{C}$ a quadratic functional. Then A generates a weakly Gaussian convolution semi-group in the sense of Bernstein.*

Proof: Let $K = \ker \varepsilon$. We can write $\mathcal{A}$ as a direct sum $\mathcal{A} = K \oplus \mathbb{C}1$, just set $a = a' + \varepsilon(a)1$, with $a' = a - \varepsilon(a)1$. The coproduct of an element $a \in K$ is of the form $\Delta(a) = a \otimes 1 + 1 \otimes a + \sum a_i^{(1)} \otimes a_i^{(2)}$, with $a_i^{(1)}, a_i^{(2)} \in K$, and therefore $S(a)$ is equal to $-a$ plus some term in K^2, more precisely,

$$S(a) = -a + \sum a_i^{(1)} S(a_i^{(2)}). \tag{8.6}$$

Set $A' = \frac{1}{2}(A + A \circ S)$, $B_1 = B_2 = \frac{1}{2}A$, then A', B_1, B_2 are also quadratic functionals. We verify $A \otimes \varepsilon + \varepsilon \otimes A' = (B_1 \otimes \varepsilon + \varepsilon \otimes B_2) \circ \Phi$ with these definitions.

On $1 \otimes 1$, $1 \otimes a$, or $a \otimes 1$, with $a \in K$, the equality is obvious. This leaves the case $a \otimes b$, $a, b \in K$, to check. In this case both sides vanish, $A(a)\varepsilon(b) + \varepsilon(a)(A(b) + A(S(b))/2 = 0$, and

$$\begin{aligned}(B_1 \otimes \varepsilon + \varepsilon \otimes B_2) \circ \Phi(a \otimes b) &= B_1(ab) + B_2(aS(b)) \\ &= \frac{1}{2}\left\{A(ab) - A(ab) + \sum A\left(ab_i^{(1)} S(b_i^{(2)})\right)\right\} \\ &= 0,\end{aligned}$$

where we used Equation (8.6) for $S(b)$. ∎

8.3.3 *The braided line*

Theorem 8.3.6 *Let $q \in \mathbb{C}\backslash\{0\}$, q not a root of unity. The weakly Gaussian convolution semi-groups in the sense of Bernstein on the braided line $\mathbb{R}_q$ are exactly the convolution semi-groups whose generator is of the form*

$$A\left(\sum_{n=0}^{\infty} c_n x^n\right) = a_0 c_0 + a_1 c_1 + a_2 c_2$$

with some constants $a_0, a_1, a_2 \in \mathbb{C}$. For $q \neq 1$ they are Gaussian in the sense of Bernstein iff $a_2 = 0$.

Remark: Using the duality of $\mathbb{R}_q$ with another copy of $\mathbb{R}_q$ we can write A as $A = a_0\varepsilon + a_1 p + \frac{a_2}{1+q}p^2$, where p denotes the generator of the dual copy.

Proof: Following Definition 8.3.2 we have to look for functionals B_1, B_2 such that we can write $(B_1 \otimes \varepsilon + \varepsilon \otimes B_2) \circ \Phi$ in the form $A \otimes \varepsilon + \varepsilon \otimes A'$.

A functional on $\mathbb{R}_q$ is characterised by its values on the monomials x^n, so that using a dual copy of $\mathbb{R}_q$ with generator p, we can write $B_i = \sum_{n=0}^{\infty} b_n^{(i)} p^n$, where $b_n^{(i)} = \frac{B_i(x^n)}{q_n!}$, $i = 1, 2$. Composing $B_1 \otimes \varepsilon + \varepsilon \otimes B_2$ with Φ is equivalent to taking $\Phi^\vee(B_1 \otimes \varepsilon + \varepsilon \otimes B_2)$, where $\Phi^\vee$ is formally identical to Φ, but the coproduct, antipode, and multiplication are now those of the dual copy of $\mathbb{R}_q$.

One calculates $\Phi^\vee(p^n \otimes 1) = \Delta p^n = \sum_{\nu=0}^{n} \begin{bmatrix} n \\ \nu \end{bmatrix}_q p^\nu \otimes p^{n-\nu}$ and $\Phi^\vee(1 \otimes p^n) = (\mathrm{id} \otimes S) \circ \Delta p^n = \sum_{\nu=0}^{n} \begin{bmatrix} n \\ \nu \end{bmatrix}_q (-1)^{n-\nu} q^{\frac{(n-\nu)(n-\nu-1)}{2}} p^\nu \otimes p^{n-\nu}$, and thus

$$\Phi^\vee(B_1 \otimes \varepsilon + \varepsilon \otimes B_2) = \sum_{n,\nu} \begin{bmatrix} n \\ \nu \end{bmatrix}_q (b_n^{(1)} + (-1)^{n-\nu} q^{\frac{(n-\nu)(n-\nu-1)}{2}} b_n^{(2)}) p^\nu \otimes p^{n-\nu}.$$

This expression has to be of the form $A \otimes \varepsilon + \varepsilon \otimes A'$, i.e. the coefficients of the terms $p^\nu \otimes p^{n-\nu}$ have to vanish for $0 < \nu < n$. This is the case if and only if all coefficients except $b_0^{(i)}$, $b_1^{(i)}$, $b_2^{(i)}$ are zero. For A this gives the form given in the theorem.

A is the generator of a Gaussian semi-group in the sense of Bernstein if we have $A = A'$. This implies that the coefficients of $p^n \otimes 1$ and $1 \otimes p^n$ have to be identical. This is the case if and only if all coefficients except $b_0^{(1)}$, $b_1^{(1)}$, and $b_0^{(2)}$ vanish. It follows that $a_2 = 0$. ∎

8.3.4 *The braided plane*

Theorem 8.3.7 *Let $q \in \mathbb{C}\backslash\{0\}$, q not a root of unity. The weakly Gaussian convolution semi-groups in the sense of Bernstein on the braided plane $\mathbb{C}_q^{2|0}$ are exactly the convolution semi-groups whose generator is of the form*

$$A\left(\sum_{n,m=0}^{\infty} c_{nm}x^n y^m\right) = a_{00}c_{00} + a_{10}c_{10} + a_{01}c_{01} + a_{20}c_{20} + a_{11}c_{11} + a_{02}c_{02},$$

with some constants $a_{00}, a_{10}, a_{01}, a_{20}, a_{11}, a_{02} \in \mathbb{C}$.

Proof: We proceed as for the braided line. This gives the following conditions:

$$b_{nm}^{(1)} + (-1)^{n-\nu+m-\mu} q^{(n-\nu+m-\mu)(n-\nu+m-\mu-1)} b_{nm}^{(2)} = 0,$$

if $1 \leq \nu \leq n-1$ or $1 \leq \mu \leq m-1$. ∎

8.3.5 *The braided q-Heisenberg-Weyl group*

Proposition 8.3.8 *Let $q \in \mathbb{C}\backslash\{0\}$, $q \neq 1$, and q not a root of unity. The generators of weakly Gaussian convolution semi-groups in the sense of Bernstein on the braided q-Heisenberg-Weyl group are of the form:*

$$A\left(\sum_{n,m,p=0}^{\infty} c_{nmp}a^n b^m c^p\right)$$
$$= a_{000}c_{000} + a_{100}c_{100} + a_{010}c_{010} + a_{001}c_{001} + a_{200}c_{200} + a_{101}c_{101} + a_{002}c_{002}$$

with some constants $a_{000}, a_{100}, a_{010}, a_{001}, a_{110}, a_{020}, a_{011}$.

Remark: Using the dual of HW_q constructed in Subsection 3.9.3, we see that the space of weakly Gaussian generators in the sense of Bernstein is spanned by $\{x, y, z, x^2, z^2, xz\}$. But for $q \neq 1$, xy, y^2, and yz do not generate weakly Gaussian convolution semi-groups, unlike in the classical case. This is due to the fact that y is no longer primitive.

Proof: By Proposition 8.3.3 we know that x, $y = xz - zx, z$, x^2, z^2, and xz are generators of weakly Gaussian semi-groups, since x and z are primitive. The hard part of the proof is to show that none of the higher order terms generate Gaussian semi-groups. This involves rather tedious calculations, and we will only outline the major steps.

We take again two general functionals $B_i = \sum_{n,m,r=0}^{\infty} b^{(i)}_{nmr}$, $i = 1,2$, and check when $\Phi^\vee(B_1 \otimes \varepsilon + \varepsilon \otimes B_2) = \Delta(B_1) + (\mathrm{id} \otimes S) \circ \Delta(B_2)$ is of the form $A \otimes \varepsilon + \varepsilon \otimes A'$, i.e. under what conditions on the coefficients $b^{(i)}_{nmr}$ the mixed terms vanish. We use the fact that $\mathrm{HW}_q^\vee$ is $\mathbb{Z}^2$-graded with $\deg(x) = (1,0)$ and $\deg(z) = (0,1)$ to control which terms in the B_i contribute to a given term $x^{n_1}y^{m_1}z^{r_1} \otimes x^{n_2}y^{m_2}z^{r_2}$.

We check that x and z are indeed primitive, and use $u = \frac{xz-qzx}{1-q}$ instead of y, so that we can use Formula (3.28). The antipode of $x^n u^m z^r$ can be written as

$$S(x^n u^m z^r) = \sum_{k=-m}^{n \wedge r} D_{nmrk} x^{n-k} u^{m+r} z^{r-k}, \tag{8.7}$$

where the coefficients D_{nmrk} are defined via certain recurrence relations. No other terms can appear since S has to respect the grading. We will only need the explicit form of the antipode for a few lower order terms: $S(x) = -x$, $S(z) = -z$, $S(x^2) = qx^2$, $S(xz) = m \circ \Psi(x \otimes z) = zx = \frac{xz}{q} + \left(1 - \frac{1}{q}\right) u$, etc.

Using Equations (3.28) and (8.7), and the values of $D_{\nu\mu\rho k}$ for small ν, μ, ρ, we can write down the conditions on the $b^{(i)}_{n,m,r}$ that are equivalent to the vanishing of the coefficient of $x^{n_1}u^{m_1}z^{r_1} \otimes x^\nu u^\mu z^\rho$ in $\Phi^\vee(B_1 \otimes \varepsilon + \varepsilon \otimes B_2)$. We get for $(\nu,\mu,\rho) = (1,0,0)$ and $n_1 + m_1 + r_1 > 0$:

$$\begin{aligned} &q_{n_1+1} q^{m_1} b^{(1)}_{n_1+1,m_1,r_1} + q_{m_1+1} b^{(1)}_{n_1,m_1+1,r_1-1} - q_{n_1+1} q^{m_1} b^{(2)}_{n_1+1,m_1,r_1} \\ &+ q_{m_1+1} b^{(2)}_{n_1,m_1+1,r_1-1} = 0, \end{aligned}$$

for $(\nu,\mu,\rho) = (0,1,0)$ and $n_1 + m_1 + r_1 > 0$:

$$\begin{aligned} &q_{m_1+1} q^{r_1} b^{(1)}_{n_1,m_1+1,r_1} - q_{m_1+1} q^{r_1-1} b^{(2)}_{n_1,m_1+1,r_1} \\ &+ \left(1 - \frac{1}{q}\right) q_{n_1+1} q_{r_1+1} q^{m_1} b^{(2)}_{n_1+1,m_1,r_1+1} = 0, \end{aligned}$$

for $(\nu,\mu,\rho) = (0,0,1)$ and $n_1 + m_1 + r_1 > 0$:

$$q_{r_1+1} b^{(1)}_{n_1,m_1,r_1+1} - q_{r_1+1} b^{(2)}_{n_1,m_1,r_1+1} = 0,$$

and for $(\nu,\mu,\rho) = (0,0,2)$ and $n_1 + m_1 + r_1 > 0$:

$$\begin{bmatrix} r_1+2 \\ 2 \end{bmatrix}_q b^{(1)}_{n_1,m_1,r_1+2} + q \begin{bmatrix} r_1+2 \\ 2 \end{bmatrix}_q b^{(2)}_{n_1,m_1,r_1+2} = 0.$$

Combining these equations we can show that $b^{(1)}_{nmr} = b^{(2)}_{nmr}$ for $n + m + r \geq 2$, $b^{(i)}_{nmr} = 0$ for $n + m + r > 2$, and, by looking at the cases for small n_1, m_1, r_1 in detail, $b^{(i)}_{110} = b^{(i)}_{020} = b^{(i)}_{011} = 0$, $i = 1,2$. This implies that A and A' have the form stated in the theorem. ∎

8.3.6 $\mathbb{R}_q$-*quantum convolution semi-groups*

We can also define the notion of (weakly) Gaussian $\mathbb{R}_q$-quantum convolution semi-groups on quantum/braided groups. Let p be the functional on $\mathbb{R}_q$ defined by $p(x^n) = \delta_{n0}$ and Φ a $\mathbb{R}_q$-quantum convolution semi-group. Then we define the generator of Φ as $A_\Phi = \Phi^*(p) = p \circ \Phi$, and call a $\mathbb{R}_q$-quantum convolution semi-group (weakly) *Gaussian in the sense of Bernstein* if its generator is the generator of a (weakly) Gaussian ccs in the sense of Bernstein. With this definition we can immediately translate the previous results into characterisations of (weakly) Gaussian $\mathbb{R}_q$-quantum convolution semi-groups.

Theorem 8.3.9 *The normed (i.e. $A_\Phi(1) = 0$) weakly Gaussian $\mathbb{R}_{q'}$-quantum convolution semigroups in the sense of Bernstein on the braided line $\mathbb{R}_q$ are of the form*

$$\Phi : x^k \mapsto \sum_{n=\lceil \frac{k}{2} \rceil}^{k} \begin{pmatrix} n \\ 2n-k \end{pmatrix} a_1^{2n-k} a_2^{k-n} \frac{q_k!}{q'_n!} (x')^n,$$

where $a_1, a_2 \in \mathbb{C}$ are constants, and x, x' denote the generators of $\mathbb{R}_q$ and $\mathbb{R}_{q'}$, respectively.

Proof: The possible generators were given in Theorem 8.3.6. This allows to construct Φ^* and thus Φ. ∎

Remark: The $\mathbb{R}_q$-qcs associated to Bozejko and Speicher's quantum Brownian motion is also weakly Gaussian in the sense of Bernstein (as a qcs!).

REFERENCES

The results of this chapter were first published in [FNS97a]. For other characterisations of Gaussian (=quadratic) generators see [Sch93, Section 5.1].

Chapter 9

Phase retrieval for probability distributions on quantum groups and braided groups

Consider the following question: Under which conditions is a probability measure (on the real line e.g.) uniquely determined (up to a shift) already by the absolute value of its Fourier transform? In other words: When is it possible to retrieve the phase (up to a constant) from the absolute value of the Fourier transform? This problem has its origin in crystallography and there exists a vast literature on it (there was even a Nobel prize given for this subject), see e.g. Carnal and Fel'dman[CF95, CF97] and the references cited there.

Let μ be the law of a random variable X (on the real line) and $\varphi(u)$ ($u \in \mathbb{R}$) its Fourier transform. Then $|\varphi(u)|^2 = \varphi(u) \cdot \overline{\varphi(u)}$ can be interpreted as the Fourier transform of the symmetrisation $\mu * \overline{\mu}$ (where $\overline{\mu}$ denotes the adjoint measure, i.e. the law of the random variable $-X$). So the above question can be reformulated in purely probabilistic terms: Under which conditions is a probability measure μ uniquely determined by its symmetrisation $\mu * \overline{\mu}$ and the first moment of μ? If one formulates the question like this, it makes sense not only on vector spaces, but also on more general convolution structures. Carnal and Fel'dman[CF97] have treated this problem on Abelian groups.

In this chapter, we will present the results of Franz, Neuenschwander, and Schott[FNS97b] for the problem of phase retrieval on nilpotent quantum groups and nilpotent braided groups: Given the symmetrisation $\mu * \overline{\mu}$ and the first moments of a unital functional μ on $\mathcal{A}$, when is it possible to retrieve the original functional μ from these data? The somewhat surprising answer is that in this framework, the retrieval is always possible (provided that the quantum or braided group is "sufficiently" non-commutative, e.g. if q is not a root of unity). By definition functionals on quantum groups have all moments and are uniquely determined by them. So it will suffice to show that the moments of the symmetrisation and the first moments of μ together allow to calculate all moments of μ recursively. Observe that one can not expect to be able to remove the condition of knowledge of the first moments of μ, since already on the classical real line, in the best possible case, μ can be determined by its symmetrisation only up to a shift. The situation on nilpotent quantum groups and nilpotent braided groups is in sharp contrast to the classical case of simply connected nilpotent Lie groups, where the moments can not be retrieved, as we see in Section 9.1.

9.1 CLASSICAL SIMPLY CONNECTED NILPOTENT LIE GROUPS

A simply connected nilpotent Lie group is a Lie group G with Lie algebra $\mathfrak{g}$ such that $\exp : \mathfrak{g} \to G$ is a diffeomorphism and that the descending central series is finite, i.e. there is some $r \in \mathbb{N}$ such that

$$\mathfrak{g}_0 \underset{\neq}{\supset} \mathfrak{g}_1 \underset{\neq}{\supset} \cdots \underset{\neq}{\supset} \mathfrak{g}_r = \{0\},$$

where

$$\mathfrak{g}_0 := \mathfrak{g}, \quad \mathfrak{g}_{k+1} := [\mathfrak{g}, \mathfrak{g}_k] \quad (0 \leq k \leq r-1).$$

G is then called step r-nilpotent. So G may be interpreted as $\mathbb{R}^d$ equipped with a Lie bracket $[.,.] : \mathbb{R}^d \times \mathbb{R}^d \to \mathbb{R}^d$ which is bilinear, skew-symmetric, and satisfies the Jacobi identity

$$[[x,y],z] + [[y,z],x] + [[z,x],y] = 0.$$

The most prominent examples are the so-called Heisenberg groups HW^d, given by $\mathbb{R}^{2d+1}$ and the Lie bracket

$$\begin{aligned} [x,y] \quad &:= \quad (0,0,\langle x',y'\rangle - \langle x'',y''\rangle) \\ & \qquad (x = (x',x'',x'''), y = (y',y'',y''') \in \mathbb{R}^d \times \mathbb{R}^d \times \mathbb{R} \cong \mathbb{R}^{2d+1}). \end{aligned}$$

Consider an adapted vector space decomposition of $G = \mathfrak{g}$ (see [GKR77]), i.e.

$$G \cong \mathfrak{g} \cong \mathbb{R}^d \cong \bigoplus_{i=1}^{r} V_i$$

such that

$$\bigoplus_{i=k}^{r} V_i = \mathfrak{g}_{k-1}.$$

Then on the "first layer" V_1, the group multiplication always reduces to Euclidean addition $+$. But on the real line, a straightforward calculation of the 3rd and the 4th moment of the symmetrisation shows that the above-mentioned moment retrieval is not possible. On the line:

$$\begin{aligned} E(X-X')^4 &= 2E(X^4) - 8E(X^3)E(X) + 6E(X^2)^2, \\ E(X-X')^3 &= 0. \end{aligned}$$

It would also be interesting to characterise classes of probability measures for which the phase retrieval is nevertheless possible on nilpotent (and even more general) Lie groups.

9.2 THE PHASE PROBLEM ON THE BRAIDED LINE

We consider first the braided line (see Subsection 3.9.1), because it turns out to be typical, but at the same time simpler since it has only one generator.

On quantum groups and braided groups the antipode plays the role of the inverse, so that the symmetrisation of a functional $\phi : \mathcal{B} \to \mathbb{C}$ is defined by

$$\phi * \overline{\phi} = (\phi \otimes (\phi \circ S)) \circ \Delta.$$

Proposition 9.2.1 *Let $q \in \mathbb{C}\backslash\{0\}$, q not a root of unity. A unital functional ϕ on $\mathbb{R}_q$ is uniquely determined by its symmetrisation $\phi * \overline{\phi}$ and its first moment $\phi(x)$.*

Proof: We have $\Delta(x^n) = \sum_{\nu=0}^{n} \begin{bmatrix} n \\ \nu \end{bmatrix}_q x^\nu (x')^{n-\nu}$, and $S(x^n) = (-1)^n q^{\frac{n(n-1)}{2}} x^n$, and thus

$$\phi * \overline{\phi}(x^n) = \phi \otimes (\phi \circ S) \circ \Delta(x^n) = \sum_{\nu=0}^{n} \begin{bmatrix} n \\ \nu \end{bmatrix}_q (-1)^{n-\nu} q^{(n-\nu)(n-\nu-1)/2} \phi(x^\nu)\phi(x^{n-\nu}).$$

If q is not a root of unity this gives a recurrence relation that allows to calculate all moments of ϕ except the first from those of the symmetrisation $\phi * \overline{\phi}$, we get $\phi(x^n) =$

$$\frac{1}{1 + (-1)^n q^{n(n-1)/2}} \left\{ (\phi * \overline{\phi})(x^n) - \sum_{\nu=1}^{n-1} \begin{bmatrix} n \\ \nu \end{bmatrix}_q q^{(n-\nu)(n-\nu-1)/2} \phi(x^\nu)\phi(x^{n-\nu}) \right\}.$$

■

9.3 THE PHASE PROBLEM ON NILPOTENT QUANTUM OR BRAIDED GROUPS

Looking at the classical case and at the braided line, we see that it is sufficient for retrieving a certain moment $\phi(x_n)$ if $S(x_n) \neq -x_n$, (if we suppose that all other moments are already known). In the classical case this is true for even moments, thus symmetrical measures can be retrieved, if they are uniquely determined by their moments, see [CD89, Proposition 2.2]. On the braided line this is true for all powers of the generator x except the first, thus all moments except the first can be retrieved (Proposition 9.2.1).

This argument can be extended to a class of quantum groups or braided groups called strongly nilpotent or S-nilpotent. In addition to the condition from Definition 8.2.1, these quantum or braided groups have to satisfy a condition on the antipode.

To be able to retrieve the values on a basis element x_n of a functional on these groups from its symmetrisation and its 'first' moments, it is sufficient, if the coefficient of x_n in the expansion of $S(x_n)$ is not equal to -1, see Theorem 9.3.2.

To formulate this rigorously, we need the following definition.

Definition 9.3.1 *A quantum group or braided group $\mathcal{B}$ is called* S-nilpotent *or* strongly nilpotent, *if there exists a basis $\{\mathbb{1}, a_\imath; \imath \in \mathcal{I}\}$, indexed by a well-ordered set $\mathcal{I}$, such that*

$$\Delta(a_\imath) - a_\imath \otimes \mathbb{1} - \mathbb{1} \otimes a_\imath \subseteq \mathcal{B}_{<\imath} \otimes \mathcal{B}_{<\imath} \tag{9.1}$$

$$S(a_\imath) \subseteq \mathcal{B}_{\leq \imath} \tag{9.2}$$

for all $\imath \in \mathcal{I}$, where $\mathcal{B}_{<\imath} = \mathrm{span}\{\mathbb{1}, a_{\imath'}; \imath' < \imath\}$, $\mathcal{B}_{\leq\imath} = \mathrm{span}\{\mathbb{1}, a_{\imath'}; \imath' \leq \imath\}$. Bases that satisfy these conditions will be called adapted.

Remark: If we require only condition (9.1), then we have the notion of nilpotency introduced in Subsection 8.2.1. It was motivated by the fact that the Hopf algebra of polynomials on a connected, simply connected nilpotent Lie group can be characterised in this way. But in that case we have $S(x_1^{k_1} \cdots x_n^{k_n}) = (-1)^{k_1+\cdots+k_n} x_1^{k_1} \cdots x_n^{k_n}$, so condition (9.2) is also satisfied.

On non-commutative nilpotent quantum groups and nilpotent braided groups there exist bases that satisfy (9.1), but not (9.2). Take for example the free algebra $\mathbb{C}\langle\langle x, y\rangle\rangle$ with two primitive generators, then the basis of monomials $\{\mathbb{1}, x, y, x^2, y^2, xy, yx, x^3, \cdots\}$ (take any order such that $a < b$ if the total degree of a is less than that of b) satisfies (9.1), but not (9.2), since $S(xy) = yx$ and $S(yx) = xy$. But we have $S^2 = \mathrm{id}$, so we can choose a basis that diagonalises S for each of the subspaces with total degree n, and the union of these bases, $\{\mathbb{1}, x, y, x^2, y^2, xy + yx, xy - yx, x^3, \cdots\}$, satisfies both (9.1) and (9.2). We do not know if their exist quantum groups or braided groups that are nilpotent, but not S-nilpotent.

Theorem 9.3.2 *Let $\mathcal{B}$ be an S-nilpotent quantum group or S-nilpotent braided group, with the adapted basis $\{a_\imath; \imath \in \mathcal{I}\}$ and let $\mathcal{K} \subseteq \mathcal{I}$. Suppose furthermore that the coefficient $s_{\imath,\imath}$ of $S(a_\imath)$ in the expansion $S(a_\imath) = \sum_{k \leq \imath} s_{\imath,k} a_k$ is not equal to -1 for all $\imath \in \mathcal{I} \backslash \mathcal{K}$. Then a normalised functional $\phi : \mathcal{B} \to \mathbb{C}$ is uniquely determined by its symmetrisation $\phi * \overline{\phi}$ and its moments on $\{a_k; k \in \mathcal{K}\}$.*

Proof: We can give an explicit procedure to calculate the moments of ϕ. Let $\imath$ be the smallest index in $\mathcal{I}$ for which we do not yet know $\phi(a_\imath)$. Then

$$(\mathrm{id} \otimes S) \circ \Delta(a_\imath) = 1 + s_{\imath,\imath} + \sum_{\imath', \imath'' < \imath} C^{(\imath)}_{\imath', \imath''} a_{\imath'} \otimes a_{\imath''}$$

and thus, since $s_{\imath,\imath} \neq -1$,

$$\phi(a_\imath) = \frac{1}{1+s_{\imath,\imath}} \left\{ \phi * \overline{\phi}(a_\imath) - \sum_{\imath',\imath''<\imath} C^{(\imath)}_{\imath',\imath''} \phi(a_{\imath'})\phi(a_{\imath''}) \right\}.$$

∎

For example on $\mathbb{R}$ the basis could be chosen as $\{x^n; n \in \mathbb{N}\}$, and the set of moments that have to be known to recover the measure is $\mathcal{K} = \{2n+1; n \in \mathbb{N}\}$. On the braided line we have only $\mathcal{K} = \{1\}$.

9.4 ON THE BRAIDED PLANE

The braided plane $\mathbb{C}_q^{2|0}$ (see Subsection 3.9.2) is S-nilpotent, and an adapted basis if given by $\{x^n y^m; n,m \in \mathbb{N}\}$, where $\mathcal{I} = \{(n,m); n,m \in \mathbb{N}\}$ is taken with the order defined by $(n_1,m_1) \leq (n_2,m_2)$ if and only if $n_1 \leq n_2$ and $m_1 \leq m_2$.

To see that this basis is really adapted, recall

$$\begin{aligned} \Delta(x^n y^m) &= \sum_{\nu=0}^{n}\sum_{\mu=0}^{m} \left[\begin{array}{c} n \\ \nu \end{array}\right]_{q^2} \left[\begin{array}{c} m \\ \mu \end{array}\right]_{q^2} q^{\mu(n-\nu)} x^\nu y^\mu \otimes x^{n-\nu} y^{m-\mu}, \\ S(x^n y^m) &= (-1)^{n+m} q^{(n+m)(n+m+1)} x^n y^m \end{aligned} \tag{9.3}$$

Proposition 9.4.1 *Let $q \in \mathbb{C}\backslash\{0\}$, q not a root of unity. A unital functional ϕ on the braided plane $\mathbb{C}_q^{2|0}$ is uniquely determined by its first moments $\phi(x), \phi(y)$, and its symmetrisation $\phi * \overline{\phi}$.*

Proof: This follows immediately from Theorem 9.3.2 with Equation (9.3), if we take $\mathcal{K} = \{(0,1),(1,0)\}$. ∎

9.5 ON THE BRAIDED HEISENBERG-WEYL GROUP

Our last example will be the braided Heisenberg-Weyl group HW_q (cf. Subsection 3.9.3).

We shall now derive some properties of the coproduct and the antipode of HW_q, in order to show that HW_q is S-nilpotent, and to determine which moments are necessary to characterise a functional on HW_q by its symmetrisatin.

This algebra is graded with $\deg(a) = (1,0)$, $\deg(c) = (0,1)$, since the relations (3.25) and (3.26) are homogeneous. Let $\mathrm{HW}_q^{(n,m)}$ be the subspace of elements of degree (n,m). The braiding of two homogeneous elements is given by

$$u'v = q^{(n_1-m_1)n_2+m_1m_2} vu', \quad \text{for } u \in \mathrm{HW}_q^{(n_1,m_1)}, \quad v \in \mathrm{HW}_q^{(n_2,m_2)}, \tag{9.4}$$

This, together with the definition of the coproduct on the generators, implies that Equation (9.1) holds for any basis of homogeneous elements ordered by their degree.

Equation (9.2) can also be satisfied, because

$$S^2|_{\mathrm{HW}_q^{(n,m)}} = q^{n(n-1)+m(m-1)-nm}\mathrm{id}_{\mathrm{HW}_q^{(n,m)}}, \tag{9.5}$$

and thus there exists a basis of homogeneous elements that diagonalises S. This basis, if ordered by the degree, is obviously adapted, and HW_q is thus S-nilpotent.

Equation (9.5) can be shown by complete induction. It is true for $\mathrm{HW}_q^{(0,0)} = \mathbb{C}\mathbb{1}$, since $S^2(\mathbb{1}) = \mathbb{1}$. Let now $u \in \mathrm{HW}_q^{(n,m)}$, then with Equation (9.4)

$$\begin{aligned} S^2(au) &= \Psi^2(a \otimes S^2(u)) = q^{2n-m}aS^2(u), \\ S^2(cu) &= \Phi^2(c \otimes S^2(u)) = q^{2m-n}cS^2(u). \end{aligned}$$

Proposition 9.5.1 *Let $q \in \mathbb{C}\backslash\{0\}$, q not a root of unity. Then a unital functional ϕ on HW_q is uniquely determined by the moments*

$$\phi(a), \quad \phi(c), \quad \phi(aca), \quad \phi(cac), \quad \phi(aacc - q^2ccaa), \quad \phi(acac - qcaca),$$

and its symmetrisation $\phi * \overline{\phi}$.

Proof: After the preceding discussion we know that there exists an adapted basis of eigenvectors of S, so that, by Theorem 9.3.2, all that remains to be done is to determine the eigenspace E_{-1} of the eigenvalue -1. By Equation (9.5) the only possible eigenvalues of $S|_{\mathrm{HW}_q^{(n,m)}}$ are $\pm q^{\frac{n(n-1)+m(m-1)-nm}{2}}$. But $n(n-1) + m(m-1) - nm \neq 0$ for all (n,m) except $(0,0),(1,0),(0,1),(2,1),(1,2),(2,2)$, so that $E_{-1} \subseteq \mathrm{HW}_q^{(0,0)} \oplus \mathrm{HW}_q^{(1,0)} \oplus \mathrm{HW}_q^{(0,1)} \oplus \mathrm{HW}_q^{(2,1)} \oplus \mathrm{HW}_q^{(1,2)} \oplus \mathrm{HW}_q^{(2,2)} =: A$. A simple calculation (note $E_{-1} = (\mathrm{id} - S)(A)$, since S is involutive on A) gives

$$E_{-1} = \mathrm{span}\{a, c, aca, cac, aacc - q^2ccaa, acac - qcaca\},$$

and this proves the proposition. ∎

REFERENCES

This chapter is taken from [FNS97b].

Chapter 10

Limit theorems on quantum groups

Limit theorems are a fundamental part of probability theory. Such theorems have been proved over the last decades for Lie groups [GKR77] and commutative hypergroups [BH95] but only few limit theorems exist in the quantum group case. In this chapter we present different aspects of that theory. In Section 10.1, the general results for limit theorems on bialgebras due to Schürmann [Sch93] is presented. Then a randomised q-central (or q-commutative) limit theorem on a family of bialgebras with one complex parameter is shown in Section 10.2. This result is due to U. Franz [Fra98]. Section 10.3 is devoted to Woronowicz results [Wor87] on convergence of convolution products of probability measures to the Haar functional on compact quantum groups. A q-central limit theorem for $U_q(su(2))$ has been proved by R. Lenczewski [Len93, Len94], this result is stated in Section 10.4 as well as a weak law of large numbers which derives easily from the central limit theorem. Then we recall D. Neuenschwander and R. Schott's (cf. [NS97]) results on domains of attraction for q-transformed random variables in Section 10.5.

10.1 ANALOGUES OF THE LAW OF LARGE NUMBERS AND THE CENTRAL LIMIT THEOREM

The most general result is probably the following theorem due to M. Schürmann.

Theorem 10.1.1 [Sch93, Theorem 6.1.1] *Let* φ_{nk}, $n \in \mathbb{N}$, $k = 1, \ldots, k_n$ ($k_n \in \mathbb{N}$), *be linear functionals on a coalgebra* $\mathcal{C}$ *satisfying*

(i) $\varphi_{n1}, \ldots, \varphi_{nk_n}$ *commute for each* $n \in \mathbb{N}$ *with respect to convolution,*

(ii) $\lim_{n\to\infty} \lim_{1\leq k\leq k_n} |(\varphi_{nk} - \varepsilon)(c)| = 0$ *for all* $c \in \mathcal{C}$,

(iii) $\sup_{n\in\mathbb{N}} \sum_{1\leq k\leq k_n} |(\varphi_{nk} - \epsilon)(c)| < \infty$ *for all* $c \in \mathcal{C}$.

Furthermore, suppose that there is a linear functional ψ on C such that for all $c \in C$

$$\lim_{n\to\infty}\left(\sum_{1\le k\le k_n}(\varphi_{nk}-\varepsilon)(c)\right)=\psi(c).$$

Then

$$\lim_{n\to\infty}\left(\prod_{1\le k\le k_n}^{\star}\varphi_{nk}\right)(c)=(\exp_\star\psi)(c)$$

for all $c \in C$ (where the product $\prod^\star$ is the convolution product).

Let us look at a corollary of this theorem for $\mathbb{N}$-graded coalgebras. Let C be an $\mathbb{N}$-graded coalgebra, i.e. $C=\bigoplus_{n\in\mathbb{N}}C^{(n)}$ and $\Delta C^{(n)}\subseteq\bigoplus_{n_1+n_2=n}C^{(n_1)}\otimes C^{(n_2)}$. Then we can define a scaling $s:\mathbb{C}\times C\to C$ by setting

$$s(z)c=z^{-\deg(c)}c,$$

for $c\in C^{(\deg(c))}$, and extending linearly in the second argument. With this definition we have $(\varphi\star\psi)\circ s(z)=(\varphi\circ s(z))\star(\psi\circ s(z))$ for all linear functionals φ,ψ on C and all $z\in\mathbb{C}$.

Corollary 10.1.2 ([Sch93, Theorem 6.1.3]) *Let C be an $\mathbb{N}$-graded coalgebra and let $\kappa\in\mathbb{N}$. If a linear functional φ on C satisfies*

(i) $\varphi|_{C^{(l)}}=0$ *for* $0<l<\kappa$,

(ii) $\varphi|_{C^{(0)}}=\varepsilon|_{C^{(0)}}$,

then for all $c\in C$

$$\lim_{n\to\infty}\left(\varphi^{\star n}\circ s(n^{1/\kappa})\right)(c)=(\exp_\star g_\varphi)(c),$$

where g_φ denotes the linear functional on C with

$$\begin{aligned}g_\varphi|_{C^{(l)}}&=0\ \text{for all}\ l\neq\kappa,\\ g_\varphi|_{C^{(\kappa)}}&=\varphi|C^{(\kappa)}.\end{aligned}$$

Remark: For $\kappa=1$ this can be considered as a generalised weak law of large numbers, for $\kappa=2$ as a generalised central limit theorem.

The braided line

The study of the 'q-convolution' and related limit theorems was initiated by Ph. Feinsilver [Fei87]. In this case the convolution can be defined via a q-characteristic function, and it is therefore possible to consider also measures that are not characterised by their moments. Thus we can look e.g. for analogues of stable laws (cf. [NS97]).

Application of Corollary 10.1.2 gives the following result.

Proposition 10.1.3 *Let* $\kappa \in \mathbb{N}$, $\varphi : \mathbb{R}_q \to \mathbb{C}$ *normalised (i.e.* $\varphi(1) = 1$*), and* $\varphi(x) = \cdots = \varphi(x^{\kappa-1}) = 0$. *Set* $\varphi_n = \varphi^{\star n} \circ s(n^{1/\kappa})$, *then*

$$\lim_{n\to\infty} \varphi_n = e^{cp^\kappa},$$

where $c = \varphi(x^\kappa)/q_\kappa!$

For $|q| < 1$, Neuenschwander and Schott's result [NS97] allows to extend the previous proposition to $\kappa \in]0, 2[$, if e^{cp^κ} is interpreted as the q-characteristic function of the limit distribution.

10.2 A MIXED QUANTUM-CLASSICAL CENTRAL LIMIT THEOREM

Let us now present the randomised central limit theorem of [Fra98]. This work was motivated by a non-commutative limit theorem due to Speicher [Spe92], in which the commutation relations of the quantum random variables are given by classical $\{-1, 1\}$-valued random variables, i.e. Bernoulli random variables.

In Franz' work it is the coproduct that depends on a sequence of i.i.d. complex-valued random variables. This also leads to q-commutation relations for the increments, but with less independence in the choice of the commutation factors. The limit distribution can be expressed as an exponential with respect to a (non-associative!) averaged convolution.

We begin by briefly recalling the results of Speicher [Spe92] and Schürmann [Sch93].

In Subsection 10.2.1, we introduce the bialgebras used to formulate the main theorem (Subsection 10.2.2), and the algebra structure of their duals.

Finally, in Section 10.2.3, we show how explicit formulas for the moments and their associated measures were obtained using dual representations.

Let $q \in \mathbb{C}$ be a fixed complex number, and $\tilde{x}_1, \tilde{x}_2, \ldots$ a sequence of quantum random variables that satisfy

$$\begin{aligned} \tilde{x}_n \tilde{x}_m &= q \tilde{x}_m \tilde{x}_n, \\ \tilde{x}_n \tilde{x}_m^* &= q \tilde{x}_m^* \tilde{x}_n, \end{aligned}$$

for $n > m$, and set

$$\tilde{S}_N = \frac{\tilde{x}_1 + \cdots + \tilde{x}_N}{\sqrt{N}}.$$

If, for $n \in \mathbb{N}$, we have e.g. $\langle \Omega, \tilde{x}_n \Omega \rangle = \langle \Omega, \tilde{x}_n^* \Omega \rangle = \langle \Omega, \tilde{x}_n^2 \Omega \rangle = \langle \Omega, (\tilde{x}_n^*)^2 \Omega \rangle = \langle \Omega, \tilde{x}_n \tilde{x}_n^* \Omega \rangle = 0$, $\langle \Omega, \tilde{x}_n^* \tilde{x}_n \Omega \rangle = g \in \mathbb{R}_+$ in the vacuum state Ω, then the moments of (S_N, S_N^*) (in the state Ω) converge for $N \to \infty$ to those of the quantum Azéma process, cf. [Sch91b].

If we take instead quantum random variables $\hat{x}_1, \hat{x}_2, \ldots$, whose commutation relations are determined by classical i.i.d. random variables $(\varepsilon_{mn})_{n,m\in\mathbb{N}}$, $P(\varepsilon_{nm} =$

$\varepsilon_{mn} = +1) = p$, $P(\varepsilon_{nm} = \varepsilon_{mn} = -1) = 1 - p$, $p \in [0,1]$,

$$\hat{x}_n \hat{x}_m = \varepsilon_{nm} \hat{x}_m \hat{x}_n,$$

(no relations between $\hat{x}_n$ and $\hat{x}_m^*$) and set again

$$\hat{S}_N = \frac{\hat{x}_1 + \cdots + \hat{x}_N}{\sqrt{N}},$$

then the moments of $(\hat{S}_N, \hat{S}_N^*)$ converge for $N \to \infty$ to those of Bozejko and Speicher's q-Brownian motion[BS91, BKS96], see [Spe92]. The parameters q and p are related via $q = 2p - 1$.

Define $\omega = \begin{pmatrix} 0 \\ 1 \end{pmatrix}$,

$$x = \begin{pmatrix} 0 & 1 \\ 0 & 0 \end{pmatrix}, \qquad y(q) = \begin{pmatrix} 1 & 0 \\ 0 & q \end{pmatrix},$$

for $q \in \mathbb{C}$, then we can realize the quantum random variables $(\tilde{x}_n)_{n\in\mathbb{N}}$ and $(\hat{x}_n)_{n\in\mathbb{N}}$ on the infinite tensor product $(\mathbb{C}^2)^{\otimes\mathbb{N},\omega}$ of $\mathbb{C}^2$ (as a Hilbert space, the product being taken with respect to the sequence $\omega_n = \omega$) as

$$\begin{aligned} \tilde{x}_n &= \underbrace{y(q) \otimes \cdots \otimes y(q)}_{n-1 \text{ times}} \otimes x \otimes 1 \otimes 1 \otimes \cdots, \\ \hat{x}_n &= y(\varepsilon_{n,1}) \otimes \cdots \otimes y(\varepsilon_{n,n-1}) \otimes x \otimes 1 \otimes 1 \otimes \cdots. \end{aligned}$$

The state is given by $\Omega = \omega^{\otimes\mathbb{N}}$.

In this realization the following amounts to replacing the parameter q in the definition of $\tilde{x}_n$ by complex-valued random variables $(q_i)_{i\in\mathbb{N}}$, i.e. we set $x_n = y(q_1) \otimes \cdots \otimes y(q_{n-1}) \otimes x \otimes 1 \otimes \cdots$. But, in order to stay in a bialgebra framework, we get less independence than in the commutation factors of the $\hat{x}_n$. It turns out that an increment x_n has to satisfy the same commutation relation either with all preceding or with all following increments, i.e. we have two possible cases

$$\begin{aligned} \text{a)} \quad x_n x_m &= q_n x_m x_n \qquad \text{for } n > m, \\ \text{b)} \quad x_n x_m &= q_m x_m n_n \qquad \text{for } n > m. \end{aligned}$$

10.2.1 *The family of bialgebras*

Let $\mathcal{A}$ be the (unital, associative) *-algebra generated by x, x^*, and $\{y_\alpha; \alpha \in \mathbb{C}\backslash\{0\}\}$ with the relations $y_\alpha^* = y_{\bar{\alpha}}$, and

$$y_\alpha x = \alpha x y_\alpha, \qquad x^* y_\alpha = \alpha y_\alpha x^*, \qquad y_\alpha y_\beta = y_{\alpha\beta} \text{ for } \alpha, \beta \in \mathbb{C}\backslash\{0\}, \qquad y_1 = 1.$$

Note that $\deg x = \deg x^* = 1$, $\deg y_\alpha = 0$ for all $\alpha \in \mathbb{C}\backslash\{0\}$ defines a grading, and so we can introduce a scaling map $s(r) : \mathcal{A} \to \mathcal{A}$, for $r \in \mathbb{R}$, by setting $s(r)a = r^{-\deg a} a$ on homogeneous elements.

A basis of $\mathcal{A}$ is given by

$$\mathcal{B} = \{y_\alpha w; w \text{ a word in the two letters } x, x^*, \text{ and } \alpha \in \mathbb{C}\backslash\{0\}\}.$$

On this algebra we can define a whole family of coalgebras, depending on one parameter $q \in \mathbb{C}\backslash\{0\}$, namely,

$$\Delta_q x = x \otimes y_q + 1 \otimes x, \quad \Delta_q x^* = x \otimes y_{\bar{q}} + 1 \otimes x^*, \quad \Delta_q(y_\alpha) = y_\alpha \otimes y_\alpha \text{ for } \alpha \in \mathbb{C}\backslash\{0\},$$

and $\varepsilon_q(x) = \varepsilon_q(x^*) = 0$, $\varepsilon_q(y_\alpha) = 1$ for all $\alpha \in \mathbb{C}\backslash\{0\}$. In fact, this is a Hopf algebra with the antipode $S_q : \mathcal{A} \to \mathcal{A}$,

$$S_q(y_\alpha) = y_{1/\alpha}, \qquad S_q(x) = -xy_{1/q}, \qquad S_q(x^*) = -x^* y_{1/\bar{q}}.$$

Using these different coalgebra structures one obtains different multiplications for functionals on $\mathcal{A}$, i.e. a one-parameter family of convolutions,

$$\varphi *_q \psi = (\varphi \otimes \psi) \circ \Delta_q$$

for linear functionals $\varphi, \psi : \mathcal{A} \to \mathbb{C}$.

For a word w in the two letters x, x^*, let $\chi(w)$ be the functional defined by

$$\chi(w)(y_\beta w') = \begin{cases} 1 & \text{if } w = w', \\ 0 & \text{else} \end{cases}$$

on the basis $\mathcal{B}$.

Then $\mathcal{U} = \operatorname{span}\{\chi(w); w \text{ a word in the two letters } x, x^*\}$ is a subalgebra of the dual of $\mathcal{A}$ (with the multiplication $m_q = \Delta_q^*$). Let $a \in \mathcal{A}$, $\Delta_q a = \sum a_i^{(1)} \otimes a_i^{(2)}$, then the product $\chi(w) *_q \chi(w')$ is defined by

$$\Big(\chi(w) *_q \chi(w')\Big)(a) = \sum \chi(w)(a_i^{(1)})\chi(w')(a_i^{(2)}).$$

We get by induction

$$\begin{aligned} \Delta_q(x^n) &= \sum_{\nu=0}^{n} \begin{bmatrix} n \\ \nu \end{bmatrix}_q x^\nu \otimes x^{n-\nu} y_{q^\nu} \\ \Delta_q\left((x^*)^n\right) &= \sum_{\nu=0}^{n} \begin{bmatrix} n \\ \nu \end{bmatrix}_q (x^*)^\nu \otimes y_{\bar{q}^\nu} (x^*)^{n-\nu}, \end{aligned}$$

where $\begin{bmatrix} n \\ \nu \end{bmatrix}_q = \frac{q_n!}{q_\nu! q_{n-\nu}!}$, $q_n! = \prod_{\nu=1}^n q_\nu$, $q_n = \sum_{\nu=1}^n q^{\nu-1} = \frac{q^n-1}{q-1}$, and from this we can calculate the coproduct of any element of $\mathcal{A}$.

To calculate the product $\chi(w) *_q \chi(w')$, we have to see what elements of $\mathcal{A}$ have a term $y_\alpha w \otimes y_\beta w'$ in their coproduct. Since the coproduct does not change the total number of x's and x^*'s, but just splits a word into two, the product has to do the inverse. We get

$$\chi(w) *_q \chi(w') = \sum_v c^v_{w,w'}(q)\chi(v), \tag{10.1}$$

where v runs over all words that can be obtained by shuffling w and w', and the $c^v_{w,w'}$ are polynomials in q and q^{-1}. To get the explicit expression, use the following procedure.

In the first term v is simply the concatenation of w and w', and the coefficient is equal to one. Then move the letters of w to the right, without changing their order, and multiply by q every time an x is moved to the right past another x, and by q^{-1} every time it is moved past an x^*. When moving an x^* to the right take the conjugate factors, i.e. a $\bar{q}$ when it is moved past an x, and a $\bar{q}^{-1}$ when it is moved past an x^*. We get e.g.

$$\begin{aligned}
\chi(x) *_q \chi(x) &= (1+q)\chi(xx), \\
\chi(x) *_q \chi(x^*) &= \chi(xx^*) + q^{-1}\chi(x^*x), \\
\chi(x^*) *_q \chi(x) &= \chi(x^*x) + \bar{q}\chi(xx^*), \\
\chi(x^*) *_q \chi(x^*) &= (1+\bar{q}^{-1})\chi(x^*x^*).
\end{aligned}$$

We will also need the averaged convolution (w.r.t. to a $\mathbb{C}\backslash\{0\}$-valued random variable q) defined by

$$u_1 \bar{*} u_2 = \mathbb{E}(u_1 *_q u_2),$$

i.e. $u_1 \bar{*} u_2$ is the functional defined by $u_1 \bar{*} u_2(a) = \int (u_1 \otimes u_2)(\Delta_q(a))\mathrm{d}\mu(q)$, where μ is the law of q. This binary operation conserves positivity, but in general it is not associative.

10.2.2 *The central limit theorem*

We will now show that the functionals

$$\begin{aligned}
\text{a)}\quad \varphi^N_{\rightarrow} &= \Big((\varphi *_{q_1} \varphi) *_{q_2} \varphi\Big) *_{q_3} \cdots *_{q_{N-1}} \varphi \circ s(\sqrt{N}) \\
&= (\varphi \otimes \cdots \otimes \varphi) \circ (\Delta_{q_1} \otimes \mathrm{id}^{\otimes N-2}) \circ \cdots \circ \Delta_{q_{N-1}} \circ s(\sqrt{N}) \\
\text{b)}\quad \varphi^N_{\leftarrow} &= \varphi *_{q_{N-1}} \cdots *_{q_3} \Big(\varphi *_{q_2} (\varphi *_{q_1} \varphi)\Big) \circ s(\sqrt{N}) \\
&= (\varphi \otimes \cdots \otimes \varphi) \circ (\mathrm{id}^{\otimes N-2} \otimes \Delta_{q_1}) \circ \cdots \circ \Delta_{q_{N-1}} \circ s(\sqrt{N})
\end{aligned}$$

converge for appropriately chosen functionals φ.

In the realization from the beginning of this section, with $\varphi(\cdot) = <\omega, \cdot\,\omega>$ and for a non-commutative polynomial $P(x, x^*)$, we can write

$$\varphi^N_{\leftarrow,\rightarrow}(P(x,x^*)) = \langle \Omega, P(S_N, S_N^*)\Omega \rangle,$$

with

$$S_N = \frac{x_1^{(N)} + \cdots + x_N^{(N)}}{\sqrt{N}}$$

and

$$\begin{aligned}
\text{a)}\quad x_n^{(N)} &= \underbrace{1 \otimes \cdots \otimes 1}_{n-1 \text{ times}} \otimes x \otimes y(q_n) \otimes \cdots \otimes y(q_{N-1}) \\
\text{b)}\quad x_n^{(N)} &= \underbrace{1 \otimes \cdots \otimes 1}_{n-1 \text{ times}} \otimes x \otimes \underbrace{y(q_{N-n}) \otimes \cdots \otimes y(q_{N-n})}_{N-n \text{ times}},
\end{aligned}$$

where, as one verifies easily, the increments $x_n^{(N)}$ satisfy commutation relations

$$\begin{array}{lrcll} \text{a)} & x_n x_m & = & q_n x_m x_n & \text{for } n > m, \\ \text{b)} & x_n x_m & = & q_m x_m n_n & \text{for } n > m. \end{array}$$

We state now our result.

Theorem 10.2.1 *Let $(q_n)_{n\in\mathbb{N}}$ be i.i.d. random variables with values in $\mathbb{C}\backslash\{0\}$ such that $\mathbb{E}(|q_1^m|) < \infty$ for all $m \in \mathbb{Z}$, and let $\varphi : \mathcal{A} \to \mathbb{C}$ be a normed functional in $\mathcal{U}$. Suppose furthermore that φ is centralised, i.e. $\varphi(x) = \varphi(x^*) = 0$.*

Then the moments of

$$\begin{aligned} \varphi_{\rightarrow}^N &= \left((\varphi *_{q_1} \varphi) *_{q_2} \varphi\right) *_{q_3} \cdots *_{q_{N-1}} \varphi \circ s(\sqrt{N}) \\ \varphi_{\leftarrow}^N &= \varphi *_{q_{N-1}} \cdots *_{q_3} \left(\varphi *_{q_2} (\varphi *_{q_1} \varphi)\right) \circ s(\sqrt{N}) \end{aligned}$$

converge for $N \to \infty$ in probability to those of the functionals

$$\begin{aligned} \varphi_{\rightarrow}^\infty &= \exp_{\bar{*},\rightarrow} g = \varepsilon + g + g\bar{*}g + (g\bar{*}g)\bar{*}g + \cdots, \\ \varphi_{\leftarrow}^\infty &= \exp_{\bar{*},\leftarrow} g = \varepsilon + g + g\bar{*}g + g\bar{*}(g\bar{*}g) + \cdots, \end{aligned}$$

where $g|_{\mathcal{A}^{(2)}} = \varphi$, and $g|_{\mathcal{A}^{(k)}} = 0$ for $k \neq 2$.

Proof: We will only prove this for $\varphi_{\rightarrow}^N$, since the convergence of $\varphi_{\leftarrow}^N$ can be shown in the same way.

We can write φ as

$$\varphi = \varepsilon + g + \tilde{\varphi},$$

where $\tilde{\varphi}$ vanishes on homogeneous elements of degree less than three.

Let us first assume $\tilde{\varphi} = 0$, i.e. $\varphi = \varepsilon + g$, $g = g_1\chi(xx) + g_2\chi(x^*x) + g_3\chi(xx^*) + g_4\chi(x^*x^*)$. To simplify the notation we assume in fact $g = \chi(x^*x)$.

Then the functional

$$\varphi_{\rightarrow}^N = \left(\cdots(\varphi *_{q_1} \varphi) *_{q_2} \cdots *_{q_{N-1}} \varphi\right) \circ S(\sqrt{N})$$

can be written as

$$\varphi_{\rightarrow}^N = \sum_v \frac{f_v^N(q_1, \ldots, q_{N-1})}{N^{\frac{|v|}{2}}} \chi(v),$$

where $|v|$ denotes the length of v. The coefficients $\frac{f_v^N(q_1,\ldots,q_{N-1})}{N^{\frac{|v|}{2}}}$ will be some rational functions of $q_1, \ldots, q_{N-1}$, as can be seen from the recurrence relations below. To prove the theorem for this case it is sufficient to show that $\mathbb{E}\left(\frac{f_v^N(q_1,\ldots,q_{N-1})}{N^{\frac{|v|}{2}}}\right)$ converges and that $\mathrm{Var}\left(\frac{f_v^N(q_1,\ldots,q_{N-1})}{N^{\frac{|v|}{2}}}\right)$ goes to zero for $N \to \infty$ (and fixed v). We will do this by induction over the length of v. The coefficient of 1 (the empty word) is constant and the coefficient of the only word with $|v| < 4$ that occurs, i.e. $v = x^*x$, is equal to N, so the induction hypothesis is satisfied for $|v| < 4$. Let

$c^v_{w,w'}(q)$ denote the coefficients of the multiplication, as in Equation (10.1). We have the following relations for $f_v^{N+1}(q_1,\ldots,q_N)$,

$$f_v^{N+1}(q_1,\ldots,q_N) = f_v^N(q_1,\ldots,q_{N-1}) + \sum_{v':\,|v'|=|v|-2} c^v_{v',x*x}(q_N) f_{v'}^N(q_1,\ldots,q_{N-1}),$$

or

$$f_v^N(q_1,\ldots,q_{N-1}) = \sum_{k=\frac{|v|}{2}}^{N-1} \sum_{v':\,|v'|=|v|-2} c^v_{v',x*x}(q_{k-1}) f_{v'}^{k-1}(q_1,\ldots,q_{k-2}),$$

and therefore

$$\begin{aligned}
&\mathbb{E}\left(\frac{f_v^N(q_1,\ldots,q_{N-1})}{N^{\frac{|v|}{2}}}\right) \\
&= \frac{1}{N}\sum_{k=\frac{|v|}{2}}^{N-1} \sum_{v':\,|v'|=|v|-2} \mathbb{E}\left(c^v_{v',x*x}(q_{k-1})\right) \mathbb{E}\left(\frac{f_{v'}^{k-1}(q_1,\ldots,q_{k-2})}{N^{\frac{|v'|}{2}}}\right) \\
&= \frac{1}{N}\sum_{v':\,|v'|=|v|-2} \mathbb{E}\left(c^v_{v',x*x}(q_{k-1})\right) \sum_{k=\frac{|v|}{2}}^{N-1} \frac{(k-1)^{\frac{|v'|}{2}}}{N^{\frac{|v'|}{2}}} \mathbb{E}\left(\frac{f_{v'}^{k-1}(q_1,\ldots,q_{k-2})}{(k-1)^{\frac{|v'|}{2}}}\right) \\
&\overset{N\to\infty}{\longrightarrow} \frac{2}{|v|} \sum_{v':\,|v'|=|v|-2} \mathbb{E}\left(c^v_{v',x*x}(q_1)\right) \lim_{N\to\infty} \mathbb{E}\left(\frac{f_{v'}^N(q_1,\ldots,q_{N-1})}{N^{\frac{|v'|}{2}}}\right)
\end{aligned} \tag{10.2}$$

The same technique works for the limit of the variance. Let us suppose that $\operatorname{Var}\left(f_v^N(q_1,\ldots,q_{N-1})\right)$ is bounded by $K_v N^{|v|-1}$, which is obviously true for $|v| = 0,1,2,3$. Then, by

$$\begin{aligned}
&\operatorname{Var}\left(f_v^N(q_1,\ldots,q_{N-1})\right) \\
&= \operatorname{Var}\left(\sum_{k=\frac{|v|}{2}}^{N-1} \sum_{v':\,|v'|=|v|-2} c^v_{v',x*x}(q_{k-1}) f_{v'}^{k-1}(q_1,\ldots,q_{k-2})\right) \\
&\leq N \sum_{k=\frac{|v|}{2}}^{N-1} \sum_{v':\,|v'|=|v|-2} \mathbb{E}\left((c^v_{v',x*x}(q_{k-1}))^2\right) \operatorname{Var}\left(f_{v'}^{k-1}(q_1,\ldots,q_{k-2})\right) \\
&\leq N^{|v|-1} \sum_{v':\,|v'|=|v|-2} \mathbb{E}\left((c^v_{v',x*x}(q_1))^2\right) K_{v'}
\end{aligned}$$

it is true for all v, and therefore $\operatorname{Var}\left(\frac{f_v^N(q_1,\ldots,q_{N-1})}{N^{\frac{|v|}{2}}}\right) \overset{N\to\infty}{\longrightarrow} 0$. So, for $N\to\infty$, the coefficients $\frac{f_v^N(q_1,\ldots,q_{N-1})}{N^{\frac{v}{2}}}$ converge to the expectation in probability, and this limit can be calculated recursively over the order of v with Equation (10.2). This shows that for $\varphi = \varepsilon + g$ the functionals $\varphi^N_{\to}$ tend to

$$\varepsilon + g + g\bar{*}g + (g\bar{*}g)\bar{*}g + \cdots = \exp_{\bar{*},\to}(g).$$

As in [Sch93, Theorem 6.1.3] the contributions of the higher order terms contained in $\tilde{\varphi}$ vanish in the limit, because they pick up higher orders of the scaling factor N^{-1}. ■

10.2.3 *Special cases*

We will now show how dual representations can be used to give explicit expressions for the moments of the limit functional. If φ is a functional on a bialgebra $\mathcal{A}$, then we can associate the operators $\rho_R(\varphi), \rho_L(\varphi) : \mathcal{A} \to \mathcal{A}$ to it by

$$\rho_R(\varphi)a = \sum a_i^{(1)} \varphi(a_i^{(2)}), \qquad \rho_L(\varphi)a = \sum \varphi(a_i^{(1)}) a_i^{(2)},$$

where $\Delta(a) = \sum a_i^{(1)} \otimes a_i^{(2)}$, i.e. by the right and left dual representation. The functional φ can be retrieved from these operators, since $\varphi = \varepsilon \circ \rho_R(\varphi) = \varepsilon \circ \rho_L(\varphi)$. Also, the map $\rho_R : \mathcal{A}^* \to \mathrm{Hom}(\mathcal{A})$ (resp. $\rho_L : \mathcal{A}^* \to \mathrm{Hom}(\mathcal{A})$) is a homomorphism (resp. anti-homomorphism), i.e.

$$\rho_R(\varphi * \psi) = \rho_R(\varphi) \circ \rho_R(\psi) \qquad \Big(\text{resp. } \rho_L(\varphi * \psi) = \rho_L(\psi) \circ \rho_L(\varphi)\Big).$$

Let $\mathcal{A}$ be as before, ρ_q the right dual representation w.r.t. Δ_q, and define

$$R_q = \rho_q(\chi(x^* x)) = (\mathrm{id} \otimes \chi(x^* x)) \circ \Delta_q, \qquad \bar{R} = \mathbb{E}(R_q).$$

These can be used to calculate the moments of $\exp_{\bar{*},\to}$, since

$$\varepsilon \circ \rho_{q_1}(\varphi_1) \circ \rho_{q_2}(\varphi_2) \circ \cdots \circ \rho_{q_n}(\varphi_n) = (\varphi_1 *_{q_2} \varphi_2) *_{q_3} \cdots *_{q_n} \varphi_n,$$

If q is real, then, for polynomials $f(z)$ in the variable $z = x + x^*$, we have

$$R_q f(z) = \frac{f(qz) - f(z) - z(q-1) f'(z)}{z^2 (q-1)^2},$$

i.e. R_q is the generator of the Azéma martingale, see Page 127 or [Fra97, Fra99] to see how this can be derived using Hopf algebra duality and a Leibniz formula.

On monomials we get $R_q z^n = k_n(q) z^{n-2}$, $k_n(q) = \sum_{\nu=1}^{n-1} \nu q^{n-\nu}$, for $n \geq 2$, $R_q z = R_q 1 = 0$. $\bar{R}$ follows by averaging over q.

Thus we have the following result.

Proposition 10.2.2 *Let q be a random variable with values in $\mathbb{R}\backslash\{0\}$ and suppose that all its moments are finite, i.e. $\mathbb{E}(|q|^n) < \infty$ for all $n \in \mathbb{N}$ and denote by $\bar{*}$ the averaged convolution with respect to q. Then*

$$\exp_{\bar{*},\to} \chi(x^* x)\Big((x + x^*)^n\Big) = \begin{cases} \frac{\bar{k}_{2m} \bar{k}_{2m-2} \cdots \bar{k}_2}{m!} & \text{if } n \text{ is even, } n = 2m, \\ 0 & \text{if } n \text{ is odd,} \end{cases}$$

where $\bar{k}_n = \sum_{\nu=1}^{n-1} \nu \mathbb{E}(q^{n-\nu})$.

Proof: Apply $\exp_{\bar{*},\rightarrow} \chi(x^*x) = \varepsilon \circ e^{\bar{R}}$ to $(x+x^*)^n$, which gives the desired expression since $\bar{R}(x+x^*)^n = \bar{k}_n(x+x^*)^{n-2}$. ∎

Remark: Similarly, the left dual representation can be used to compute $\exp_{\bar{*},\leftarrow}(g)$.

Examples:

- In the deterministic case, i.e. if $P(q = q_0) = 1$ for some $q_0 \in \mathbb{R}\backslash\{0\}$, we get the marginal distribution of the Azéma martingale with parameter q_0.
- For Bernoulli random variables, i.e. if $P(q = 1) = p$, $P(q = -1) = 1 - p$, for some $p \in [0, 1]$, we obtain

$$\varphi_\infty\left((x+x^*)^{2m}\right) = \prod_{\mu=0}^{m-1}(1+2\mu p)$$

i.e. the moments of $\mu = c_p|x|^{\frac{1}{p}-1}e^{-x^2/2p}dx$.

10.3 CONVERGENCE TO THE HAAR MEASURE ON COMPACT QUANTUM GROUPS

We know that on a compact topological group G (with countable basis) the convolution powers of an adapted probability measure converge weakly to the Haar measure, i.e. to the unique bi-invariant probability measure on G, see, e.g., [GKR77]. A probability measure is called adapted, if the closed subgroup generated by the support of the measure is equal to G. Woronowicz has shown a similar result for compact quantum groups, in this case the condition of adaptedness is replaced by a condition of faithfulness. A state ϕ on a $*$-algebra A is called faithful, if $\phi(a^*a) = 0$ implies $a = 0$ for all $a \in A$.

Woronowicz' result is as follows:

Theorem 10.3.1 *[Wor87, Theorem 4.2] Let (A, Δ) be a compact quantum group (see Definition 3.11.1). Then the iterated convolution of every faithful state on A converge to the Haar state in the pointwise Cesaro-limit, i.e. we have for any faithful state ϕ on A,*

$$h(a) = \mathrm{C}{-}\lim_{n\to\infty} \phi^{\star n}(a) \qquad \textit{for all } a \in A,$$

where h is the Haar state (see Theorem 3.11.3 for existence of the Haar state).

10.4 q-CENTRAL LIMIT THEOREM FOR $U_q(su(2))$

It has been proved by R. Lenczewski [Len93] that more precise statements for limit theorems can be given for specific quantum groups (for example for $U_q(su(2))$). We present these results in this section.

10.4.1 *Introduction*

Let $q > 0, q \neq 1$. By C_q we denote a unital $*$-algebra over $\mathbb{C}$ generated by the set

$$\{e, f, t, t^{-1}\}$$

where $e^* = f$ and t, t^{-1} are hermitian, i.e. $t^* = t$, $(t^{-1})^* = t^{-1}$, subject to relations

$$te = q^2 et, \quad tf = q^{-2} ft, \quad tt^{-1} = t^{-1}t = 1,$$

with the coproduct

$$\Delta : C_q \to C_q \otimes C_q$$

$$\Delta(w) = t^{-1} \otimes w + w \otimes t, \quad \Delta(t^\sigma) = t^\sigma \otimes t^\sigma, \quad \Delta(1) = 1 \otimes 1,$$

where $\sigma \in \{1, -1\}$, $w \in \{e, f\}$, the counit

$$\epsilon : C_q \to \mathbb{C}$$

$$\epsilon(w) = 0, \quad \epsilon(t^\sigma) = 1,$$

and the coinverse

$$S : C_q \to C_q$$

$$S(w) = -twt^{-1}, \quad S(t) = t^{-1}, \quad S(t^{-1}) = t.$$

It can be easily checked that C_q is a $*$-Hopf algebra. Note that if we add the relation

$$[e, f] = \frac{t^2 - t^{-2}}{q^2 - q^{-2}}$$

then we obtain $U_q(su(2))$, the q-deformed universal enveloping algebra of $su(2)$ (note that Δ, ϵ and S preserve this relation).

On C_q we define two gradations:

(i) $d(e) = d(f) = 1$, $d(t) = d(t^{-1}) = d(1) = 0$, with the direct sum decomposition

$$C_q = (C_q)_{(0)} \oplus (C_q)_{(1)} \oplus \dots$$

(ii) $b(f) = -1$, $b(e) = 1$, $b(t) = b(t^{-1}) = b(1) = 0$ with the direct sum decomposition

$$C_q = \dots \oplus C_q^{(-1)} \oplus C_q^{(0)} \oplus C_q^{(1)} \dots .$$

Definition 10.4.1 *We will say that the functional $\phi \in C_q^*$ is even if ϕ vanishes outside $C_q^{(0)}$. This means that the only correlations which don't vanish are those for which the number of e's is equal to the number of f's.*

The N-th iteration of the coproduct Δ is defined as $\Delta_N = (\Delta_{N-1} \otimes \mathrm{id}) \circ \Delta$ with $\Delta_1 = \Delta$. It can be seen that in the case of C_q and $U_q(su(2))$ it takes the form

$$\Delta_{N-1}(t^\sigma) = (t^\sigma)^{\otimes N}, \quad \Delta_{N-1}(w) = \sum_{i=1}^{N} j_{i,N}(w),$$

where

$$j_{i,N}(w) = (t^{-1})^{\otimes(i-1)} \otimes w \otimes t^{\otimes(N-i)}.$$

Thus, $\Delta_{N-1}(e)$ and $\Delta_{N-1}(f)$ can be viewed as sums of "q-independent" random variables. They satisfy the q-commutation relations

$$j_i(u)j_k(w) = q(u,w)j_k(w)j_i(u)$$

where $u, w \in \{e, f\}$ and $q(u,w)$ is the commutation factor given by

$$q(u,w) = q^{2b(w)+2b(u)}$$

for $i < k$, i.e. $q(e,f) = q(f,e) = 1$, $q(e,e) = q^4$, $q(f,f) = q^{-4}$.

For given functionals ϕ, ψ on a bialgebra C^*, their convolution is the functional

$$\phi \star \psi : C \to \mathbb{C}$$

defined as

$$\phi \star \psi = (\phi \otimes \psi) \circ \Delta.$$

The N-th convolution power of $\phi \in C^*$ will be denoted by $\phi_N^\star$, i.e.

$$\phi_N^\star = \phi \star \ldots \star \phi \quad (N \text{ times}).$$

It is well-known that it can be expressed in terms of the $N-1$-th coproduct iteration as $\phi_N^\star = \phi^{\otimes N} \circ \Delta_{N-1}$.

In convolution-type limit theorems for C_q and $U_q(su(2))$ we need to evaluate

$$\phi_N^\star(v_1 \ldots v_n),$$

where $v_1, \ldots, v_n \in \{e, f, t, t^{-1}\}$, rescale the variables $v_i \to \tau_N(v_i)$ and their products in an appropriate way, then evaluate the pointwise limit

$$\lim_{N\to\infty} \phi_N^\star \circ \tau_N,$$

and finally, find the GNS realization of the limit law.

By an *ordered partition* of the set $I = \{1, \ldots, p\}$ we understand the r-tuple $S = (S_1, \ldots, S_r)$ of mutually disjoint and nonempty subsets of I, for which

$$I = S_1 \cup \ldots \cup S_r.$$

By the signature of S we understand the r-tuple $(\gamma_{S_1}, \ldots, \gamma_{S_r})$, where $\gamma_{S_k} = \#S_k$. The signature is even if γ_{S_k} is even for each k. Then S is also called even. The set of all (all even) ordered partitions of $I = \{1, \ldots, p\}$ is denoted $\mathcal{P}_p$ ($\mathcal{P}_p^e$). An inversion of S is a pair (i,j), such that $i \in S_k$, $j \in S_m$, $i > j$, $k < m$. The set of all inversions of S is denoted W_S.

10.4.2 *A q-Central Limit Theorem*

A q-analogue of the convolution limit theorem was derived for C_q and $U_q(su(2))$ in [Len93] (extended in [Len94] to Jimbo-Drinfeld quantum groups $U_q(g)$ and the associated $*$-Hopf algebras) under the assumption that the state ϕ is a homomorphism on $\mathbb{C}[t,t^{-1}]$. We summarise these results here.

Definition 10.4.2 *Let $\mathcal{B} \subset \mathcal{A}$ be associative unital $*$-algebras and let ϕ be a state on $\mathcal{A}$. We say that ϕ is left or right $\mathcal{B}$-independent if*

$$\phi(ba) = \phi(b)\phi(a) \quad \text{or} \quad \phi(ab) = \phi(a)\phi(b),$$

respectively, for all $a \in \mathcal{A}$, $b \in \mathcal{B}$.

Proposition 10.4.3 *Let $\mathcal{B} \subset \mathcal{A}$ be associative unital $*$-algebras and let ϕ be a state on $\mathcal{A}$. Then ϕ is left and right $\mathcal{B}$-independent iff the restriction of ϕ to $\mathcal{B}$ is a unital $*$-homomorphism.*

Proof: Clearly, if ϕ is left and right $\mathcal{B}$-independent, then it is a unital $*$-homomorphism on $\mathcal{B}$. Conversely, if ϕ is a unital $*$-homomorphism on $\mathcal{B}$, then we use the Cauchy-Schwarz inequality to obtain

$$|\phi((b-\phi(b))a)| \leq \sqrt{\phi(a^*a)\phi((b^*-\phi(b^*))(b-\phi(b)))} = 0$$

for any $b \in \mathcal{B}$ and $a \in \mathcal{A}$. Thus,

$$\phi(ba) = \phi(b)\phi(a)$$

and ϕ is left $\mathcal{B}$-independent. In a similar fashion we show that it is right $\mathcal{B}$-independent. ∎

Corollary 10.4.4 *If $\phi \in C_q$ is a non-trivial $*$-homomorphism on $\mathbb{C}[t,t^{-1}]$, then it is even.*

Proof: If $c \in (C_q)^{(r)}$ and $n \in \mathbb{Z}$, then

$$\phi(t^n c) = q^{2nr}\phi(ct^n).$$

By Proposition 2.2, ϕ is right and left $\mathbb{C}[t,t^{-1}]$-independent and thus we obtain

$$\phi(t^n)\phi(c) = q^{2nr}\phi(t^n)\phi(c),$$

and therefore, since $\phi(t) \neq 0$ and $0 < q < 1$, we must have $\phi(c) = 0$, i.e. ϕ vanishes outside $(\mathcal{U}_q)^{(0)}$, i.e. ϕ is even. ∎

Definition 10.4.5 *Let $\mathcal{A}$ denote the free $*$-algebra generated by $\bar{e}$ (denote $\bar{e}^* = \bar{f}$). Let $0 < a < 1$, $0 < q < 1$. For a given state ϕ on C_q, define a functional $\Psi_{a,q}$ on $\mathcal{A}$ as the linear extension of $\Psi_{a,q}(1) = 1$ and*

$$\Psi_{a,q}(\bar{w}_1 \ldots \bar{w}_p) = \sum_{S \in \mathcal{P}_p^c} D(S)\phi(w_{S_1}) \ldots \phi(w_{S_r})$$

where $w_1, \ldots, w_p \in \{e, f\}$, $w_{S_j} = \prod_{k \in S_j} w_k$ *(in the natural order) and*

$$D(S) = Q(S) \frac{(a^{-4} - 1)^{p/2}}{(a^{-2\gamma s_1} - 1)(a^{-2\gamma s_1 - 2\gamma s_2} - 1) \ldots (a^{-2\gamma s_1 - \ldots - 2\gamma s_r} - 1)}$$

where $Q(S)$ *is a* q*-dependent "commutation factor" of the form*

$$Q(S) = \prod_{W_S} q(b_i, b_j)$$

(recall that W_S *is the set of inversions of* S*).*

In the central limit theorem the q-normalisation was given by

$$w \to \tau_N(w) = w / \sqrt{[N]_{a^2}},$$

where $[N]_{a^2} = (a^{2N} - a^{-2N})/(a^2 - a^{-2})$, which was obtained from the calculation of the variance

$$\phi_N^\star(vw) = \phi(vw)[N]_{a^2}$$

where $v, w \in \{e, f\}$.

Remark. The origin of Definition 10.4.5 is the q-central limit theorem for C_q and $U_q(su(2))$ in which $\Psi_{a,q}$ appears as the limit functional. Thus Definition 10.4.5 may be viewed as the combinatorial form of the central limit state. More information on this state is given in the sequel.

Theorem 10.4.6 *Let* $0 < a < 1$, $0 < q < 1$. *If* ϕ *is a state on* C_q *and a* $*$-*homomorphism on* $\mathbb{C}[t, t^{-1}]$ *with* $\phi(t) = a$, *then*

$$\lim_{N \to \infty} \phi_N^\star \circ \tau_N(w_1 \ldots w_p) = \Psi_{a,q}(\bar{w}_1 \ldots \bar{w}_p)$$

for p *even. If* p *is odd, then the limit vanishes. Thus,* $\Psi_{a,q}$ *is the limit state for a convolution* q*-analogue of the central limit theorem for* C_q.

Remark. One can extend $\Psi_{a,q}$ to a state $\tilde{\Psi}_{a,q}$ on $\tilde{\mathcal{A}} = \mathcal{A} * \mathbb{C}[r]$, where r is a hermitian generator, by assuming that $\tilde{\Psi}_{a,q}$ is a homomorphism on $\mathbb{C}[r]$ and the two-sided ideal generated by the relations

$$\bar{e}\bar{f} - \bar{f}\bar{e} = \frac{a^2 - a^{-2}}{q^2 - q^{-2}} r^2,$$

$$\bar{e}r = q^2 r\bar{e}, \quad \bar{f}r = q^{-2} r\bar{f}$$

is contained in ker $\tilde{\Psi}_{a,q}$. Then it can be seen (or shown) that $\tilde{\Psi}_{a,q}$ is the vacuum state for the canonical Fock space representation of a q-harmonic oscillator with scale c given by the relations

$$b_q b_q^\dagger - b_q^\dagger b_q = c r_q^2$$

$$r_q b_q = q^{-2} b_q r_q, \quad r_q b_q^\dagger = q^2 b_q^\dagger r_q$$

where $c = (a^2 - a^{-2})/(q^2 - q^{-2})$ (thus $c = 1$ if $a = q$). Note that one can write $r_q = q^{2\hat{N}}$, where $\hat{N}$ is the number operator.

Corollary 10.4.7 *Let $0 < a < 1$, $0 < q < 1$. If ϕ is a state on $U_q(su(2))$ and a homomorphism on $\mathbb{C}[t, t^{-1}]$, then*

$$\lim_{N\to\infty} \phi_N^\star \circ \tau_N(w_1 \dots w_p) = \tilde{\Psi}_{a,q}(\bar{w}_1 \dots \bar{w}_p),$$

for p even. For p odd, the above limit vanishes. Thus, the limit state is the vacuum state for the q-oscillator with scale $c = (a^2 - a^{-2})(q^2 - q^{-2})$.

Proposition 10.4.8 *The limit law of $b_q + b_q^\dagger$ is q^4-Gaussian.*

Proof: It can be shown that the moments $m_k = \Psi((b_q + b_q^\dagger)^k)$ coincide with the moments $m'_k = \Psi'((c_q + c_q^\dagger)^k)$, $k \in \mathbb{Z}^+$, respectively, where $c_q c_q^\dagger - q^4 c_q^\dagger c_q = 1$. ∎

The explicit formula for the density of the q-Gaussian law can be found in the works of Bozejko and Speicher[BS91].

10.4.3 *A q-Weak Law of Large Numbers*

To get the normalisation in the weak law of large numbers, we evaluate

$$\phi_N^\star(v) = \phi(v)[N]_a,$$

where $v \in \{e, f\}$, for any state ϕ on C_q or $U_q(su(2))$. Therefore, we will use the following normalisation:

$$\tau'_N(v) = \frac{1}{[N]_a} v$$

and extend it multiplicatively to the free algebra generated by e, f (appropriate normalisation for t and t^{-1} is obtained easily). Note that when $a \to 1$, $[N]_a \to N$, therefore this normalisation may be viewed as the canonical q-analogue of that in the usual weak law of large numbers.

The simplest case is when we assume that the state ϕ is a homomorphism on $\mathbb{C}[t, t^{-1}]$. Let me discuss this case first and show that it boils down to the central limit theorem. This is because ϕ must be even then and all that the new normalisation changes is the scale of the q-oscillator.

Theorem 10.4.9 *Let $0 < a < 1$, $0 < q < 1$. If ϕ is a state on C_q and a homomorphism on $\mathbb{C}[t, t^{-1}]$ with $\phi(t) = a$, then*

$$\lim_{N\to\infty} \phi_N^\star \circ \tau'_N(w_1 \dots w_p) = \left(\frac{a - a^{-1}}{\sqrt{a^2 - a^{-2}}}\right)^p \Psi_{a,q}(\bar{w}_1 \dots \bar{w}_p)$$

for p even. For p odd, the above limit vanishes. Thus, the limit state is a rescaled central limit state.

Proof: Obvious. ∎

Corollary 10.4.10 *Let $0 < a < 1$, $0 < q < 1$. If ϕ is a state on $U_q(su(2))$ and a homomorphism on $\mathbb{C}[t, t^{-1}]$ with $\phi(t) = a$, then*

$$\lim_{N\to\infty} \phi_N^\star \circ \tau_N'(w_1 \dots w_p) = \left(\frac{a - a^{-1}}{\sqrt{a^2 - a^{-2}}}\right)^p \tilde{\Psi}_{a,q}(\bar{w}_1 \dots \bar{w}_p)$$

for p even. For p odd, the above limit vanishes. Thus, the limit state is the vacuum state of a q-oscillator with scale $c' = (a - a^{-1})^2/(q^2 - q^{-2})$.

Proof: Obvious. ∎

Therefore, assuming that ϕ is a homomorphism on $\mathbb{C}[t, t^{-1}]$, we get a modification of the q-central limit theorem. The general case, i.e. when ϕ is not a homomorphism on $\mathbb{C}[t, t^{-1}]$, seems of interest.

Example: The simplest example is furnished by the states associated with the two- dimensional fundamental representation of $U_q(su(2))$. Take pure states $\psi(.) = \langle x, .x\rangle$ with $x = \alpha x_1 + \beta x_2$, where x_1, x_2 are canonical basis vectors in $\mathbb{C}^2$ and $|\alpha|^2 + |\beta|^2 = 1$. In particular, then we have $\phi(t^n) = q^n|\alpha|^2 + q^{-n}|\beta|^2$, $n \in \mathbb{Z}$. Note that if $\alpha \notin \{0, 1\}$, then ψ does not satisfy the homomorphism assumption.

It seems desirable to be able to prove a q-analogue of the weak law for such states (for which $\psi(e) \neq 0 \neq \psi(f)$). However, we'll show that if one wants the limits of all correlations (moments) to exist, then using a natural normalisation one has to assume that ϕ is a homomorphism on $\mathbb{C}[t, t^{-1}]$.

Some additional notation is needed (this is taken from [Len94]). For a given ordered partition $S = (S_1, \dots, S_r)$ and $j \in \{1, \dots, r\}$ define $S_{j)} = S_1 \cup \dots \cup S_{j-1}$, $S_{(j} = S_{j+1} \cup \dots \cup S_r$, $S_{j]} = S_1 \cup \dots \cup S_j$, $S_{[j} = S_j \cup \dots \cup S_r$, and put $S_{0]} = S_{[r+1} = S_{(r} = S_{1)} = \emptyset$. Define the following partition-dependent mappings:

$$\tau_j^S(v_k) = \begin{cases} t & if \quad k \in S_{j)} \\ v_k & if \quad k \in S_j \\ t^{-1} & if \quad k \in S_{(j} \end{cases}$$

where $1 \leq j \leq r$. Another family of partition-dependent mappings is given by

$$\sigma_j^S(v_k) = \begin{cases} t & if \quad k \in S_{j)} \\ t^{-1} & if \quad k \in S_{[j} \end{cases}$$

for $v_k \in \{e, f\}$. Both maps are extended multiplicatively to all products $v = v_1 \dots v_k$, $v_i \in \{e, f\}$, $1 \leq i \leq k$. Using those maps we can evaluate convolutions of functionals. The lemma below gives a formula for the N-th convolution power (this is Lemma 1 of [Len94]).

Lemma 10.4.11 *Let ϕ be any functional on C_q or $U_q(su(2))$ and let $v = v_1 \dots v_p$ where $v_1, \dots, v_p \in \{e, f\}$. It holds*

$$\phi_N^\star(v_1 \dots v_p)$$
$$= \sum_{r=1}^{p} \sum_{1\leq i_1 < \dots < i_r \leq N} \sum_{S=(S_1,\dots,S_r)\in\mathcal{P}_p} \prod_{m=1}^{r+1} \phi(\sigma_m^S(v))^{(i_m - i_{m-1} - 1)} \prod_{j=1}^{r} \phi(\tau_j^S(v))$$

where we put $i_0 = 0$, $i_{r+1} = N+1$.

Let us find the natural normalisation in this case. For any state ϕ on $\mathcal{C}_q$ or $U_q(su(2))$, we obtain

$$\phi_N^\star(e) = \phi(e)[N]_{\phi(t),\phi(t^{-1})}$$

where

$$[N]_{a,b} = \frac{a^N - b^N}{a-b}$$

and $\phi(t) \neq \phi(t^{-1})$. Therefore, let us normalise the variables e, f as follows:

$$\tau_N''(e) = \frac{1}{[N]_{\phi(t),\phi(t^{-1})}}\, e, \quad \tau_N''(f) = \frac{1}{[N]_{\phi(t),\phi(t^{-1})}}\, f.$$

Let us calculate the convolution powers of ϕ. We have

$$\sigma_j^S(v_1 \dots v_p) = t^{|S_{j)}|-|S_{(j}|} = t^{\eta_j^S}$$

where

$$\eta_j^S = \gamma_1^S + \dots + \gamma_{j-1}^S - \gamma_j^S - \dots \gamma_r^S.$$

We put $\phi(t^{\eta_j^S}) = q^{\alpha_j^S}$ and evaluate

$$\sum_{1 \le i_1 < \dots < i_r \le N} q^{\alpha_1^S(i_1-1)+\alpha_2^S(i_2-i_1-1)+\dots+\alpha_{r+1}^S(N-i_r)}$$

$$= q^{-\alpha_1^S-\dots-\alpha_r^S+\alpha_{r+1}^S N} G_N(\alpha_1^S - \alpha_2^S, \alpha_2^S - \alpha_3^S, \dots, \alpha_r^S - \alpha_{r+1}^S | q)$$

where

$$G_N(\beta_1, \dots, \beta_r | q) = \sum_{1 \le k_1 < \dots < k_r \le N} q^{\beta_1 k_1 + \dots + \beta_r k_r}$$

for $q, \beta_1, \dots, \beta_r \in \mathbb{R}$. It holds

$$G_N(\beta_1, \dots, \beta_r | q) = \sum_{i=0}^{r} q^{N(\beta_{i+1}+\dots+\beta_r)} g_i(\beta_1, \dots, \beta_r | q),$$

where g_i's do not depend on N (their explicit form is given in [Len98]). Note that for $q = 1$,

$$G_N(\beta_1, \dots, \beta_r | q) = \binom{N}{r}.$$

Corollary 10.4.12 *Under the assumptions of Lemma 10.4.11 we have*

$$\phi_N^\star(v_1 \dots v_p)$$
$$= \sum_{S=(S_1,\dots,S_r) \in \mathcal{P}_p} q^{-\sum_{i=1}^r \alpha_i^S + N\alpha_{r+1}^S} G_N(\alpha_1^S - \alpha_2^S, \dots, \alpha_r^S - \alpha_{r+1}^S | q) \prod_{j=1}^r \phi(\tau_j^S(v))$$

Corollary 10.4.13 *Under the assumptions of Lemma 10.4.11 we have*

$$\phi_N^* \circ \tau_N''(v_1 \dots v_p) =$$

$$\sum_{S=(S_1,\dots,S_r)\in\mathcal{P}_p} q^{-\sum_{i=1}^r \alpha_i^S} \sum_{j=0}^r \left(\frac{q^{N\alpha_{j+1}^S}}{[N]^p_{\phi(t),\phi(t^{-1})}} \right) g_j(\alpha_1^S - \alpha_2^S, .., \alpha_r^S - \alpha_{r+1}^S | q) \prod_{m=1}^r \phi(\tau_m^S(v)).$$

Proof: Immediate from Corollary 10.4.12 and the form of τ_N''. ∎

It remains to evaluate the limit of the above expressions as $N \to \infty$. In order to do that one needs to compare the growth rate of $q^{N\alpha_{j+1}^S}$ with that of $[N]^p_{\phi(t),\phi(t^{-1})}$ for each j. Note that in particular

$$q^{\alpha_1^S} = \phi(t^{-p}), \quad q^{\alpha_{r+1}^S} = \phi(t^p),$$

which implies that one needs to compare

$$\max\{\phi(t^{-p})^N, \dots, \phi(t^p)^N\}$$

with

$$(\phi(t)^N - \phi(t^{-1})^N)^p$$

This boils down to comparing

$$\max\{\phi(t^p), \dots, \phi(t^{-p})\}$$

with $\phi(t)^p$ or $\phi(t^{-1})^p$, depending which one is larger. But Hölder's inequality gives

$$\phi(t)^p \leq \phi(t^p).$$

If we had a strict inequality for some positive p, then the limits of correlations of order p would not exist since the normalisation would not grow fast enough. Therefore, we must have an equality which means that ϕ must be a homomorphism on $\mathbb{C}[t]$. Finally, it is not hard to see (using Cauchy Schwarz Inequality) that this implies that ϕ must be a homomorphism on $\mathbb{C}[t, t^{-1}]$. This is the case considered in the beginning. Therefore, one can conclude that if we want the limits of all moments to exist (which seems standard in QP), then we must assume that ϕ is a homomorphism on $\mathbb{C}[t, t^{-1}]$ and thus an even state.

10.5 DOMAINS OF ATTRACTION FOR q-TRANSFORMED RANDOM VARIABLES

q-algebra and q-analysis ($0 < q < 1$) on the real line may be interpreted as a generalisation of ordinary addition, which (roughly speaking) corresponds to the case $q = 1$. However, it is not possible to define q-addition directly on the real

line itself, but rather indirectly on the space of measures on $\mathbb{R}$ in the sense that the q-convolution of two Dirac measures is in general not a Dirac measure (in the ordinary sense). So the whole theory is somewhat similar to hypergroups (cf. [BH95]), but it does not fit exactly into this context.
Feinsilver [Fei87] began a probabilistic study on q-added random variables. In the last part of his paper, he initiated an investigation of limit theorems for q-sums of random variables. The result of Neuenschwander-Schott [NS97], which we recall below, states that a random variable X lies in the strict domain of attraction of a non-degenerate strictly stable random variable Z with exponent $\alpha \in]0,2[$ iff the q-transform of X lies in the strict domain of attraction of mZ for some constant m depending on q and α with the same norming sequence. The proof consists essentially of getting rid of the centering constants appearing in the case of ordinary addition and of a desintegration procedure for the "only if"-direction.

10.5.1 *q-addition*

Let $0 < q < 1$. We define q-addition (indirectly) by defining a "Dirac measure" $\delta_{x\oplus y})$ by

$$\delta_{x\oplus y}(x^k) = (\delta_x * \delta_y)(x^k) = \sum_{j=0}^{k} \begin{bmatrix} k \\ j \end{bmatrix}_q x^{k-j}y^j.$$

The q-convolution of measures $\mu \in M^b(\mathbb{R})$ is then extended from the q-convolution of Dirac probability measures as indicated above also in the natural way by linearity and weak continuity.
The q-characteristic function of $\mathcal{L}(X)$ for the random variable X is defined by

$$\psi_X(u) = E(e(iuX)) \qquad (u \in \mathbb{R}).$$

The symbol $\varphi_X(u)$ will be used for the ordinary characteristic function. It has been shown by [Fei87, Theorem 3] that $\mathcal{L}(X)$ is uniquely determined by its q-characteristic function ψ_X. Furthermore, if X_1, X_2 are independent random variables, then the q-convolution of X_1 and X_2 is a random variable Z whose q-characteristic function is the product of the q-characteristic functions of X_1 and X_2: $\psi_Z = \psi_{X_1}\psi_{X_2}$ (cf. [Fei87]). Define the random variable Y by

$$Y = \sum_{k=0}^{\infty} T_k, \tag{10.3}$$

where the T_k are independent random variables, T_k obeying to an exponential law with mean q^k. Assume X is any random variable on $\mathbb{R}$, independent of Y. Then the q-characteristic function of X is the ordinary characteristic function of XY: $\psi_X = \varphi_{XY}$ (cf. [Fei87], Proposition 4). We will call XY the q-transform of X. If $F(x) = P(X \le x)$ is the law of the random variable X, then the law of the q-transform is given by the mixture

$$G(x) = P(XY \le x) = \int_0^{\infty} F(\frac{x}{y})\mathcal{L}(Y)(dy). \tag{10.4}$$

So what we have to study are sums of the type

$$\sum_{k=1}^{n} X_k Y_k,$$

where $X_1, X_2, \dots$ are any random variables and $Y_1, Y_2, \dots$ are i.i.d., as in (10.3), and independent of $X_1, X_2, \dots$.

10.5.2 *Domains of attraction*

First, we recall some facts on stable laws with respect to ordinary addition. As references, see e.g. [GK54] or [Bre68]. A random variable Z is called stable if for every $n \geq 1$ and i.i.d. copies $Z_1, Z_2, \dots, Z_n$ of Z there are $c_n > 0, d_n \in \mathbb{R}$ such that

$$X \stackrel{\mathcal{L}}{=} c_n \sum_{k=1}^{n} (Z_k + d_n).$$

Equivalently, Z is stable iff there are i.i.d. random variables $X_1, X_2, \dots$ and $a_n > 0, b_n \in \mathbb{R}$ such that

$$a_n \sum_{k=1}^{n} (X_k + b_n) \stackrel{w}{\to} Z \tag{10.5}$$

(where $\stackrel{w}{\to}$ denotes weak convergence). If $X \stackrel{\mathcal{L}}{=} X_1$, then X is said to lie in the domain of attraction of Z. The sequence $\{(a_n, b_n)\}_{n \geq 1}$ is called norming sequence. We will use the term "strictly stable" and "strict domain of attraction" if $d_n = 0$ resp. $b_n = 0$. It can be shown that there exists $\alpha \in]0, 2]$ such that

$$c_n = n^{-1/\alpha}.$$

The number α is called the exponent of stability. The case $\alpha = 2$ corresponds to the case where Z obeys to a normal distribution. Z is non-degenerate and stable with exponent $\alpha \in]0, 2[$ iff its characteristic function has the form

$$\begin{aligned} \varphi_Z(u) &= \exp\{i\gamma u + (v \int_{-\infty}^{0} + w \int_{0}^{\infty})(e^{iux} - 1 - \frac{iux}{1+x^2})\frac{dx}{|x|^{1+\alpha}}\} \\ & (v, w \geq 0, v + w > 0). \end{aligned}$$

For short, we write $Z \stackrel{\mathcal{L}}{=} (\alpha, \gamma, v, w)$. It follows that for $k \in \mathbb{N}$, we have $(\alpha, k\gamma, kv, kw) \stackrel{\mathcal{L}}{=} k^{1/\alpha} Z$.
For the following lemma see [PWZ81], Remark 3 on p.628.

Lemma 10.5.1 *In the case* $0 < \alpha < 1$ *we have*

$$n a_n b_n \to b$$

for some $b \in \mathbb{R}$.

The next lemma follows also from [PWZ81], Remark 3 on p.628 (see also [GK54], Theorem 35.3).

Lemma 10.5.2 *In case* $1 < \alpha < 2$ *we have*

$$b_n = -E(X_1) + \frac{b + o(1)}{na_n} \quad (n \to \infty)$$

for some $b \in \mathbb{R}$.

What remains, is the case $\alpha = 1$. This situation is somewhat special in the following sense. If $\alpha \in]0,1[\cup]1,2[$ and if Z is α-stable, then it is always possible to center Z so that it becomes strictly stable (see e.g. [Sha69], Theorem 6). However the following property follows at once by considering the characteristic function for $\alpha = 1$ in the "explicit form"

$$\begin{aligned} \varphi_Z(u) &= \exp\{i\beta u - \rho|u|(1 + i\theta \frac{u}{|u|}\frac{2}{\pi}\log|u|)\} \\ & (\beta \in \mathbb{R}, \rho > 0, \theta = \frac{v-w}{v+w}); \end{aligned}$$

the centering statement then follows from [Fel71], Theorem VII.5.3.

Lemma 10.5.3 *The only non-degenerate strictly stable laws* μ *with exponent* $\alpha = 1$ *are the shifted (with shift* $b \in \mathbb{R}$*) symmetric (Cauchy) ones (i.e. those with* $v = w$*). In this case, we have*

$$a_n b_n = -E(\sin(a_n X_1)) + \frac{b + o(1)}{n} \quad (n \to \infty),$$

where $b \in \mathbb{R}$ *is the afore-mentioned shift of* μ.

The domains of attraction of stable laws with exponent $0 < \alpha < 2$ may be characterised as follows (cf. [Mee86], p.344):

Proposition 10.5.4 *For a non-degenerate stable random variable* Z *with exponent* $0 < \alpha < 2$ *(10.5) holds iff*

$$nP(a_n X_1 < x) \to v \int_{-\infty}^{x} \frac{dt}{|t|^{1+\alpha}} \quad (n \to \infty) \quad (x < 0)$$

and

$$nP(a_n X_1 > x) \to w \int_{x}^{\infty} \frac{dt}{t^{1+\alpha}} \quad (n \to \infty) \quad (x > 0).$$

Let X, Z be random variables. The q-transform of X lies in the strict domain of attraction of Z with norming sequence $\{a_n\}$, if for i.i.d. copies $X_1, X_2, \ldots$ of X and i.i.d. random variables $Y_1, Y_2, \ldots$ as in (10.3) and independent of $X_1, X_2, \ldots$ there exist $a_n > 0$ such that

$$a_n \sum_{k=1}^{n} X_k Y_k \xrightarrow{w} Z. \tag{10.6}$$

Lemma 10.5.5 *The characteristic function of* $\log Y$ *is analytic in a neighbourhood of the real axis.*

Proof: By [Fei87], the density of Y is given by

$$g(x) = C\sum_{j=0}^{\infty} \frac{(-1)^j}{q_j!} q^{\binom{j}{2}} e^{-q^{-j}x} \quad (x \geq 0),$$

hence the density of $\log Y$ is

$$h(x) = e^x g(e^x) = C\sum_{j=0}^{\infty} \rho_j e^{-q^{-j}e^x + x}, \tag{10.7}$$

where

$$\rho_j = \frac{(-1)^j}{q_j!} q^{\binom{j}{2}}.$$

Clearly

$$\sum_{j=0}^{\infty} |\rho_j| < \infty. \tag{10.8}$$

Since

$$-q^{-j}e^x + x \leq -e^x + x \leq K \quad (x \in \mathbb{R}), \tag{10.9}$$

it follows from (10.8) and (10.9) that the series in (10.7) converges uniformly for $x \in \mathbb{R}$, so we get, with the aid of (10.7)-(10.9),

$$\begin{aligned} P(|\log Y| > x) &= \int_{\mathbb{R}\setminus[-x,x]} h(t)dt \\ &\leq C(\sum_{j=0}^{\infty} |\rho_j|) \int_{\mathbb{R}\setminus[-x,x]} e^{-e^t+t}dt \\ &= O(1 - e^{-e^{-x}} + e^{-e^x}) \\ &= O(e^{-x}) \quad (x \to \infty). \end{aligned}$$

Now the assertion follows from [Luk70], Theorem 7.2.1. ■
For fixed α and Y as in (1), define the constant

$$m = (EY^\alpha)^{1/\alpha}.$$

Theorem 10.5.6 *Let* Z *be a non-degenerate strictly stable random variable with exponent* $0 < \alpha < 2$ *and let* X *be any random variable. Then the q-transform of* X *lies in the strict domain of attraction of* mZ *with norming sequence* $\{a_n\}_{n\geq 1}$ *iff* X *lies in the strict domain of attraction of* Z *with norming sequence* $\{a_n\}_{n\geq 1}$.

Proof: 1. "If"-direction: Assume X lies in the strict domain of attraction of Z with norming sequence $\{a_n\}_{n\geq 1}$. Let $Y_1, Y_2, \ldots$ be as in (10.6). By [GK54], Theorem 25.1 and the Remark on p.121 it follows that the conditions (i)-(iii) mentioned before Proposition 8 in [Fei87] are indeed fulfilled. Hence by [Fei87], Proposition 8 the condition of our Proposition 10.5.4 carries over to the q-transforms (2) of $X_1, X_2, \ldots$ in the sense that XY lies in the domain of attraction of mZ with some norming sequence $\{a_n, b_n\}_{n\geq 1}$ for certain $b_n \in \mathbb{R}$.

1.1 Case $0 < \alpha < 1$**:** By Lemma 10.5.1 it follows that XY lies in the strict domain of attraction of $mZ - b$ for some $b \in \mathbb{R}$. By the convergence of types theorem (see e.g. [Bre68], Theorem 8.32) it follows that $mZ - b$ is also strictly α-stable, hence $b = 0$.

1.2 Case $1 < \alpha < 2$**:** By Lemma 10.5.2 it follows that

$$na_n E(X) \to b \quad (n \to \infty)$$

for some $b \in \mathbb{R}$. Hence also

$$na_n E(XY) \to bE(Y) \quad (n \to \infty).$$

So it follows from Lemma 10.5.2 that XY lies in the strict domain of attraction of $mZ - b' + bE(Y)$ for some $b, b' \in \mathbb{R}$ and the rest of the proof is as under 1.1.

1.3 Case $\alpha = 1$**:** By Lemma 10.5.3 it follows that

$$nE(\sin(a_n X)) \to b \quad (n \to \infty),$$

where $b \in \mathbb{R}$ is the shift of Z as in Lemma 10.5.3. Hence by the dominated convergence theorem, Proposition 10.5.4, and the stability property, we get

$$\begin{aligned}
\lim_{n\to\infty} nE(\sin(a_n XY)) &= \lim_{n\to\infty} \int_0^\infty nE(\sin((a_n y)X))\mathcal{L}(Y)(dy) \\
&= \lim_{n\to\infty} \int_0^\infty y\frac{n}{y}E(\sin(a_{n/y}X))\mathcal{L}(Y)(dy) \\
&= bE(Y) \\
&= mb.
\end{aligned}$$

So by Lemma 10.5.3 it follows that XY lies in the strict domain of attraction of mZ.

"Only if"-direction: By Proposition 10.5.4 it follows that

$$\begin{aligned}
& n\int_0^\infty \int_J \mathcal{L}(a_n y X_1)(dt)\mathcal{L}(Y)(dy) \\
\to\ & m^\alpha \int_0^\infty \int_J (v \cdot \mathbf{1}\{t < 0\} + w \cdot \mathbf{1}\{t > 0\}) \frac{dt}{|t|^{1+\alpha}} \\
=\ & \int_0^\infty \int_J (v \cdot \mathbf{1}\{t < 0\} + w \cdot \mathbf{1}\{t > 0\}) \frac{dt}{|t/y|^{1+\alpha}} \mathcal{L}(Y)(dy) \quad (n \to \infty)
\end{aligned}$$

and thus

$$n\int_0^\infty\int_J \frac{(t/y)^2}{1+(t/y)^2}\mathcal{L}(a_n y X_1)(dt)\mathcal{L}(Y)(dy)$$

$$\to \int_0^\infty\int_J \frac{(t/y)^2}{1+(t/y)^2}$$

$$(v\cdot \mathbf{1}\{t<0\}+w\cdot\mathbf{1}\{t>0\})\frac{dt}{|t/y|^{1+\alpha}}\mathcal{L}(Y)(dy) \qquad (n\to\infty), \tag{10.10}$$

where J is an interval of the form $]-\infty,x]$ $(x<0)$ or $[x,\infty[$ $(x>0)$. Let η and η_n denote the finite measures on $]0,\infty[$ given by

$$\eta(B) = w\int_B \frac{x^2}{1+x^2}\frac{dx}{x^{1/\alpha}}$$

and

$$\eta_n(B) = n\int_B \frac{x^2}{1+x^2}\mathcal{L}(a_n X_1)(dx)$$

for Borel subsets $B\subset]0,\infty[$. Define λ and λ_n to be the finite measures on $\mathbb{R}$ given by

$$\int_{-\infty}^{\infty} f(x)\lambda(dx) = \int_0^\infty f(\log y)\eta(dy)$$

and

$$\int_{-\infty}^{\infty} f(x)\lambda_n(dx) = \int_0^\infty f(\log y)\eta_n(dy)$$

for bounded real-valued continuous functions on $\mathbb{R}$. Let H be the distribution function of $\log Y$. By the "basic estimate" in [Fei87] and (10.10) it follows (similarly as in the proof of [Fei87], Proposition 8) that

$$\lambda_n * H \overset{w}{\to} \lambda * H \qquad (n\to\infty) \tag{10.11}$$

Let $\zeta(u)$, $\zeta_n(u)$, resp. $\xi(u)$ be the characteristic function (in the ordinary sense) of λ, λ_n resp. H. Then (10.11) may be rewritten as

$$\zeta_n(u)\cdot\xi(u)\to\zeta(u)\cdot\xi(u) \qquad (n\to\infty) \qquad (u\in\mathbb{R}). \tag{10.12}$$

By Lemma 10.5.5, ξ is analytic in a neighbourhood of the real axis, hence it has only isolated zeroes there. So we may divide (10.12) by $\xi(u)$ for all real u with the exception of isolated points, and of course in a neighbourhood of $u_0=0$. Thus it follows from the Lévy continuity theorem and the continuity of ζ that

$$\zeta_n(u)\to\zeta(u) \qquad (n\to\infty) \qquad (u\in\mathbb{R}),$$

hence

$$\lambda_n \xrightarrow{w} \lambda \quad (n \to \infty)$$

and thus

$$\eta_n \xrightarrow{w} \eta \quad (n \to \infty).$$

An analogous argument holds also for the negative real axis, hence by Proposition 10.5.4 it follows that X lies in the domain of attraction of Z with norming sequence $\{a_n, b_n\}_{n \geq 1}$ for certain $b_n \in \mathbb{R}$. Now the same type of argument as in the proof of the "if"-direction shows that one may replace b_n by 0 by the strictness of the domain of attraction. ∎

REFERENCES

This chapter is based on [Sch91b, Sch93, Fra98, Len93, Len94, NS97, Wor87]

Bibliography

[Abe80] E. Abe. *Hopf Algebras.* Cambridge Unversity Press, Cambridge, 1980.

[AD89] J. Aczél and J. Dhombres. *Functional equations in several variables.* Cambridge University Press, Cambridge, 1989.

[ALV94] L. Accardi, Y.G. Lu, and I. Volovich. Non-commutative (quantum) probability, master fields and stochastic bosonization. *Preprint Centro V. Volterra 198-94*, 1994.

[And69] R.F.V. Anderson. The Weyl functional calculus. *J. Funct. Anal.*, 4:240–267, 1969.

[ASW88] L. Accardi, M. Schürmann, and W.v. Waldenfels. Quantum independent increment processes on superalgebras. *Math. Z.*, 198:451–477, 1988.

[Azé85] J. Azéma. Sur les fermés aléatoires. In J. Azéma and M. Yor, editors, *Séminaire de Probabilités XIX, Strasbourg 1983/1984*, volume 1223 of *Lecture Notes in Math.* Springer-Verlag, 1985.

[Bal85] P. Baldi. Unicité du plongement d'une mesure de probabilité dans un semi-groupe de convolution gaussien. cas non-abélien. *Math. Z.*, 188:411–417, 1985.

[BCG+92] F. Bonechi, E. Celeghini, R. Giachetti, E. Sorace, and M. Tarlini. Heisenberg XXY model and quantum Galilei group. *J. Phys. A: Math. Gen.*, 25:L939–943, 1992.

[BCG+94] F. Bonechi, E. Celeghini, R. Giachetti, C. M. Pereña, E. Sorace, and M. Tarlini. Exponential mapping for non-semisimple quantum groups. *J. Phys. A: Math. Gen.*, 27:1307–1315, 1994.

[Bel98] V.P. Belavkin. On quantum Itô algebras. *Math. Phys. Lett.*, 7:1–16, 1998.

[BGS99] A. Ben Ghorbal and M. Schürmann. On the algebraic foundations of a non-commutative probability theory. Prépublication 99/17, Institut E. Cartan, Nancy, 1999.

[BH95] W.R. Bloom and H. Heyer. *Harmonic Analysis of Probability Measures on Hypergroups*, volume 20 of *de Gruyter Studies in Mathematics*. de Gruyter, Berlin, 1995.

[Bia93] P. Biane. *Ecole d'été de Probabilités de Saint-Flour*, volume 1608 of *Lecture Notes in Math.*, chapter Calcul stochastique non-commutatif. Springer-Verlag, Berlin, 1993.

[Bia98] P. Biane. Processes with free increments. *Math. Z.*, 227(1):143–174, 1998.

[BKS96] M. Bozejko, B. Kümmerer, and R. Speicher. q-Gaussian processes: Non-commutative and classical aspects. *Commun. Math. Phys.*, 185(1):129–154, 1996.

[BM93] W. K. Baskerville and S. Majid. The braided Heisenberg group. *J. Math. Phys.*, 34:3588–3606, 1993.

[Bre68] L. Breiman. *Probability*. Addison-Wesley, Reading (Mass.), 1968.

[BS91] M. Bozejko and R. Speicher. An example of generalized Brownian motion. *Commun. Math. Phys.*, 137:519–531, 1991.

[CD89] H. Carnal and M. Dozzi. On a decomposition problem for multivariate probability measures. *J. Multivariate Anal.*, 31:165–177, 1989.

[CF95] H. Carnal and G.M. Fel'dman. Phase retrieval for probability measures on abelian groups. *J. Theor. Prob*, 8(3):717–725, 1995.

[CF97] H. Carnal and G.M. Fel'dman. Phase retrieval for probability measures on abelian groups ii. *J. Theor. Prob.*, 10(4):1065–1074, 1997.

[CGST91] E. Celeghini, R. Giachetti, E. Sorace, and M. Tarlini. The quantum Heisenberg group $H(1)_q$. *J. Math. Phys.*, 32:1155–1158, 1991.

[CP95] V. Chari and A. Pressley. *A guide to quantum groups*. Cambridge University Press, 1995.

[DHL91] H. D. Doebner, J. D. Hennig, and W. Lücke. Mathematical guide to quantum groups. In H. D. Doebner and J. D. Hennig, editors, *Proceedings of Quantum Groups Conference held in Clausthal in 1989, Springer Lecture Notes in Phys., Vol. 370*, 1991.

[Dob92] V. K. Dobrev. Duality for the matrix quantum group $GL_{p,q}(2, \mathbb{C})$. *J. Math. Phys.*, 33:3419–3430, 1992.

[DP93a] V. K. Dobrev and Preeti Pashar. Duality for a Lorentz quantum group. *Lett. Math. Phys.*, 29:259–269, 1993.

[DP93b] V. K. Dobrev and Preeti Pashar. Duality for multiparametric quantum GL(n). *J. Phys. A: Math. Gen.*, 26:6991–7002, 1993.

[Dri87] V. G. Drinfeld. Quantum groups. In *Proceedings of the International Congress of Mathematicians, Berkeley 1986*. American Mathematical Society, 1987.

[Eme89] M. Emery. On the Azéma martingales. In *Séminaire de Probabilités XXIII*, volume 1372 of *Lecture Notes in Math.* Springer-Verlag, Berlin, 1989.

[Fei78] P. Feinsilver. Processes with independent increments on a Lie group. *Trans. AMS*, 242:73–121, 1978.

[Fei87] P. Feinsilver. Generalized convolutions and probability theory based on the q-Heisenberg algebra. In *Proc. First World Congr. Bernoulli Soc.*, pages 729–741. VNU Science Press, 1987.

[Fel71] W. Feller. *An introduction to probability theory and its applications, Vol. II.* Wiley, New York, 2nd edition, 1971.

[FF] P. Feinsilver and U. Franz. Brownian motion on the affine group and generalized Gegenbauer polynomials. in preparation.

[FFS95] P. Feinsilver, U. Franz, and R. Schott. Duality and multiplicative processes on quantum groups. In H.D. Doebner, V. Dobrev, and P. Nattermann, editors, *Proceedings of the International Conference on "Nonlinear, Deformed and Irreversible Quantum Systems"*, pages 462–468. World Scientific, 1995.

[FFS97] P. Feinsilver, U. Franz, and R. Schott. Duality and multiplicative processes on quantum groups. *J. Theor. Prob.*, 10(3):795–818, 1997.

[FG93] C. Fronsdal and A. Galindo. The dual of a quantum group. *Lett. Math. Phys.*, 27:59–71, 1993.

[FK93] J. Fröhlich and T. Kerler. *Quantum groups, quantum categories and quantum field theory*, volume 1542 of *Lecture Notes Math.* Springer-Verlag, 1993.

[FKS93] P. Feinsilver, J. Kocik, and R. Schott. Representations and stochastic processes on groups of type-H. *J. Funct. Anal.*, 115:144–165, 1993.

[FNS97a] U. Franz, D. Neuenschwander, and R. Schott. Gauss laws in the sense of bernstein and uniqueness of embedding into convolution semigroups on quantum groups and braided groups. *Probability Theory and Related Fields*, 109:101–127, 1997.

[FNS97b] U. Franz, D. Neuenschwander, and R. Schott. Phase retrieval for probability distributions on quantum groups and braided groups. Prépublication 97/n° 4, Institut Elie Cartan, Nancy, 1997. To appear in J. Theor. Prob.

[Fra94] U. Franz. Stochastic processes on Lie groups and quantum groups. Diplomarbeit/ mémoire de DEA, jointly prepared at TU Clausthal and Université H. Poincaré-Nancy 1, 1994.

[Fra97] U. Franz. *Contribution á l'étude des processus stochastiques sur les groupes quantiques.* PhD thesis, Université H. Poincaré-Nancy 1, 1997.

[Fra98] U. Franz. A mixed quantum-classical central limit theorem. *Banach Center Publ.*, 43:183–189, 1998.

[Fra99] U. Franz. Classical markov processes from quantum Lévy processes. *Inf. Dim. Anal., Quantum Prob. and Rel. Topics*, 2(1):105–129, 1999.

[FS89a] P. Feinsilver and R. Schott. An operator approach to processes on Lie groups. In *Lecture Notes in Mathematics, Vol. 1391*, pages 59–65, New York, 1989. Conference Proceedings, 'Probability Theory on Vector Spaces', Lancut 1987, Springer-Verlag.

[FS89b] P. Feinsilver and R. Schott. Operators, stochastic processes, and Lie groups. In *Lecture Notes in Mathematics, Vol. 1379*, pages 75–85, New York, 1989. Oberwolfach Conference, 'Probability Measures on Groups', Springer-Verlag.

[FS90] P. Feinsilver and R. Schott. Special functions and infinite-dimensional representations of Lie groups. *Math. Z.*, 203:173–191, 1990.

[FS92] P. Feinsilver and R. Schott. Appell systems on Lie groups. *J. Theor. Probability*, 5:251–281, 1992.

[FS93] P. Feinsilver and R. Schott. *Algebraic Structures and Operator Calculus, Vol. I: Representations and Probability Theory.* Kluwer Academic Publishers, Dordrecht, 1993.

[FS98a] U. Franz and R. Schott. Diffusions on braided spaces. *J. Math. Phys.*, 39(5):2748–2762, 1998.

[FS98b] U. Franz and R. Schott. Evolution equations and Lévy processes on quantum groups. *J. Phys. A: Math. and Gen.*, 31:1395–1404, 1998.

[FSS98] U. Franz, R. Schott, and M. Schürmann. Braided independence and Lévy processes on braided spaces. Prépublication 98/n° 32, Institut Elie Cartan, Nancy, 1998.

[Fuc92] J. Fuchs. *Affine Lie Algebras and Quantum Groups, an Introduction with Applications in Conformal Field Theory.* Cambridge University Press, Cambridge, 1992.

[GDF82] B. Gruber, H. D. Doebner, and P. Feinsilver. Representations of the Heisenberg-Weyl algebra and group. *Kinam*, 4:241–278, 1982.

[GK54] B.V. Gnedenko and A.N. Kolmogorov. *Limit distributions for sums of independent random variables.* Addison-Wesley, Reading (Mass.), 1954.

[GKR77] Y. Guivarc'h, M. Keane, and B. Roynette. *Marches Aléatoires sur les Groupes de Lie*, volume 624 of *Lecture Notes in Math.* Springer-Verlag, Berlin, 1977.

[GP89] A. Ch. Ganchev and V. B. Petkova. $U_q(sl(2))$ invariant operators and minimal theories fusion matrices. *Phys. Lett. B*, 233:374 – 382, 1989.

[Gui95] A. Guichardet. *Groupes quantiques.* InterÉdition/CNRS Éditions, Paris, 1995.

[Haz77] W. Hazod. *Stetige Faltungshalbgruppen von Wahrscheinlickeitsmaßen und erzeugende Distributionen*, volume 595 of *Lecture Notes in Math.* Springer-Verlag, Berlin, 1977.

[HDL86] M. Hakim-Dowek and D. Lépingle. L'exponentielle stochastique des groupes de Lie. In *Lecture Notes in Math., Vol. 1204*, pages 352–374, Berlin, 1986. Springer-Verlag.

[Hey77] H. Heyer. *Probability measures on locally compact groups.* Springer-Verlag, Berlin, 1977.

[HKK98] J. Hellwig, C. Köstler, and B. Kümmerer. Stationary quantum Markov processes as solutions of stochastic differential equations. *Banach Center Publ.*, 43:217–229, 1998.

[HN91] J. Hilgert and K.-H. Neeb. *Lie-Gruppen und Lie-Algebren.* Vieweg, Braunschweig, 1991.

[Hum87] J.E. Humphreys. *Introduction to Lie algebras and representation theory.* Springer-Verlag, Berlin, 1987.

[Jim85] M. Jimbo. A q-difference analogue of $U(g)$ and the Yang-Baxter equation. *Lett. Math. Phys.*, 10:63–69, 1985.

[Jim86] M. Jimbo. A q-analogue of $U(gl(N+1))$, Hecke algebra, and the Yang-Baxter equation. *Lett. Math. Phys*, 11:247–252, 1986.

[JS91a] A. Joyal and R. Street. The geometry of tensor calculus I. *Adv. Math.*, 88(1):55–112, 1991.

[JS91b] A. Joyal and R. Street. An introduction to Tannaka duality and quantum groups. In A. Carboni, M.C. Pedicchio, and G. Rosolini, editors, *Category theory, Proc. Int. Conf., Como/Italy 1990*, pages 413–492. Springer-Verlag, 1991. Lecture Notes in Math., Vol. 1488.

[JS93] A. Joyal and R. Street. Braided tensor categories. *Adv. Math.*, 102(1):20–78, 1993.

[Kas95] Ch. Kassel. *Quantum Groups*, volume 155 of *Graduate texts in mathematics*. Springer-Verlag, Berlin, 1995.

[KM94] A. Kempf and S. Majid. Algebraic q-integration and Fourier theory on quantum and braided spaces. *J. Math. Phys.*, 35:6801–6837, 1994.

[Koo91] T.H. Koornwinder. Positive convolution structures associated with quantum groups. In *10th Oberwolfach Conference, 'Probability Measures on Groups'*, New York, 1991. Plenum Press.

[Koo95] T.H. Koornwinder. Minicourse "special functions and q-commuting variables". In *Workshop on "Special functions, q-series and related topics"*. Fields Institute, Toronto, Canada, 12-23 June 199, 1995.

[Küm88] B. Kümmerer. Survey on a theory of non-commutativ stationary Markov processes. In L. Accardi and W.v. Waldenfels, editors, *Quantum Probability and Applications III*, pages 228–244. Springer-Verlag, 1988.

[Küm90] B. Kümmerer. Markov dilations and non-commutative poisson processes. preprint, 1990.

[Len93] R. Lenczewski. On sums of q-independent $su_q(2)$ quantum variables. *Comm. Math. Phys*, 154:127–134, 1993.

[Len94] R. Lenczewski. Addition of independent variables in quantum groups. *Rev. Math. Phys.*, 6:135–147, 1994.

[Len98] R. Lenczewski. A noncommutative limit theorem for homogeneous correlations. *Studia Math.*, 129:225–252, 1998.

[Luk70] E. Lukacs. *Characteristic functions*. Griffin, London, 2nd edition, 1970.

[Lus93] G. Lusztig. *Introduction to Quantum Groups*, volume 110 of *Progress in mathematics*. Birkhauser, Boston, 1993.

[Mac71] S. MacLane. *Categories for the working mathematician*, volume 5 of *Graduate texts in mathematics*. Springer-Verlag, Berlin, 1971.

[Maj93a] S. Majid. Beyond supersymmetry and quantum symmetry (an introduction to braided groups and braided matrices). In M.L. Ge and H.J. Vega, editors, *Quantum Integrable Statistical Models and Knot Theory*, pages 231–282. World Scientific, 1993.

[Maj93b] S. Majid. Braided momentum in the q-Poincaré group. *J. Math. Phys.*, 34:2045–2058, 1993.

[Maj93c] S. Majid. Free braided differential calculus, braided binomial theorem, and the braided exponential map. *J. Math. Phys.*, 34:4843–4856, 1993.

[Maj93d] S. Majid. Quantum random walks and time-reversal. *Int. J. Mod. Phys.*, 8:4521–4545, 1993.

[Maj94] S. Majid. Quantum double as quantum mechanics. *J. Geom. Phys.*, 13:169–202, 1994.

[Maj95a] S. Majid. *-Structures on braided spaces. *J. Math. Phys.*, 36:4436–4449, 1995.

[Maj95b] S. Majid. *Foundations of quantum group theory*. Cambridge University Press, 1995.

[Maj96] S. Majid. Introduction to braided geometry and q-Minkowski space. In *Proc. School 'Enrico Fermi' CXXVII*, pages 267–348. IOS Press, 1996.

[Maj97] S. Majid. Quasi-* structure on q-Poincaré algebras. *J. Geom. Phys.*, 22:14–58, 1997.

[McK69] H.P. McKean. *Stochastic integrals*. Academic Press, 1969.

[Mee86] M. Meerschaert. Domains of attraction of nonnormal operator-stable laws. *J. Multivariate Anal.*, 19:342–347, 1986.

[Mey93] P.-A. Meyer. *Quantum Probability for Probabilists*, volume 1538 of *Lecture Notes in Math*. Springer-Verlag, Berlin, 1993.

[MMN+90a] T. Masuda, K. Mimachi, Y. Nakagami, M. Noumi, Y. Saburi, and K. Ueno. Unitary representations of the quantum group $SU_q(1,1)$: Structure of the dual space of $U_q(sl(2))$. *Lett. Math. Phys.*, 19:187–194, 1990.

[MMN+90b] T. Masuda, K. Mimachi, Y. Nakagami, M. Noumi, Y. Saburi, and K. Ueno. Unitary representations of the quantum group $SU_q(1,1)$: II - matrix elements of unitary representations and the basic hypergeometric functions. *Lett. Math. Phys.*, 19:195–204, 1990.

[Mon93] S. Montgomery, editor. *Hopf algebras and their actions on rings*, volume 082 of *CBMS-NSF-Regional conference series in mathematics*, Providence, RI, 1993. AMS.

[MRP94] S. Majid and M. J. Rodríguez-Plaza. Random walk and the heat equation on superspace and anyspace. *J. Math. Phys.*, 35:3753–3760, 1994.

[MV94] A. Morozov and L. Vinet. Free-field representation of group element for simple quantum groups. preprint htp-th/9409093, 1994.

[Nel66] E. Nelson. Derivation of the Schrödinger equation from Newtonian mechanics. *Phys. Review*, 150:1079–1085, 1966.

[Nel88] E. Nelson. Stochastic mechanics and random fields. In *Calcul des probabilites, Ec. d'Ete, Saint-Flour/Fr. 1985-87*, volume 1362 of *Lecture Notes in Math.*, pages 427–450. Springer-Verlag, 1988.

[Neu93] D. Neuenschwander. Gauss measures in the sense of Bernstein on the Heisenberg group. *Prob. Math. Stat.*, 14:253–256, 1993.

[Neu96] D. Neuenschwander. On the uniqueness problem for continuous convolution semigroups of probability measures on simply connected nilpotent Lie groups. submitted, 1996.

[NRS97] D. Neuenschwander, B. Roynette, and R. Schott. Characterizations of Gaussian distributions on simply connected nilpotent Lie groups and symmetric spaces. *C. R. Acad. Sci. Paris*, 324:87–92, 1997.

[NS97] D. Neuenschwander and R. Schott. A note on domains of attraction for q-transformed random variables. *Prob. Math. Stat.*, 17(2):387–394, 1997.

[Pap94] G. Pap. Uniqueness of embedding into a Gaussian semigroup on a nilpotent Lie group. *Arch. Math.*, 62:282–288, 1994.

[Par90] K.R. Parthasarathy. Azéma martingales and quantum stochastic calculus. In R.R. Bahadur, editor, *Proc. R.C. Bose Memorial Symposium*. Wiley Eastern, 1990.

[Par92] K.R. Parthasarathy. *An Introduction to Quantum Stochastic Calculus*. Birkhäuser, 1992.

[PS91] K.R. Parthasarathy and K.B. Sinha. Unification of quantum noise processes in fock spaces. In L. Accardi, editor, *Quantum Probability and Related Topics VI*. World Scientific, 1991.

[PWZ81] R. Le Page, M. Woodroofe, and J. Zinn. Convergence to a stable distribution via order statistics. *Ann. Prob.*, 9(4):624–632, 1981.

[RTF90] N. Yu. Reshetikhin, L. A. Takhtadzhyan, and L. D. Faddeev. Quantization of Lie groups and Lie algebras. *Leningrad Math. J.*, 1(1):193–224, 1990.

[Sch53] M. P. Schützenberger. Une interprétation de certaines solutions de l'équation fonctionnelle: $f(x+y) = f(x)f(y)$. *C. R. Acad. Sci. Paris*, 236:352–353, 1953.

[Sch91a] M. Schürmann. The Azéma martingales as components of quantum independent increment processes. In J. Azéma, P.A. Meyer, and M. Yor, editors, *Séminaire de Probabilités XXV*, volume 1485 of *Lecture Notes in Math.* Springer-Verlag, 1991.

[Sch91b] M. Schürmann. Quantum q-white noise and a q-central limit theorem. *Commun. Math. Phys.*, 140:589–615, 1991.

[Sch93] M. Schürmann. *White Noise on Bialgebras*, volume 1544 of *Lecture Notes in Mathematics*. Springer-Verlag, Berlin, 1993.

[Sch95a] M. Schürmann. Direct sums of tensor products and non-commutative independence. *J. Funct. Anal.*, 1995.

[Sch95b] M. Schürmann. Non-commutative probability on algebraic structures. In H. Heyer, editor, *Proceedings of XI Oberwolfach Conference on Probability Measures on Groups and Related Structures*, pages 332–356. World Scientific, 1995.

[Ser92] J.-P. Serre. *Lie algebras and Lie groups : 1964 lecture given at Harvard*, volume 1500 of *Lecture Notes in Math.* Springer-Verlag, Berlin, 2nd edition, 1992.

[Sha69] M. Sharpe. Operator-stable probability distributions on vector groups. *Trans. Amer. Math. Soc.*, 136, 1969.

[Sie73] E. Siebert. Über die Erzeugung von Faltungshalbgruppen auf beliebigen lokalkompakten Gruppen. *Math. Z.*, 131:313–333, 1973.

[Sie81] E. Siebert. Fourier analysis and limit theorems for convolution semigroups on a locally compact group. *Adv. Math.*, 39:111–154, 1981.

[Ske94] M. Skeide. *The Levy-Khintchine formula for the quantum group $SU_q(2)$*. PhD thesis, Univ. Heidelberg, Naturwiss.-Math. Gesamtfak., 1994.

[Spe92] R. Speicher. A non-commutative central limit theorem. *Math. Z.*, 209:55–66, 1992.

[Spe97] R. Speicher. Universal products. In D. Voiculescu, editor, *Free probability theory. Papers from a workshop on random matrices and operator algebra free products, Toronto, Canada, Mars 1995*, volume 12 of *Fields Inst. Commun.*, pages 257–266. American Mathematical Society, Providence, RI, 1997.

[SS93] S. Shnider and S. Sternberg. *Quantum Groups: from Coalgebras to Drinfeld Algebras, a Guided Tour*, volume 002 of *Graduate texts in mathematical physics*. International Press, 1993.

[SS98] M. Schürmann and M. Skeide. Infinitesimal generators of the quantum group $SU_q(2)$. *Inf. Dim. Anal., Quantum Prob. and Rel. Topics*, 1(4):573–598, 1998.

[SW86] D. H. Sattinger and O. L. Weaver. *Lie Groups and Algebras with Applications to Physics, Geometry, and Mechanics*. Springer-Verlag, New York, 1986.

[SW97] R. Speicher and R. Woroudi. Boolean convolution. In D. Voiculescu, editor, *Free probability theory. Papers from a workshop on random matrices and operator algebra free products, Toronto, Canada, Mars 1995*, volume 12 of *Fields Inst. Commun.*, pages 267–279. American Mathematical Society, Providence, RI, 1997.

[Swe69] M. E. Sweedler. *Hopf Algebras.* Benjamin, New York, 1969.

[SWZ92] P. Schupp, P. Watts, and B. Zumino. The two-dimensional quantum Euclidean algebra. *Lett. Math. Phys.*, 24:141–145, 1992.

[TV91] P. Truini and V. S. Varadarajan. The concept of a quantum semisimple group. *Lett. Math. Phys.*, 21:287–292, 1991.

[TV93] P. Truini and V. S. Varadarajan. Quantization of reductive Lie algebras: Construction and universality. *Rev. Math. Phys.*, 5(2):363–415, 1993.

[Var84] V.S. Varadarajan. *Lie groups, Lie algebras and their representations.* Springer-Verlag, Berlin, 1984.

[VDN92] D. Voiculescu, K. Dykema, and A. Nica. *Free Random Variables.* AMS, 1992.

[VS88] L. L. Vaksman and Ya. S. Soibel'man. Algebra of functions on the quantum group $SU(2)$. *Funkt. Anal. Appl.*, 22:170–181, 1988.

[Wal73] W.v. Waldenfels. An approach to the theory of pressure broadening of spectral lines. In M. Behara, K. Krickeberg, and J. Wolfowitz, editors, *Probability and Information Theory II*, volume 296 of *Lecture Notes in Math.* Springer-Verlag, Berlin, 1973.

[Wey31] H. Weyl. *The Theory of Groups and Quantum Mechanics.* Methuen, London, 1931. reprinted by Dover Publications, 1950.

[Wig32] E. Wigner. On the quantum corrections for thermodynamic equilibrium. *Phys. Rev.*, 40:749–759, 1932.

[Wor87] S. L. Woronowicz. Compact matrix pseudogroups. *Commun. Math. Phys.*, 111:613–665, 1987.

[Wor91] S. L. Woronowicz. Quantum $E(2)$ group and its Pontryagin dual. *Lett. Math. Phys.*, 23:251–263, 1991.

[Wor98] S.L. Woronowicz. Compact quantum groups. In A. Connes, K. Gawedzki, and J. Zinn-Justin, editors, *Symétries Quantiques, Les Houches, Session LXIV, 1995*, pages 845–884. Elsevier Science, 1998.

Index

Other *Mathematics and Its Applications* titles of interest:

P.M. Alberti and A. Uhlmann: *Stochasticity and Partial Order. Doubly Stochastic Maps and Unitary Mixing*. 1982, 128 pp. ISBN 90-277-1350-2

A.V. Skorohod: *Random Linear Operators*. 1983, 216 pp. ISBN 90-277-1669-2

I.M. Stancu-Minasian: *Stochastic Programming with Multiple Objective Functions*. 1985, 352 pp. ISBN 90-277-1714-1

L. Arnold and P. Kotelenez (eds.): *Stochastic Space-Time Models and Limit Theorems*. 1985, 280 pp. ISBN 90-277-2038-X

Y. Ben-Haim: *The Assay of Spatially Random Material*. 1985, 336 pp. ISBN 90-277-2066-5

A. Pazman: *Foundations of Optimum Experimental Design*. 1986, 248 pp. ISBN 90-277-1865-2

P. Kree and C. Soize: *Mathematics of Random Phenomena. Random Vibrations of Mechanical Structures*. 1986, 456 pp. ISBN 90-277-2355-9

Y. Sakamoto, M. Ishiguro and G. Kitagawa: *Akaike Information Criterion Statistics*. 1986, 312 pp. ISBN 90-277-2253-6

G.J. Szekely: *Paradoxes in Probability Theory and Mathematical Statistics*. 1987, 264 pp. ISBN 90-277-1899-7

O.I. Aven, E.G. Coffman (Jr.) and Y.A. Kogan: *Stochastic Analysis of Computer Storage*. 1987, 264 pp. ISBN 90-277-2515-2

N.N. Vakhania, V.I. Tarieladze and S.A. Chobanyan: *Probability Distributions on Banach Spaces*. 1987, 512 pp. ISBN 90-277-2496-2

A.V. Skorohod: *Stochastic Equations for Complex Systems*. 1987, 196 pp. ISBN 90-277-2408-3

S. Albeverio, Ph. Blanchard, M. Hazewinkel and L. Streit (eds.): *Stochastic Processes in Physics and Engineering*. 1988, 430 pp. ISBN 90-277-2659-0

A. Liemant, K. Matthes and A. Wakolbinger: *Equilibrium Distributions of Branching Processes*. 1988, 240 pp. ISBN 90-277-2774-0

G. Adomian: *Nonlinear Stochastic Systems Theory and Applications to Physics*. 1988, 244 pp. ISBN 90-277-2525-X

J. Stoyanov, O. Mirazchiiski, Z. Ignatov and M. Tanushev: *Exercise Manual in Probability Theory*. 1988, 368 pp. ISBN 90-277-2687-6

E.A. Nadaraya: *Nonparametric Estimation of Probability Densities and Regression Curves*. 1988, 224 pp. ISBN 90-277-2757-0

H. Akaike and T. Nakagawa: *Statistical Analysis and Control of Dynamic Systems*. 1998, 224 pp. ISBN 90-277-2786-4

A.V. Ivanov and N.N. Leonenko: *Statistical Analysis of Random Fields*. 1989, 256 pp. ISBN 90-277-2800-3

Other *Mathematics and Its Applications* titles of interest:

V. Paulauskas and A. Rackauskas: *Approximation Theory in the Central Limit Theorem. Exact Results in Banach Spaces.* 1989, 176 pp. ISBN 90-277-2825-9

R.Sh. Liptser and A.N. Shiryayev: *Theory of Martingales.* 1989, 808 pp. ISBN 0-7923-0395-4

S.M. Ermakov, V.V. Nekrutkin and A.S. Sipin: *Random Processes for Classical Equations of Mathematical Physics.* 1989, 304 pp. ISBN 0-7923-0036-X

G. Constantin and I. Istratescu: *Elements of Probabilistic Analysis and Applications.* 1989, 488 pp. ISBN 90-277-2838-0

S. Albeverio, Ph. Blanchard and D. Testard (eds.): *Stochastics, Algebra and Analysis in Classical and Quantum Dynamics.* 1990, 264 pp. ISBN 0-7923-0637-6

Ya.I. Belopolskaya and Yu.L. Dalecky: *Stochastic Equations and Differential Geometry.* 1990, 288 pp. ISBN 90-277-2807-0

A.V. Gheorghe: *Decision Processes in Dynamic Probabilistic Systems.* 1990, 372 pp. ISBN 0-7923-0544-2

V.L. Girko: *Theory of Random Determinants.* 1990, 702 pp. ISBN 0-7923-0233-8

S. Albeverio, PH. Blanchard and L. Streit: *Stochastic Processes and their Applications in Mathematics and Physics.* 1990, 416 pp. ISBN 0-9023-0894-8

B.L. Rozovskii: *Stochastic Evolution Systems. Linear Theory and Applications to Non-linear Filtering.* 1990, 330 pp. ISBN 0-7923-0037-8

A.D. Wentzell: *Limit Theorems on Large Deviations for Markov Stochastic Process.* 1990, 192 pp. ISBN 0-7923-0143-9

K. Sobczyk: *Stochastic Differential Equations. Applications in Physics, Engineering and Mechanics.* 1991, 410 pp. ISBN 0-7923-0339-3

G. Dallaglio, S. Kotz and G. Salinetti: *Distributions with Given Marginals.* 1991, 300 pp. ISBN 0-7923-1156-6

A.V. Skorohod: *Random Processes with Independent Increments.* 1991, 280 pp. ISBN 0-7923-0340-7

L. Saulis and V.A. Statulevicius: *Limit Theorems for Large Deviations.* 1991, 232 pp. ISBN 0-7923-1475-1

A.N. Shiryaev (ed.): *Selected Works of A.N. Kolmogorov, Vol. 2: Probability Theory and Mathematical Statistics.* 1992, 598 pp. ISBN 90-277-2795-X

Yu.I. Neimark and P.S. Landa: *Stochastic and Chaotic Oscillations.* 1992, 502 pp. ISBN 0-7923-1530-8

Y. Sakamoto: *Categorical Data Analysis by AIC.* 1992, 260 pp. ISBN 0-7923-1429-8

Lin Zhengyan and Lu Zhuarong: *Strong Limit Theorems.* 1992, 200 pp. ISBN 0-7923-1798-0

Other *Mathematics and Its Applications* titles of interest:

J. Galambos and I. Katai (eds.): *Probability Theory and Applications.* 1992, 350 pp. ISBN 0-7923-1922-2

N. Bellomo, Z. Brzezniak and L.M. de Socio: *Nonlinear Stochastic Evolution Problems in Applied Sciences.* 1992, 220 pp. ISBN 0-7923-2042-5

A.K. Gupta and T. Varga: *Elliptically Contoured Models in Statistics.* 1993, 328 pp. ISBN 0-7923-2115-4

B.E. Brodsky and B.S. Darkhovsky: *Nonparametric Methods in Change-Point Problems.* 1993, 210 pp. ISBN 0-7923-2122-7

V.G. Voinov and M.S. Nikulin: *Unbiased Estimators and Their Applications. Volume 1: Univariate Case.* 1993, 522 pp. ISBN 0-7923-2382-3

V.S. Koroljuk and Yu.V. Borovskich: *Theory of* U-*Statistics.* 1993, 552 pp. ISBN 0-7923-2608-3

A.P. Godbole and S.G. Papastavridis (eds.): *Runs and Patterns in Probability: Selected Papers.* 1994, 358 pp. ISBN 0-7923-2834-5

Yu. Kutoyants: *Identification of Dynamical Systems with Small Noise.* 1994, 298 pp. ISBN 0-7923-3053-6

M.A. Lifshits: *Gaussian Random Functions.* 1995, 346 pp. ISBN 0-7923-3385-3

M.M. Rao: *Stochastic Processes: General Theory.* 1995, 635 pp. ISBN 0-7923-3725-5

Yu.A. Rozanov: *Probability Theory, Random Processes and Mathematical Statistics.* 1995, 267 pp. ISBN 0-7923-3764-6

L. Zhengyan and L. Chuanrong: *Limit Theory for Mixing Dependent Random Variables.* 1996, 352 pp. ISBN 0-7923-4219-4

A.V. Ivanov: *Asymptotic Theory of Nonlinear Regression.* 1997, 333 pp. ISBN 0-7923-4335-2

E.M.J. Bertin†, I. Cuculescu and R. Theodorescu: *Unimodality of Probability Measures.* 1997, 264 pp. ISBN 0-7923-4318-2

A. Swishchuk: *Random Evolutions and Their Applications.* 1997, 215 pp. ISBN 0-7923-4533-9

V. Kalashnikov: *Geometric Sums: Bounds for Rare Events with Applications.* Risk Analysis, Reliability, Queueing. 1997, 283 pp. ISBN 0-7923-4616-5

V. Buldygin and S. Solntsev: *Asymptotic Behaviour of Linearly Transformed Sums of Random Variables.* 1997, 516 pp. ISBN 0-7923-4632-7

Yu.A. Rozanov: *Random Fields and Stochastic Partial Differential Equations.* 1998, 238 pp. ISBN 0-7923-4984-9

U. Franz and René Schott: *Stochastic Processes and Operator Calculus on Quantum Groups.* 1999, 236pp. ISBN 0-7923-5883-X

